OXFORD

GCSE Maths

For Edexcel

HIGHER

Marguerite Appleton
Dave Capewell
Derek Huby
Jayne Kranat
Peter Mullarkey

OXFORD
UNIVERSITY PRESS

OXFORD
UNIVERSITY PRESS

Great Clarendon Street, Oxford OX2 6DP
Oxford University Press is a department of the University of Oxford.
It furthers the University's objective of excellence in research, scholarship,
and education by publishing worldwide in

Oxford New York
Auckland Cape Town Dar es Salaam Hong Kong Karachi
Kuala Lumpur Madrid Melbourne Mexico City Nairobi
New Delhi Shanghai Taipei Toronto

With offices in

Argentina Austria Brazil Chile Czech Republic France Greece
Guatemala Hungary Italy Japan South Korea Poland Portugal
Singapore Switzerland Thailand Turkey Ukraine Vietnam

Oxford is a registered trade mark of Oxford University Press
in the UK and in certain other countries

ISBN 9780199139484

10 9 8 7 6 5 4 3 2 1

Printed in Spain by Cayfosa (Impresia Iberica)

Paper used in the production of this book is a natural, recyclable product made from wood
grown in sustainable forests. The manufacturing process conforms to the environmental
regulations of the country of origin.

Acknowledgements
The Publisher would like to thank Edexcel for their kind permission to reproduce past exam
questions.
Edexcel Ltd, accepts no responsibility whatsoever for the accuracy or method of working in the answers given

The Publisher would like to thank the following for permission to reproduce photographs:
p2-3: Ivan Kmit/Dreamstime; **p20-21:** Martin Fischer/Shutterstock; **p36-37:** David Martyn Hughes/Dreamstime.com; **p56-57:** Odyssei/
Dreamstime.com; **p70-71:** Godrick/Dreamstime.com; **p86-87:** SSPL via Getty Images; **p93:** OUP; **p106-107:** Jakub Krechowicz/Dreamstime;
p122-123: Saniphoto/Dreamstime.com; **p134:** Phil Date/Dreamstime.com; **p140-141:** OUP/Photodisc; **p158-159:** Alex Segre/Alamy;
p172-173: Diman Oshchepkov/Dreamstime.com; **p186-187:** Rex Features; **p202-203:** Slobodan Djajic/Dreamstime.com; **p216-217:** Monkey
Business Images/Dreamstime.com; **p228-229:** Dgareri/Dreamstime.com; **p246-247:** Thor Jorgen Udvang/Dreamstime.com; **p260-261:** Giovanni
Benintende/Shutterstock; **p278-279:** Wrangler/Dreamstime.com; **p298-299:** Bobbigmac/Dreamstime.com; **p314-315:** Creative Commons/
Wikipedia; **p321:** OUP; **p322:** Raja Rc/Dreamstime.com; **p328-329:** Bbbar/Dreamstime.com; **p348-349:** Dreamstime Agency/Dreamstime.com;
p362-363: Marian Mocanu/Shutterstock; **p376-377:** Pablo H Caridad/Shutterstock; **p398-399:** Idrutu/Dreamstime.com.

The Publisher would also like to thank Anna Cox for her work in creating the case studies.
Figurative artwork is by Peter Donnelly

About this book

This book has been specifically written to help you get the best possible grade in your Edexcel GCSE Mathematics examinations. It is designed for students who have achieved a secure level 6 at Key Stage 3 and are looking to progress to a grade B at GCSE, Higher tier.

The authors are experienced teachers and examiners who have an excellent understanding of the Edexcel specification and so are well qualified to help you successfully meet your objectives.

The book is made up of chapters that are based on Edexcel specification A (linear) and is organised clearly into a suggested teaching order.

Functional maths and **problem solving** are flagged in the exercises throughout.

- In particular there are **case studies,** which allow you apply your GCSE knowledge in a variety of engaging contexts.

- There are also **rich tasks,** which provide an investigative lead-in to the chapter – you may need to study some of the techniques in the chapter in order to be able to complete them properly.

Also built into this book are the new **assessment objectives:**

AO1   recall knowledge of prescribed content.

AO2   select and apply mathematical methods in a range of contexts.

AO3   interpret and analyse problems and select strategies to solve them.

AO2 and AO3 are flagged throughout, particularly in the regular **summary assessments,** as these make up around 50% of your examination.

Finally, you will notice an icon that looks like this:

This shows opportunities for **Quality of Written Communication,** which you will also be assessed on in your exams.

Best wishes with your GCSE Maths – we hope you enjoy your course and achieve success!

Contents

About this book

This book has been specifically written to help you get the best possible grade in your Edexcel GCSE Mathematics examinations. It is designed for students who have achieved a secure level 6 at Key Stage 3 and are looking to progress to a grade B at GCSE, Higher tier.

The authors are experienced teachers and examiners who have an excellent understanding of the Edexcel specification and so are well qualified to help you successfully meet your objectives.

The book is made up of chapters that are based on Edexcel specification A (linear) and is organised clearly into a suggested teaching order.

Functional maths and **problem solving** are flagged in the exercises throughout.

- In particular there are **case studies,** which allow you apply your GCSE knowledge in a variety of engaging contexts.

- There are also **rich tasks,** which provide an investigative lead-in to the chapter – you may need to study some of the techniques in the chapter in order to be able to complete them properly.

Also built into this book are the new **assessment objectives:**

AO1   recall knowledge of prescribed content.

AO2   select and apply mathematical methods in a range of contexts.

AO3   interpret and analyse problems and select strategies to solve them.

AO2 and AO3 are flagged throughout, particularly in the regular **summary assessments,** as these make up around 50% of your examination.

Finally, you will notice an icon that looks like this:

This shows opportunities for **Quality of Written Communication,** which you will also be assessed on in your exams.

Best wishes with your GCSE Maths – we hope you enjoy your course and achieve success!

Contents

Finding your way around this book

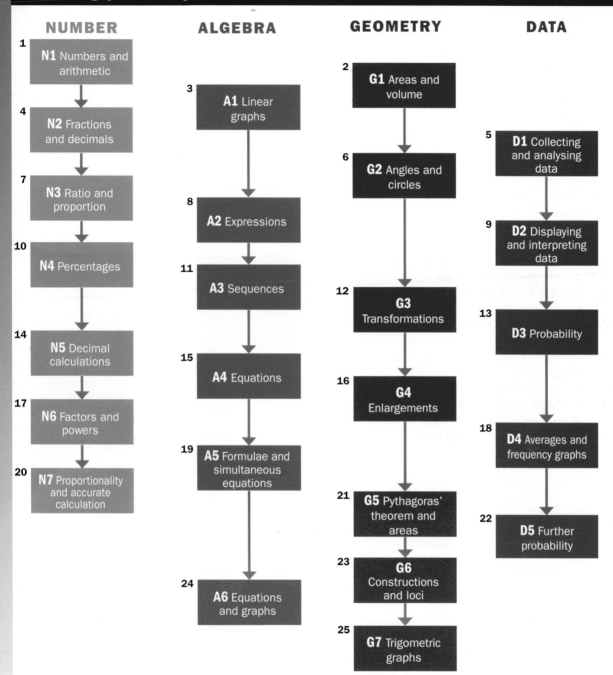

NUMBER

1 **N1** Numbers and arithmetic

4 **N2** Fractions and decimals

7 **N3** Ratio and proportion

10 **N4** Percentages

14 **N5** Decimal calculations

17 **N6** Factors and powers

20 **N7** Proportionality and accurate calculation

ALGEBRA

3 **A1** Linear graphs

8 **A2** Expressions

11 **A3** Sequences

15 **A4** Equations

19 **A5** Formulae and simultaneous equations

24 **A6** Equations and graphs

GEOMETRY

2 **G1** Areas and volume

6 **G2** Angles and circles

12 **G3** Transformations

16 **G4** Enlargements

21 **G5** Pythagoras' theorem and areas

23 **G6** Constructions and loci

25 **G7** Trigometric graphs

DATA

5 **D1** Collecting and analysing data

9 **D2** Displaying and interpreting data

13 **D3** Probability

18 **D4** Averages and frequency graphs

22 **D5** Further probability

Numbers and arithmetic

Today we take numbers for granted but throughout history mathematicians have struggled to understand and accept numbers beyond the integers. Decimal numbers only became popular in the late Renaissance period. Whilst negative numbers were mistrusted up until the eighteenth century and numbers like the square root of two were only fully accepted in the nineteenth century!

What's the point?

The rules of negative numbers are now agreed and they find many uses. For example, as temperatures, losses in financial records, as coordinates or whenever we need to show something is in an opposite direction.

Check in

You should be able to

■ **understand place value**

1 Write in words the value of the digit 4 in each of these numbers.

    a 4506   **b** 23 409   **c** 200.45   **d** 13.054

■ **use negative numbers**

2 a Sketch a number line showing values from -5 to $+5$ and mark these numbers on it.

    i -3 **ii** $+4$ **iii** -2.5 **iv** 0

    b Write this set of directed numbers in ascending order.

        $+5, -2.4, -3, +6, 0, +1.5, -1.8$

What I need to know

KS3 Understand place
value
Do basic arithmetic

What I will learn

- Do arithmetic with
  negative numbers
  and decimals
- Round numbers
  and approximate
  calculations
- Use the correct
  order of arithmetic
  operations

What this leads to

N2 Calculations with
fractions
N5 Written calculations
with decimals
N7 Approximate and
exact calculations

How many steps would you need to take to walk to the moon?
How many steps would a beetle need to walk the same distance?
What if you had to walk to the Sun, the nearest star, the nearest galaxy...
Investigate.

Place value and ordering numbers

This spread will show you how to:

- Use place value to round and order whole numbers and decimals
- Multiply and divide any number by powers of 10

Keywords
Ascending
Descending
Digit
Integer
Order
Round

- You can **round** large numbers to the nearest hundred, thousand or any other power of ten and round decimal numbers to any number of decimal places.
  - Identify the final **digit** required
  - Round it up if the following digit is a 5 or more
  - Write the rounded number, including any zeros needed to make the place value correct.

Example

Round 72 456 to the nearest   **a** 10   **b** 100   **c** 1000.
....
a 72 456 = 72 460 to the nearest 10   **b** 72 456 = 72 500 to the nearest 100
c 72 456 = 72 000 to the nearest 1000

Example

Round 6.0374 to the nearest   **a** tenth   **b** hundredth   **c** thousandth.
....
a 6.0374 = 6.0 to the nearest tenth   **b** 6.0374 = 6.04 to the nearest hundredth
c 6.0374 = 6.037 to the nearest thousandth

To **order** decimals, look at the tenths digit first, then the hundredths digit, then the thousandths and so on.

Example

Write these numbers in **ascending** order.
0.3   0.275   0.28   0.3005   0.269997
....
0.275, 0.28 and 0.269997 are smaller than 0.3 and 0.3005.
0.269 997 is smallest.
0.275 is smaller than 0.28.
0.3 is smaller than 0.3005.
In ascending order, the numbers are 0.269997, 0.275, 0.28. 0.3, 0.3005.

Do not be misled by the number of digits. 0.28 is equal to 0.280, and is larger than 0.275.

- Multiplying a number by 10 moves the digits one place to the left. Multiplying by 100 moves the digits two places to the left.
- Dividing a number by 10 moves the digits one place to the right. Dividing by 100 moves the digits two places to the right.

Example

Work out   **a** 3.72 ÷ 100   **b** 0.0349 × 10 000   **c** 17.3 ÷ 1000
....
a 3.72 ÷ 100 = 0.0372   Move the digits 2 places right.

b 0.0349 × 10 000 = 349   Move the digits 4 places left.

c 17.3 ÷ 1000 = 0.0173   Move the digits 3 places right.

1   Write these sets of numbers in ascending order.
    a 0.3, 3.1, 1.3, 2, 1, 0.1   **b** 607, 77.2, 27.6, 7.06, 6.07

2   Write these sets of numbers in descending order.
    a 6008, 682.8, 862.6, 6000.8, 8000.6   **b** 47.9, 94.7, 49.7, 79.4, 74.9, 97.4

3   Multiply these numbers by 10.
    a 16.7   **b** 24.8   **c** 0.716   **d** 1.095 **e** 243   **f** 281.3

4   Divide these numbers by 10.
    a 214   **b** 67.3   **c** 4106   **d** 200.7 **e** 6.025   **f** 86

5   Decide which number in each pair is bigger.
    Explain your answers.
    a 4.52 and 4.05   **b** 5.5 and 5.05   **c** 16.8 and 16.75   **d** 16.8 and 16.15

6   Write these sets of numbers in ascending order.
    a 7.83, 7.3, 7.8, 7.08, 7.03, 7.38   **b** 4.2, 8.24, 8.4, 4.18, 2.18, 2.4

7   Write these sets of numbers in descending order.
    a 16.7, 18.16, 16.18, 17.16, 18.7, 17.6   **b** 1.06, 13.145, 1.1, 2.38, 13.2, 2.5

8   Round these numbers to the nearest   **i** 10   **ii** 100.
    a 3048   **b** 1763 **c** 294   **d** 51   **e** 43   **f** 743

9   Round these numbers to the nearest 1000.
    a 2964   **b** 1453 **c** 17   **d** 24 598   **e** 16 344   **f** 167 733

10   Round these numbers to   **i** 1 decimal place   **ii** 2 decimal places.
    a 39.114   **b** 7.068   **c** 5.915   **d** 512.715
    e 4.259   **f** 12.007   **g** 0.833   **h** 26.8813

11   Round these numbers to the nearest
    i tenth   **ii** hundredth   **iii** thousandth.
    a 0.07   **b** 15.9184   **c** 127.9984
    d 887.172   **e** 55.144 55   **f** 0.007 49

12   Calculate
    a 13.06×100 **b** $208.5 \div 100$ **c** 1.085×1000
    d $2487 \div 1000$ **e** $0.008 \div 10$ **f** $0.006\ 19 \times 1000$
    g $45.13 \div 1000$ **h** $0.000\ 045 \times 100$

This spread will show you how to:

- Add and subtract with positive and negative numbers

Keywords
Directed number
Negative
Number line
Positive

A number with a plus or minus sign is a **directed number**.
You can extend the basic rules of addition and subtraction to
include negative numbers.

- Adding a **positive** number counts as addition. Move right along the **number line**.
- Subtracting a positive number counts as subtraction. Move left along the number line.
- Adding a **negative** number counts as subtraction. Move left along the number line.
- Subtracting a negative number counts as addition. Move right along the number line.

For subtracting a
negative number,
think of reducing
an overdraft, or
taking ice cubes
out of a cold
drink.

Example

Calculate **a** $-5 + -6$ **b** $+4 - -2$ **c** $-7 - +2$ **d** $-5 + +8$

a Start at -5 on the number line and
move 6 places to the left.
The answer is -11.

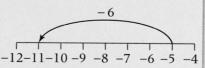

b Start at $+4$ on the number line and
move 2 places to the right. The answer is $+6$.

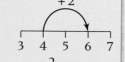

c Start at -7 on the number line and
move 2 places to the left. The answer is -9.

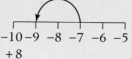

d Start at -5 on the number
line and move 8 places to the
right. The answer is $+3$.

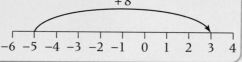

Example

Ben writes: $-5 + -2 = +7$
Is he correct?

> Two minuses make a plus.
> I've got –5 and –2, so the
> answer must be positive.

No.
Using the number line, you start at -5 and move
2 places to the left.
The correct answer is $-5 + -2 = -7$.

Examiner's tip
Avoid simple
rules like 'two
minuses make a
plus', which can
be misleading.
Use the number
line.

1 Calculate
  a $+7 - +9$ **b** $+5 - +6$ **c** $+8 - +10$
  d $-7 + +5$ **e** $-11 + +6$ **f** $-7 + +2$

2 Calculate
  a $-7 + +8$ **b** $-9 + +12$ **c** $-6 + +10$
  d $+3 - -6$ **e** $+5 - -7$ **f** $+2 - -3$

3 Calculate
  a $-9 - -4$ **b** $-8 - -6$ **c** $-5 - -1$
  d $-6 - -8$ **e** $-5 - -9$ **f** $-3 - -10$

4 Copy and complete these calculations by replacing the boxes with
  the correct number or sign.
  a $-3 + \boxed{} = -5$ **b** $+7 \boxed{} -5 = +2$
  c $\boxed{}8 + \boxed{}5 = +3$ **d** $\boxed{}2 + \boxed{}11 = -13$

5 Calculate
  a $+8 - -14$ **b** $-1 + -11$ **c** $-9 - -7$
  d $+3 + -17$ **e** $+8 - -4$ **f** $+13 + -1$
  g $+48 - +29$ **h** $-19 + +4$ **i** $+34 + -23$
  j $-104 + +43$ **k** $+208 - -136$ **l** $+347 + -298$

6 Calculate
  a $-4.5 + -6.3$ **b** $-2.8 - -3.5$ **c** $+5.6 - -7.9$
  d $-9.4 + +8.7$ **e** $-26.5 + -11.7$ **f** $+45.9 - -66.8$

A02 Functional Maths

7 Find the balance in these bank accounts after the transactions
  shown.
  a Opening balance £133.45. Deposits of £45.55 and £63.99,
    followed by withdrawals of £17.50 and £220.
  b Opening balance is −£459.77. Deposit of £650, followed by a
    withdrawal of £17.85.

A negative
number
represents an
overdraft.

8 Find the final temperatures in these science experiments.
  a Starting temperature 55 °C. It goes up 32°, then down 100°.
  b Starting temperature −15 °C. It goes down 28°, increases by
    75°, and then goes down 17°.
  c Starting temperature −22 °C. It goes up 12°, then down 2°,
    then increases by 53°.

Multiplying and dividing negative numbers

This spread will show you how to:

● Multiply and divide with negative numbers

For multiplication and division, these simple rules tell you the sign of the answer when negative numbers are multiplied or divided.

- positive number × positive number = positive number
- positive number × negative number = negative number
- negative number × negative number = positive number

The same rules apply to division.

Example

Calculate   **a** $+4 \times +3$ **b** $-5 \times +4$ **c** $+7 \times -2$
               d -6×-2 **e** $-5 \times +7$

....

     a $+4 \times +3 = +12$ **b** $-5 \times +4 = -20$
     c $+7 \times -2 = -14$ **d** $-6 \times -2 = +12$
     e $-5 \times +7 = -35$

Example

Calculate   **a** $-12 \div +2$ **b** $+50 \div +2$ **c** $+24 \div -8$
               d $-18 \div -3$ **e** $-48 \div +4$

....

     a $-12 \div +2 = -6$ **b** $+50 \div +2 = +25$
     c $+24 \div -8 = -3$ **d** $-18 \div -3 = +6$
     e $-48 \div +4 = -12$

These examples use both rules.

p.16

Example

Calculate   **a** $+4 - +3 \times -2$ **b** $\dfrac{-2 \times +3}{+2 + -4}$

               c $+5 - -2 \times +3$ **d** $2(3 \times -4) \div 4(-5 \times 2)$

....

   a $(+4) - (+3) \times (-2)$ **b** $\dfrac{-2 \times +3}{+2 + -4}$

      $= (+4) - (-6)$ $= \dfrac{-6}{-2}$

      $= +10$ $= +3$

   c $(+5) - (-2) \times (+3)$ **d** $2(3 \times -4) \div 4(-5 \times 2)$

      $= (+5) - (-6)$ $= 2(-12) \div 4(-10)$

      $= +11$ $= \dfrac{-24}{-40}$

                                 $= \dfrac{3}{5}$ Cancel by -8.

Carry out multiplication and division before addition and subtraction.

When multiplying or dividing negative numbers, the combination of the signs gives the sign of the answer.

1 Calculate
  a $+5 \times -3$　　**b** $+2 \times -9$　　　**c** $+7 \times -3$
  d $-8 \times +7$　　**e** $-4 \times +9$　　　**f** $-6 \times +2$

2 Calculate
  a -4×-4　　**b** -2×-8　　　**c** -3×-5
  d -6×-7　　**e** -7×-8　　　**f** -9×-9

3 Calculate
  a $+5 \times -5$　　**b** $+4 \times -8$　　　**c** $-8 \times +9$　　　**d** $-4 \times +5$
  e -3×-10　**f** -7×-7　　　**g** $+8 \times +2$　　　**h** $+5 \times -4$
  i $-2 \times +9$　　**j** -13×-2　　**k** $-7 \times +6$　　　**l** $+12 \times -4$

4 Calculate
  a $-36 \div +12$ **b** $-16 \div +4$　　　**c** $+28 \div -4$　　　**d** $+18 \div -9$
  e $-38 \div -2$ **f** $-80 \div -16$

5 Calculate
  a $-18 \div +9$ **b** $-20 \div +4$　　　**c** $-30 \div -6$　　　**d** $-12 \div -3$
  e $-66 \div +3$ **f** $+47 \div -47$　　　**g** $-80 \div -2$　　　**h** $+24 \div +6$
  i $-45 \div -9$ **j** $-51 \div +3$　　　**k** $+57 \div -19$　　**l** $-81 \div -3$

6 Copy and complete these calculations, replacing the boxes with the
  correct positive or negative number or sign.
  a $-7 \times \boxed{} 8 = -56$　　　　**b** $+48 \div \boxed{} = -8$
  c $\boxed{} \div +45 = +1$　　　　　**d** $+108 \div \boxed{} = -9$

7 Multiply these numbers by 10.
  a $+45$　　　　**b** -15　**c** $+6.3$　　**d** -2.5 **e** -0.073 **f** $+0.0092$

8 Multiply these numbers by -10.
  a $+4.9$　　　　**b** -6.3 **c** -0.377 **d** $+61.97$　　　**e** -14.09　　　**f** -0.009

9 Divide these numbers by $+10$.
  a -360　　　　**b** $+1$　**c** -9.8　　**d** -0.087　　　**e** $+0.073$　　　**f** -0.0006

10 Divide these numbers by -10.
  a $+550$　　　　**b** -4.8 **c** -52.66 **d** $+1560$　　　**e** -0.082　　　**f** $+5.0005$

11 Calculate
  a $+18 \div +100$　　　**b** $+9 \times -3$　　　**c** $-14 \div +2$　　　**d** $-3.8 \times +100$

Approximation and rounding

This spread will show you how to:
- Round numbers to a given power of ten or number of decimal places
- Round numbers to any number of significant figures

Keywords
Decimal
Power
Round
Significant

Numbers are **rounded** when it is not appropriate to give an answer that is too precise.

- Numbers can be rounded:

  to **decimal places**

  to the nearest unit, 10, 100, 1000

  $4.16 = 4.2$ to 1 dp, and
  $5.663 = 5.66$ to 2 dp

  $32\,559 = 33\,000$ to the nearest thousand.

Always check the digit after the one you're rounding to: if it is a 5 or more, round up your final digit.

p.322

- When rounding to **significant figures**, count from the first non-zero digit

  to 2 sf: $712.4 = 710$ and $0.00405 = 0.0041$.

  to 3 sf: $6.339 = 6.34$ and $0.000\,000\,338\,754 = 0.000\,000\,339$.

dp and sf are abbreviations for 'decimal places' and 'significant figures'.

Example

a Round these numbers to 2 dp.

  i 34.567   ii 3.887 126   iii 215.587 54

b Round 323 754.885 to the nearest:

  i unit   ii 10   iii 100   iv 1000 v 10 000

c Round these numbers to 2 sf.

  i 39.54   ii 217   iii 0.000 455   iv 12 019 v 25.505

. . . .

a i 34.57   ii 3.89   iii 215.59
b i 323 755 ii 323 750   iii 323 800   iv 324 000
    v 320 000
c i 40   ii 220 iii 0.000 46   iv 12 000 v 26

Unit, 10, 100, ... are **powers** of ten.

You should always round the original value.

Example

Round 3.447 to   a 2 dp   b 1 dp.

. . . .

a $3.447 = 3.45$ to 2 dp   b $3.447 = 3.4$ to 1 dp

Do not round the 2 dp answer to get the 1 dp answer: use the original value 3.447.

Example

Find approximate answers to:

a $12.3 - 8.9$ b $76.5 + 184.2$ c $20 - 14.53$

. . . .

a $12.3 - 8.9 \approx 12 - 9 = 3$
b $76.5 + 184.2 \approx 80 + 200 = 280$
c $20 - 14.53 \approx 20 - 15 = 5$

Rounding to 1 sf is a useful way of finding a quick approximate answer to a calculation.

1 Round these numbers to the nearest 10.
   a 28   b 32   c 50   d 209 e 776   f 23 775

2 Round these decimal numbers to the nearest whole number.
   a 5.8   b 4.4   c 21.67   d 39.175
   e 18.405   f 453.66

3 Round these numbers to the nearest 100.
   a 205   b 173   c 52   d 734   e 1389   f 134 545

4 Round these numbers to the nearest 1000.
   a 2239   b 12 563   c 7500   d 11 452
   e 78 466   f 155 669

5 Round these numbers to one decimal place (nearest tenth).
   a 0.31   b 0.73   c 0.25   d 0.205 e 4.55   f 105.449

6 Round these whole numbers to two decimal places (nearest hundredth).
   a 0.317   b 0.455   c 15.304   d 104.675
   e 16.445   f 0.0036

7 Round these whole numbers to two significant figures.
   a 483   b 1206   c 488   d 13 562
   e 533   f 14 511

8 Round these numbers to two significant figures.
   a 0.355   b 0.421   c 0.0566   d 0.004 673
   e 1.357   f 0.000 004 152

9 Round these numbers to one significant figure.
   a 157   b 2488   c 4.66   d 13.77
   e 0.000 453 f 121 450

10 Use a calculator to work these out.
   Write your answers correct to two significant figures.
   a $8 \div 13$ b $4 \div 7$ c $5 \div 9$ d 24×16
   e 7.8×71 f 2093×3493

11 By rounding all of the numbers to one significant figure,
   write a calculation that you could carry out mentally to
   estimate the answers to these calculations.
   a $355 \div 21$ b 39×43 c $1053 \div 92$
   d $4385 + 11\ 655$ e $108 + (2360 \div 52)$

12 Use mental calculations to work out the value of each
   of the estimates that you wrote for question **11**.

13 Use a calculator to work out an exact answer for each of the
   calculations from question **11**.
   For each one, write a sentence to say how well the calculator result
   agrees with the estimated answer that you wrote in question **12**.

Mental calculations

This spread will show you how to:

p.222

- Use mental and written methods to multiply and divide with decimals
- Understand place value and where to place the decimal point

Keywords
Inverse
Place value

- Multiplying a positive number by a number between 0 and 1 makes it smaller.

  $6 \times 0.5 = 3$

- Dividing a positive number by a number between 0 and 1 makes it bigger.

  $6 \div 0.5 = 12$

Always start calculations with an estimate.

- Work out a mental calculation using the significant digits from the question; for example, for $12.5 \div 0.05$, work out $125 \div 5$.
- Finally, use your initial estimate to check your answer and adjust the **place value**.
- Use **inverses** where possible.
  $10 \div 0.2 = 10 \times 5$ and $10 \times 0.25 = 10 \div 4$

$12.5 \div 0.05$
$\approx 10 \div 0.05$
$= 200$

$125 \div 5 = 25$

$12.5 \div 0.05$
$= 250$

Example

Use a mental method to work out
a 320×0.4 **b** $320 \div 0.4$ **c** $3.2 \div 0.4$

....

a $320 \times 0.4 \approx 300 \times 0.4$
$= 120$
$32 \times 4 = 64 \times 2$
$= 128$
$320 \times 0.4 = 128$

b $320 \div 0.4 \approx 320 \div \frac{1}{2} = 640$
$32 \div 4 = 8$ so
$320 \div 0.4 = 800$
c $3.2 \div 0.4 \approx 3 \div \frac{1}{2} = 6$
$32 \div 4 = 8$ so
$3.2 \div 0.4 = 8$

Estimate first.

Compare with estimate to get final answer.

Example

Write a multiplication that is equivalent to $34.5 \div 0.25$.

....

Dividing by a quarter is the same as multiplying by 4.
$34.5 \div 0.25 = 34.5 \times 4$

$34.5 \div 0.25 = 138$
$34.5 \times 4 = 138$

Example

Calculate mentally   **a** $72.5 \div 0.05$ **b** 340×0.3 **c** $8.46 \div 0.2$

....

a $72.5 \div 0.05 \approx 70 \times 200 = 1400$
    $725 \div 5 = 145$ so
    $72.5 \div 0.05 = 1450$

b $340 \times 0.3 \approx 300 \div 3 = 100$
    $340 \times 3 = 1020$ so
    $340 \times 0.3 = 102$

c $8.46 \div 0.2 \approx 9 \times 5 = 45$
    $8.46 \div 2 = 4.23$ so
    $8.46 \div 0.2 = 42.3$

$0.05 = \frac{1}{200}$

$0.3 \approx \frac{1}{3}$

$0.2 = \frac{1}{5}$

1 Use a mental method to work out
   a 5×0.2　**b** 4×0.3　**c** 0.5×3　**d** 16×0.5
   e 7×0.2　**f** 30×0.25　**g** 40×0.4　**h** 0.6×25

2 Use a mental method to work out
   a $8 \div 0.2$　**b** $4 \div 0.4$　**c** $6 \div 0.3$　**d** $32 \div 0.4$
   e $0.8 \div 4$　**f** $0.3 \div 0.03$　**g** $0.4 \div 0.04$　**h** $50 \div 0.01$

3 Use a mental method to work out these calculations. Start with an
   estimate, and show your working.
   a 2×0.4　**b** 20×0.04　**c** 3×7　**d** 0.3×0.7
   e 12×0.4　**f** $12 \div 0.3$　**g** $3.6 \div 4$　**h** $3.6 \div 0.9$

4 Use a calculator to check your answers to question **3**.

5 Write a division that is equivalent to each of these multiplications.
   a 4×0.5　**b** 6×0.2　**c** 12×0.2　**d** 2×0.001

6 Write a multiplication that is equivalent to each of these divisions.
   a $4 \div 0.5$　**b** $6 \div 0.25$　**c** $16 \div 0.01$　**d** $15 \div 0.05$

7 Use a mental method to find these calculations. Show your method.
   a $8 \div 0.25$　**b** $15 \div 0.5$　**c** $7 \div 0.2$　**d** 4×0.25
   e 16×0.02　**f** $24 \div 0.12$　**g** 20×0.05　**h** $18 \div 0.025$
   i 3×0.125　**j** $7 \div 0.25$　**k** 40×0.025　**l** 8×0.875

8 Use a mental method to find an estimate for each of these
   calculations.
   a $37 \div 0.47$　**b** $319 \div 0.3$　**c** 3.8×134　**d** $17 \div 0.031$

9 Use a calculator to find the exact answers to question **8**.

10 Use a mental method to work out these calculations. Start each one
   with an estimate, and show your method.
   a 31×0.3　**b** $49 \div 0.07$　**c** $3.66 \div 0.3$　**d** $4.24 \div 0.4$
   e $13.9 \div 0.03$ **f** $3.9 \div 0.03$　**g** $171 \div 0.3$　**h** 5.2×0.125

11 Use a calculator to check your answers to question **10**.

This spread will show you how to:
- Use mental and written methods to multiply and divide with decimals
- Estimate answers to calculations, using these to check the solution

Keywords
Check
Estimate
Place value

For column methods of addition and subtraction, use the full decimal numbers as given in the question.

Use the decimal point to ensure that the columns are aligned correctly.

Example

Calculate
a $135.23 + 27.8$ **b** $34.56 - 18.729$

p.218

a Estimate $140 + 30 = 170$

```
  1 3 5 . 2 3
+     2 7 . 8
  1 6 3 . 0 3
        1 1
```

b Estimate $30 - 20 = 10$

```
  ²3̶ ¹³4̶ . ¹5 ⁵6̶ ¹0
- 1  8 . 7  2  9
  1  5 . 8  3  1
```

Use your estimates to **check** your answers.

For standard methods of multiplication and division, work with the significant digits from the numbers in the question.

Use an estimate to adjust the place value correctly.

Example

Calculate
a 18.5×7.9 **b** $47.592 \div 1.8$

p.222

a Estimate $20 \times 8 = 160$

```
        1 8 5
      ×   7 9
    1 6₇6₄5
  1 2₅9₃5 0
  1 4₁6₁1 5
```

b Estimate $50 \div 2 = 25$
This could be done by long division or by short division. Using short division you get:

```
        2  6  4  4
18)4 7¹¹5 ⁷9 ⁷2
```

$20 \times 8 = 160$

$50 \div 2 = 25$

Use your **estimate** to check your answer.

$18.5 \times 7.9 = 146.15$

146.15 is close to 160, so this is about right.

Use your estimate to adjust the **place values**.

So $47.592 \div 1.8 = 26.44$

You should show your working for the questions in the exercise.

1 Use a written method to calculate
   a $24.72 - 14.04$ **b** $1.52 - 1.09$
   c $6.149 - 2.052$ **d** $6.64 - 15.88$

2 Use a written method to calculate
   a $5.23 - 3.11$ **b** $17.45 - 13.26$
   c $6.41 - 4.37$ **d** $23.6 - 17.9$

3 Use a written method to calculate
   a $1.09 + 154$ **b** $0.09 + 0.36$
   c $14.52 + 9.8$ **d** $13.92 + 0.8$

4 Use a written method to calculate
   a $4.5 - 0.53$ **b** $3.085 - 2.99$
   c $16.3 - 3.86$ **d** $112.14 - 53.8$

5 Use a written method to calculate
   a $11.1 - 8.29$ **b** $2.09 - 1.333$
   c $102.8 - 14.79$ **d** $978 + 148.72$

6 Use a calculator to check your answers to questions **1** to **5**.

7 Use a written method to work out these multiplications.
   a 15.9×4 **b** 17.9×0.3
   c 16.9×0.8 **d** 0.048×0.07

8 Calculate these divisions, using a written method.
   a $1.36 \div 0.8$ **b** $3.01 \div 7$
   c $19.2 \div 0.4$ **d** $13.45 \div 0.05$

9 Use long multiplication (or an equivalent written method) to evaluate
   a 8.8×1.9 **b** 190×0.054
   c 189×4.2 **d** 214×0.037

10 Use long division (or an equivalent written method) to evaluate
   a $211.68 \div 24$ **b** $133.98 \div 0.66$
   c $292.38 \div 0.33$ **d** $5.913 \div 0.27$

11 Use a calculator to check your answers to questions **7** to **10**.

This spread will show you how to:

- Perform the operations within a calculation in the correct order
- Estimate answers to calculations, using these to check the solution
- Round numbers to sensible degrees of accuracy

Keywords
BIDMAS
Operation
Order

When a calculation involves a number of steps, or **operations**, you need to do them in the right order.

The order in which operations are carried out is:

- Brackets – start by working out the contents of any brackets
- Powers (indices) – for example, squares, cubes or square roots – come next
- Multiplication and division are done next
- Addition and subtraction are done last.

$(3 + 2) \times 4^2 - 6$
$= 5 \times 4^2 - 6$
$= 5 \times 16 - 6$
$= 80 - 6$
$= 74$

BIDMAS (**B**rackets, **I**ndices or powers, **D**ivision, **M**ultiplication, **A**ddition, **S**ubtraction) will help you to remember this.

Example

Evaluate **a** $4 + 3 \times 2$ **b** $5 + 3^2$ **c** $\sqrt{(5 + 4 \times 11)}$ **d** $\sqrt{(5^2 - 4^2)}$

a $4 + 3 \times 2 = 4 + 6 = 10$
b $5 + 3^2 = 5 + 9 = 14$
c $\sqrt{(5 + 4 \times 11)} = \sqrt{(5 + 44)} = \sqrt{49} = 7$
d $\sqrt{(5^2 - 4^2)} = \sqrt{(25 - 16)} = \sqrt{9} = 3$

Examiner's tip
Using brackets in parts **c** and **d** shows that **all** the values are contained in the square root.

Example

Estimate the answer to $\dfrac{6.3 + \sqrt{9.7^2 - 17}}{149}$
Round all the numbers to a sensible amount.

$\dfrac{6.3 + \sqrt{9.7^2 - 17}}{149} \approx \dfrac{6 + \sqrt{10^2 - 20}}{150} = \dfrac{6 + \sqrt{80}}{150} \approx \dfrac{6 + 9}{150} = \dfrac{15}{150} = 0.1$

Deal with the numerator and denominator separately.

Example

Adam explained how he would calculate $\dfrac{5 + \sqrt{9}}{11}$
What is the problem here?

There is a root, an addition and a division. I'll do the root first, then divide, and then add.

Even though there are no brackets in this expression, the whole of the 'top line' is divided by 11, so you need to find the square root, then add, then divide. The expression could be written as $(5 + \sqrt{9}) \div 11$.

1 Evaluate
a $5 + 6 \times 9$ **b** $4 \times 9 + 1$ **c** $8 - 3 - 3$ **d** $4 \times 3 + 7 \times 2$

2 Evaluate
a $9 - 4 \times 2$ **b** $(9 - 4) \times 2$ **c** $5 \times 7 + 4 \times 2$ **d** $5 \times (7 + 4) \times 2$

3 Copy and complete these equations, replacing the ● with the correct operation.
a $5 \bullet 3 \times 7 = 26$ **b** $4 \times 6 \bullet 2 = 22$
c $4 \bullet 7 + 1 = 29$ **d** $17 \bullet 2 + 2 = 6^2$

4 Copy these calculations, inserting brackets to make the answers correct.
a $11 - 1 \times 5 = 50$ **b** $12 + 3 \div 3 = 5$
c $12 - 4 - 1 = 9$ **d** $8 \div 4 + 4 + 1 = 2$

5 Copy and complete these equations, replacing the ☐ with the correct number.
a $(\square + 2) \times 9 = 36$ **b** $64 \div (\square + 3) = 8$
c $\sqrt{(\square - 10)} = 5 \times 2$ **d** $\sqrt{\square - (5 \times 2)} = 2$

6 Find the values of these expressions.
a $(5^2 + 3) \times 7$ **b** $(9 - 7)^2$
c $(5 - 3) \times (4^2 - 7)$ **d** $(5^2 - 8)^2$

7 Calculate the values of these expressions.
a $(4 + 7)^2$ **b** $(6 + 7) \times 9 \div 3$
c $\dfrac{6 + (5^2 - 13)}{4}$ **d** $\sqrt{100 - 2 \times 6^2}$

8 Find the values of these expressions.
a $\frac{28}{4} + \sqrt{100 - (9^2 + 5 \times 2)}$ **b** $\sqrt{28 + 4^2 - (10 - 2)} + 4 \times 3$

9 Estimate the value of each of these expressions without using a calculator. Show all of your working.
a $\dfrac{5.2 \times (4.8^2 - 12)}{3.9}$ **b** $73 \times \dfrac{202 - 11}{38 \times 5} + 29$
c $18.4^2 - \dfrac{592}{11.4}$ **d** $\sqrt{26.4 \times \dfrac{12.5 + 7.4}{5.36}}$

Summary

Check out

You should now be able to:

- Multiply or divide any number by powers of 10
- Round numbers to a given power of 10, number of decimal places and significant figures
- Multiply and divide negative numbers
- Check and estimate answers to calculations
- Use the correct order of operations, including brackets, in a calculation

Worked exam question

Work out an estimate for $\dfrac{412 \times 5.904}{0.195}$

(3)

(Edexcel Limited 2006)

$412 \longrightarrow 400$

$5.904 \longrightarrow 6$

$0.195 \longrightarrow 0.2$

> Write an approximation for each number.

$$\frac{412 \times 5.904}{0.195} \approx \frac{400 \times 6}{0.2} = \frac{2400}{0.2} = 12000$$

OR

$$\frac{412 \times 5.904}{0.195} \approx \frac{400 \times 6}{0.2} = 2000 \times 6 = 12000$$

OR

> You should show your working whichever method you choose.

$$\frac{412 \times 5.904}{0.195} \approx \frac{400 \times 6}{0.2} = 400 \times 6 \times 5 = 12000$$

Exam questions

1 Work out an estimate for $\dfrac{497 \times 10.05}{24.8}$ (3)

2 Work out an estimate for $\dfrac{805 \times 11.03}{0.39}$ (3)

3 Work out an estimate for

   a $\sqrt{6.8^2 + 29.4}$ (3)

   b $\dfrac{21.7^2 - 186}{190 \times 21.3}$ (3)

4 Isaac says $\sqrt{20^2 - 16^2}$ is 4
   Pam says $\sqrt{20^2 - 16^2}$ is 12

   a Who is right?
   Give a reason for your answer. (2)

   b Work out $\dfrac{\sqrt{(120.1 + 0.9)}}{4}$ (1)

5 Find the value of

   a $\dfrac{-14 \times -3}{-14 + -3}$ (1)

   b $\dfrac{4 \times -3}{4 - 7}$ (1)

   c $\dfrac{-5 \times (-2 + 8)}{-3 \times 4}$ (1)

Area and volume

In modern manufacturing and architecture designers uses computer aided design (CAD) packages to create virtual, 3-dimensional models of their products. They build up complex shapes and surfaces by putting together simpler shapes like triangles and cuboids. After developing their design the computer can analyse the structure and produce detailed plans.

What's the point?

Since complex shapes can be broken down into simpler shapes, understanding the properties of a few basic shapes allows a mathematician to work out the properties of any shape.

You should be able to

■ **Find simple areas and perimeters**

1 These shapes are drawn on a centimetre square grid.

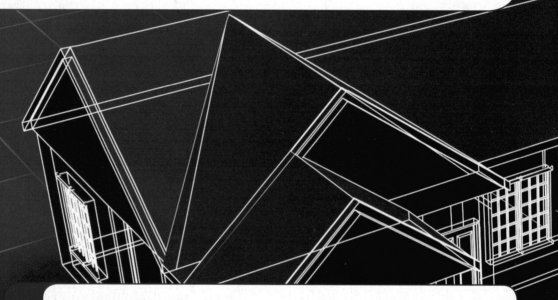

    a   b   c

    For shape **a–c**
        write the length of each side in cm.
        Hence find the perimeter in **i** cm, **ii** mm.
    ii calculate the area of the shape in cm^2.

■ **Draw nets and find surface areas and volumes**

2 **a** Draw a net of this cube
   b Hence calculate to surface area
   c Calculate its volume.

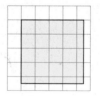
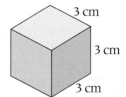

Orientation

What I need to know	What I will learn	What this leads to
KS3 Draw the net of a 3D shape	■ Find the area of triangles and quadrilaterals ■ Find the volume and surface area of prisms ■ Use metric and compound measures	**G5** Areas and volumes of more complex shapes **N3** Use compound measures Design, Engineering

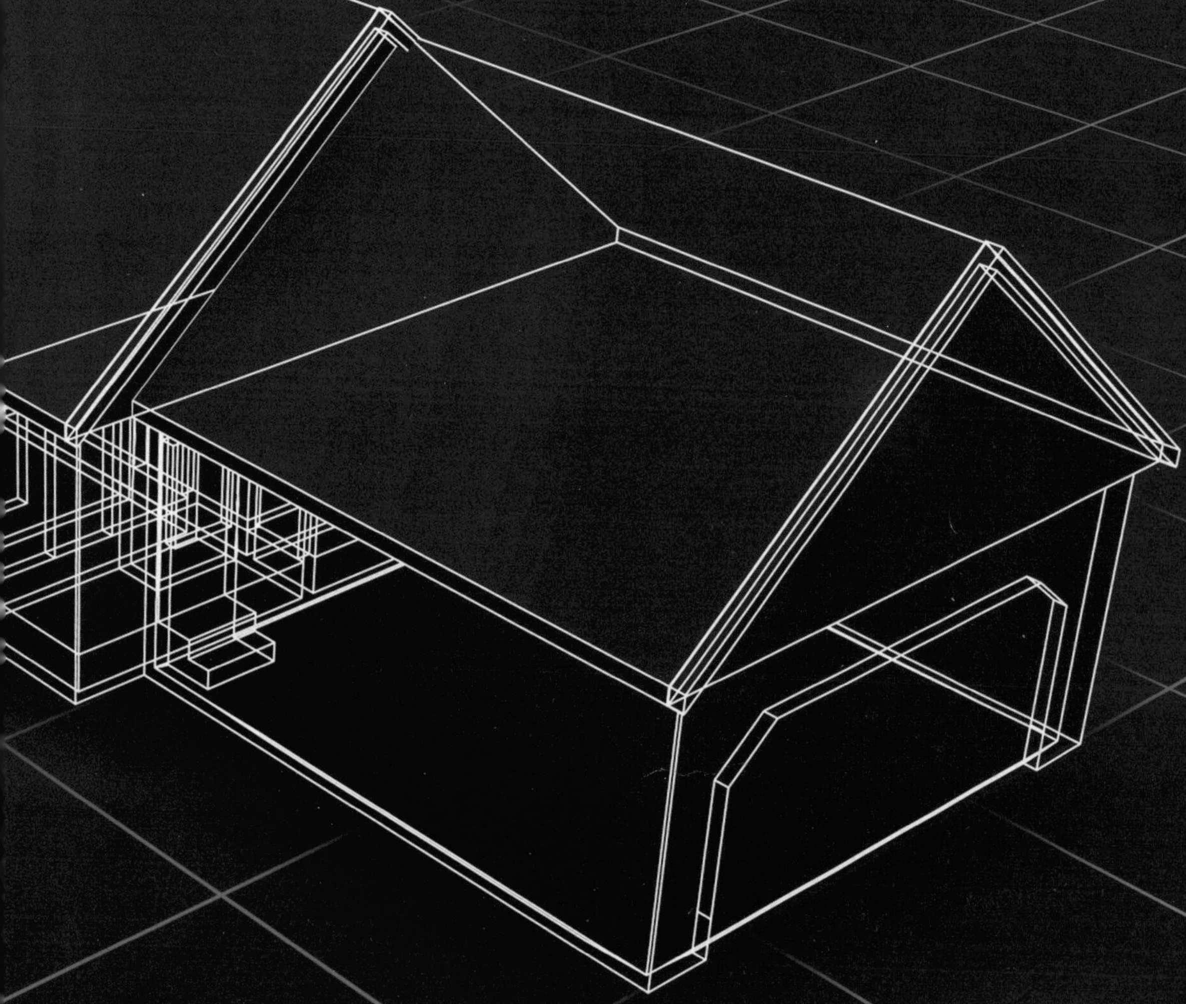

Rich task

You will need some square dotty paper for this investigation

This diagram shows a quadrilateral with an area of 4 cm^2. There are 6 dots on the perimeter of the quadrilateral and 2 dots inside the perimeter.

Investigate.

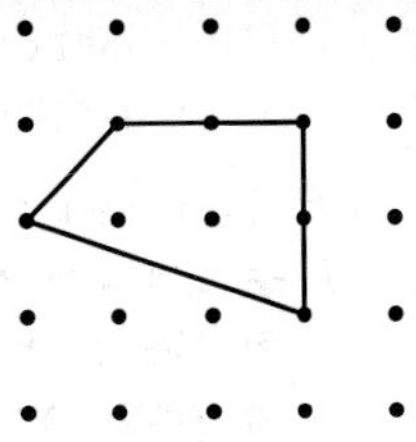

This spread will show you how to:
- Calculate the perimeter and area of shapes made from rectangles and triangles

The **area** of a shape is the amount of space it covers.
- You can use formulae to find the areas of **rectangles** and **triangles**.

Rectangle

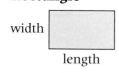

width

length

- Area = length × width

The area of the triangle is half the rectangle.

Triangle

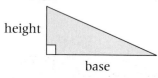

height

base

- Area = $\frac{1}{2}$ × base × height

The height of a triangle is always at right angles to the base.

height

base

Example

Find the area of each shape.

a   9 cm

2 cm

b   50 mm

22 mm

a Area = length × width
= 9 × 2 = 18 cm²

b Area = $\frac{1}{2}$ base × height
= $\frac{1}{2}$ × 50 × 22 = 550 mm²

Remember to write the units. The ² shows the measurement is an area.

- You can split compound shapes into rectangles and triangles.

Example

Find   **a** the perimeter
b the area of this shape.

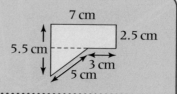

7 cm

2.5 cm

5.5 cm

3 cm

5 cm

a Perimeter = 5.5 + 7 + 2.5 + 3 + 5
= 23 cm
b Area = area of rectangle + area of triangle
= (7 × 2.5) + ($\frac{1}{2}$ × 4 × 3)
= 17.5 + 6
= 23.5 cm²

Triangle:
height = 7 − 3 = 4 cm
base = 5.5 − 2.5 = 3 cm

height

base

Perimeter is the distance around a shape.

1 Calculate the areas of these rectangles.

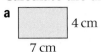
a 4 cm
7 cm

b 5.3 cm
4.2 cm

c 8.7 cm
3 cm

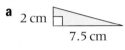
d 62 mm
120 mm

e 49 mm
210 mm

2 Find the areas of these triangles.

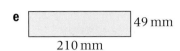
a 2 cm
7.5 cm

b 4.2 cm
5.6 cm

c 12 mm
21 mm

d 4 cm
3 cm

e 3.5 cm
8 cm

f 2.4 cm
6 cm

g 3 cm
7.2 cm

3 Split these shapes into rectangles and triangles to work out
i the perimeter   **ii** the area.

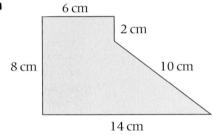
a 6 cm
2 cm
8 cm
10 cm
14 cm

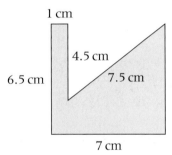
b 1 cm
4.5 cm
6.5 cm
7.5 cm
7 cm

c 4 cm   12 cm
2 cm
5 cm
5 cm   13 cm

4 Pete is making a mobile out of shapes
like this:
He cuts the shape out of a piece of card
that is 30 cm × 20 cm.
What is the area of the card left over?

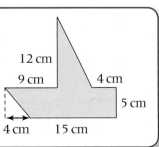
12 cm
9 cm   4 cm
5 cm
4 cm   15 cm

Area of a parallelogram and a trapezium

This spread will show you how to:

- Calculate the area of parallelograms and trapeziums

Keywords
Area
Parallelogram
Trapezium

You can use the formula for the **area** of a rectangle to find the formula for the area of a **parallelogram**.

This parallelogram can be made into a rectangle.

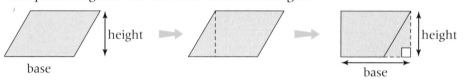

You can cut a triangle from one side and stick it on the other.

The height is at right angles to the base.

- Area of parallelogram = base × height.

Two congruent **trapeziums** make a parallelogram.

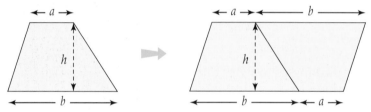

Area of the parallelogram = base × height
$$= (a + b) \times h$$

- Area of trapezium = $\frac{1}{2}(a + b) \times h$.

The area of a trapezium is half the sum of the parallel sides times the distance-between them.

Example

Find the areas of these shapes.

a 3 cm
7 cm

b 4 cm
3 cm
6 cm

a Area of parallelogram
$$= \text{base} \times \text{height}$$
$$= 7 \times 3$$
$$= 21 \, \text{cm}^2$$

b Area of trapezium
$$= \frac{1}{2} \times (a + b) \times h$$
$$= \frac{1}{2} \times (4 + 6) \times 3$$
$$= 5 \times 3$$
$$= 15 \, \text{cm}^2$$

1 Find the area of each parallelogram.

a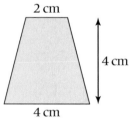
6 cm
2.5 cm

b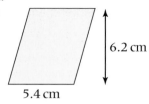
5.4 cm
6.2 cm

c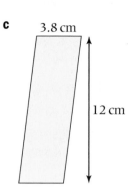
3.8 cm
12 cm

d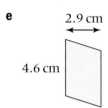
4.2 cm
7.5 cm

e
2.9 cm
4.6 cm

2 Find the area of each trapezium.

a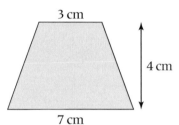
2 cm
4 cm
4 cm

b
3 cm
4 cm
7 cm

c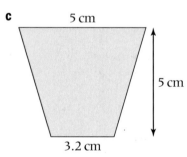
5 cm
5 cm
3.2 cm

d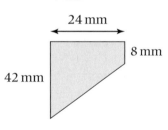
24 mm
8 mm
42 mm

e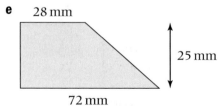
28 mm
25 mm
72 mm

DID YOU KNOW?

The word 'trapezium' originates from the shape made by the ropes and bar of an old-fashioned flying trapeze.

3 Caroline has drawn a sandcastle.
What is the area of her castle and flag?
Start by dividing the shape into parts.

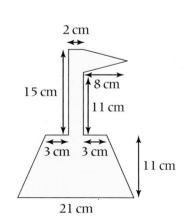
2 cm
15 cm
8 cm
11 cm
3 cm   3 cm
11 cm
21 cm

This spread will show you how to:

- Analyse 3-D shapes through 2-D projections and cross-sections, including plans and elevations

Keywords
Elevation
Plan
Projection
2-D
3-D

- A **plan** of a solid is the view from directly overhead (bird's eye view).

- An **elevation** is the view from the front or the side of the solid.

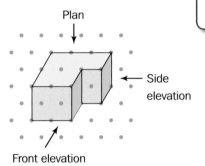

You draw solids on isometric paper.

Plans and elevations are **projections** of a **3-D** solid onto a **2-D** surface.

Example

For this solid, draw **a** the plan **b** the front **c** side elevation.

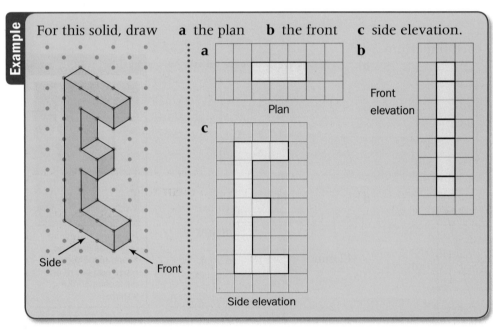

You draw plans and elevations on squared paper.

Example

Here are the plan and front elevation of a prism.
The front elevation shows the cross-section of the prism.

A prism has the same cross-section throughout its length.

Draw a 3-D sketch of the prism.

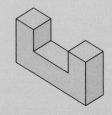

You do not need isometric paper for a sketch.

1 For these solids, draw
  i the plan
  ii the front elevation from the direction marked with an arrow.

a

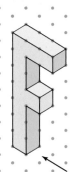

b

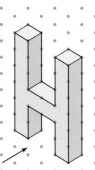

c

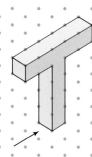

d

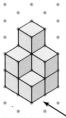

e

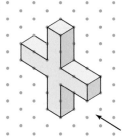

f

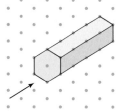

A03 Problem

2 The diagrams show the plan and the front elevation of different solids.
  Draw a sketch of each solid.

The numbers on the plan tell you the number of cubes in each column.

a

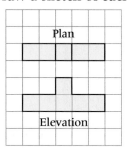

b

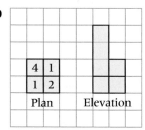

4	1
1	2
Plan   Elevation

c

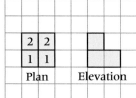

2	2
1	1
Plan   Elevation

d

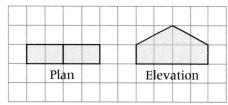

Plan   Elevation

e

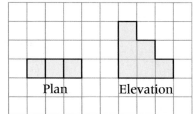

Plan   Elevation

f

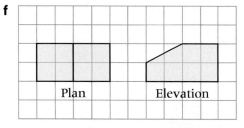

Plan   Elevation

This spread will show you how to:

● Find the surface area of simple 3-D shapes

Keywords
Cuboid
Prism
Surface area

● **Surface area** is the total area of all the faces of a 3-D shape.

To find the surface area, first imagine the net of the shape.

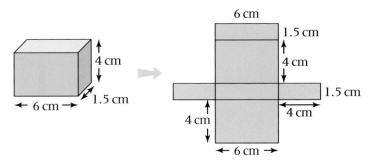

The faces of this **cuboid** are in pairs.

Front/back:	Area = $6 \times 4 = 24$
Side:	Area = $4 \times 1.5 = 6$
Top/bottom:	Area = $6 \times 1.5 = 9$

$$\begin{aligned} \text{Surface area} \ &= 2 \times (24 + 6 + 9) \\ &= 2 \times 39 \\ &= 78\,\text{cm}^2 \end{aligned}$$

Example

Find the surface area of this triangular **prism**.

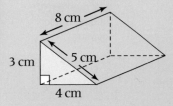

The two end faces of a prism are identical.

Triangle: Area = $\frac{1}{2} \times 4 \times 3 = 6$

Triangle: Area = $\frac{1}{2} \times 4 \times 3 = 6$

Side: Area = $3 \times 8 = 24$

Bottom: Area = $4 \times 8 = 32$

Sloping
side: Area = $5 \times 8 = 40$

Surface area
$= 6 + 6 + 24 + 32 + 40 = 108\,\text{cm}^2$

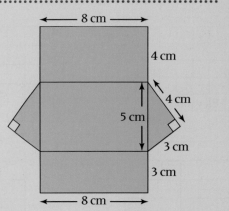

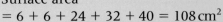

1 Work out the surface areas of these cuboids.

Give your answers to 1 dp.

a
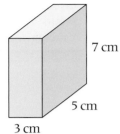
7 cm
5 cm
3 cm

b
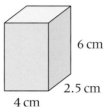
6 cm
2.5 cm
4 cm

c
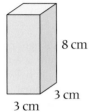
8 cm
3 cm
3 cm

d
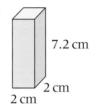
7.2 cm
2 cm
2 cm

e
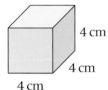
4 cm
4 cm
4 cm

f
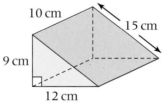
2 mm
9 mm
3 mm

2 Work out the surface areas of these prisms.

a
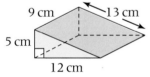
9 cm
13 cm
5 cm
12 cm

b
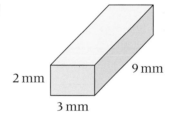
10 cm
15 cm
9 cm
12 cm

3 A scout troop is making a tent out of canvas and a groundsheet out of PVC.
   a What area of PVC do they need for the groundsheet?
   b What area of canvas do they need for the cover (including front and back flaps)?

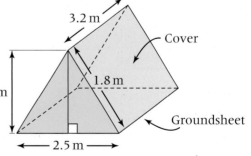
3.2 m
Cover
1.8 m
1.3 m
Groundsheet
2.5 m

This spread will show you how to:

- Calculate volumes of right prisms

Keywords
Cross-section
Prism
Volume

- A **prism** is an object with constant **cross-section**.

- **Volume** of a prism = area of cross-section × length.
  = A × l

In a right prism there is a right angle between the-length and the-base.

Example

Work out the volume of this cuboid.

4 cm

1.5 cm

7 cm

··

Area of cross-section = 4 × 1.5
= 6 cm²
Volume = 6 × 7
= 42 cm³

Area of triangle = $\frac{1}{2}bh$.

Work out the volume of this prism.

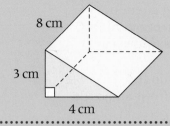

8 cm

3 cm

4 cm

··

Area of cross-section = $\frac{1}{2}$ × 4 × 3
= 6 cm²
Volume = 6 × 8
= 48 cm³

1 Find the volume of each cuboid.

a
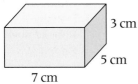
3 cm
5 cm
7 cm

b
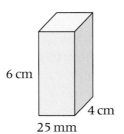
6 cm
4 cm
25 mm

c
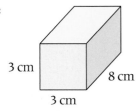
3 cm
8 cm
3 cm

d
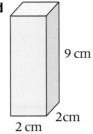
9 cm
2 cm
2 cm

e
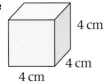
4 cm
4 cm
4 cm

f
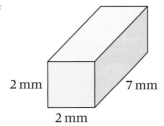
2 mm
7 mm
2 mm

2 Find the volume of each prism.

a
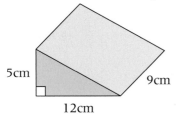
5 cm
9 cm
12 cm

b
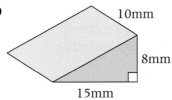
10 mm
8 mm
15 mm

Problem
A03

3 Find the volume of this shape

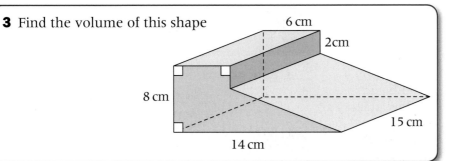
6 cm
2 cm
8 cm
14 cm
15 cm

Compound measures 1

This spread will show you how to:

● Understand and use compound measures, including speed and density

p.118

Keywords
Capacity
Density
Rate of flow
Speed

Compound measures describe one quantity in relation to another. These are examples of compound measures.

● **Speed** $= \dfrac{\text{total distance travelled}}{\text{total time taken}}$ Units such as m/s; km/h

● **Density** $= \dfrac{\text{mass}}{\text{volume}}$ Units such as g/cm³

● Population density $= \dfrac{\text{population}}{\text{area}}$ Units such as number of people/km²

Use the triangle to work out which calculation to use.

Cover D (for distance)
You multiply
S(speed) × T(time)

Example

Kerry jogs at an average speed of 5 km/h for $1\frac{1}{2}$ hours. What distance does she jog?

Distance $= 5 \times 1\frac{1}{2}$
$= 7.5\,\text{km}$

Example

Find the density of a piece of wood with cross-section area 42 cm², length 12 cm and mass 693 g.

Volume $= 42 \times 12 = 504\,\text{cm}^3$
Density $= 693 \div 504$
$= 1.375\,\text{g/cm}^3$

Density $= \dfrac{\text{mass}}{\text{volume}}$

Mass in grams
Volume in cm³
So Density in g/cm³.

● **Capacity** is the volume of liquid that a container can hold. Metric units of capacity are litre, centilitre, millilitre.

Rate of flow is a compound measure. It is the volume of liquid that passes through a container in a unit of time.

● **Rate of flow** $= \dfrac{\text{volume}}{\text{time}}$ Units such as litres/s

Example

a 12 litres of water flows from a hosepipe in 15 seconds. What is the rate of flow in litres/s?

b Sand was falling from the back of a lorry at a rate of 0.4 kg/s. It took 20 minutes for all the sand to fall from the lorry. How much sand was the lorry carrying?

a Rate of flow $= \dfrac{\text{volume}}{\text{time}}$
$= 12 \div 15$
$= 0.8$ litres/s

b 20 minutes $= 20 \times 60$ s
$= 1200$ s
Rate $= \dfrac{m}{t}$
$0.4 = \dfrac{m}{1200}$
$m = 1200 \times 0.4 = 480\,\text{kg}$

Here sand is flowing, not a liquid.
Rate of flow $= \dfrac{\text{mass}}{\text{time}}$
Units kg/s
You could use the triangle:

1 Find the rate of flow for pipes A and B in litres/s.
  a Pipe A: 20 litres of water in 8 seconds.
  b Pipe B: 48 litres of water in 30 seconds.

2 Water empties from a tank at a rate of 2 litres/s.
It takes 10-minutes to empty the tank.
How much water was in the tank?

3 An engine uses oil at a rate of 0.3 ml/km.
How much oil will it use on a journey of
  a 100 km   **b** 80 km   **c** 42 km?

4 A car travelled at an average speed of 48 km/h.
  a How far did it travel in
    i 2 hours   **ii** 15 minutes   **iii** 20 minutes?
  b How long did it take to travel
    i 144 km   **ii** 72 km   **iii** 8 km?

5 The table shows information about some journeys Shaun made in one week.
Copy and complete the table.
Remember to show the units.

Distance	Time taken	Average speed
120 km	$1\frac{1}{2}$ hours	
250 miles	4 hours	
4 km		16 km/hour
	24 seconds	5 m/s
300 m	15 seconds	
0.4 km	160 seconds	
3 km		24 m/s
	20 minutes	60 km/h

6 The table shows the densities of different metals.
Use the information in the table to find
  a the mass of 0.8 m³ of zinc
  b the mass of 0.5 m³ of cast iron
  c the mass of 3.2 m³ of gold
  d the volume of 910 g of tin
  e the volume of 220 g of nickel
  f the volume of a brass statue that has mass 17 kg.

Metal	Density
Zinc	7130 kg/m³
Cast iron	6800 kg/m³
Gold	19 320 kg/m³
Tin	7280 kg/m³
Nickel	88 kg/m³
Brass	8500 kg/m³

7 In this question, give your answers in kg/m³.
  a The volume of 24 g of silver is 3 cm³.
    Work out the density of silver.
  b The volume of 18 g of titanium is 4 cm³.
    Work out the density of titanium.
  c A sheet of aluminium foil has volume 0.4 cm³ and
    mass 1.08 g. Work out the density of aluminium foil.

33

Summary

Check out

You should now be able to:

- Calculate the perimeter and area of shapes made from rectangles and triangles
- Calculate the area of a parallelogram and a trapezium
- Use 2-D representations of 3-D shapes, including plans and elevations
- Calculate the surface area and volume of right prisms
- Understand and use compound measures, including speed and density

Worked exam question

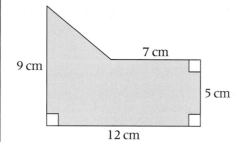

Diagram NOT accurately drawn.

Work out the area of the shape.

(4)

(Edexcel Limited 2008)

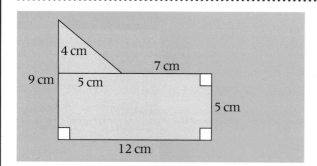

Draw the extra line on the diagram and write the extra measurements on the diagram.

Area of triangle $= \frac{1}{2} \times 5 \times 4$
$= 10 \text{ cm}^2$

Area of rectangle $= 12 \times 5$
$= 60 \text{ cm}^2$

Area of whole shape $= 10 + 60$
$= 70 \text{ cm}^2$

Show the working out for the area of each shape.

Exam questions

1 Here are the front elevation, side elevation and the plan of a 3-D shape.

Front elevation   Side elevation

Plan

Draw a sketch of the 3-D shape.

(2)

(Edexcel Limited 2008)

2

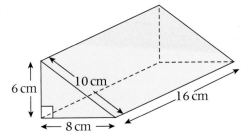

Diagram NOT
accurately drawn.

6 cm   10 cm   16 cm   8 cm

Work out the total surface area of the triangular prism.
Give the units with your answer.

(4)

A02

3 A plane flies 1400 kilometres in 2 hours 20 minutes.
Calculate the average speed, in km/h, of the plane.

(3)

(Edexcel Limited 2007)

Linear graphs

When you hire a car the price increases in equal amounts for each extra day of hiring. When you empty a bath the volume of water going down the plug hole is the same every second. These are examples of linear functions. The quantities involved change at a constant rate.

What's the point?

Any linear function can be written using algebra and represented as a straight line graph. The graphs of linear functions will always be straight because of the constant rate of change.

Check in

You should be able to

■ **plot points to draw a straight line**

1 Plot these coordinate groups on one set of axes, joining them to produce a line in each case.

  i (3, 1), (3, 2), (3, 3)   **ii** (1, −2), (2, −2), (3, −2)

  iii (1, 3), (2, 5), (3, 7)   **iv** (1, 4), (2, 3), (3, 2)

■ **write fractions in different forms**

2 Convert these top heavy fractions to mixed numbers and vice versa.

  a $\frac{13}{2}$ **b** $\frac{21}{4}$ **c** $\frac{33}{7}$ **d** $-\frac{100}{11}$

  e $2\frac{1}{3}$ **f** $7\frac{1}{2}$ **g** $5\frac{3}{4}$ **h** $-4\frac{3}{8}$

■ **solve linear equations**

3 Find the value of the unknown in each equation.

  a $9 = 3n + 3$ **b** $6 = 1 + 5m$ **c** $10p - 4 = 1$

What I need to know

What I will learn

What this leads to

KS3 Plot lines given
their equations
Basic algebra

→

Plot straight line graphs
Find gradients and
intercepts from graphs
and equations
Use the form
$y = mx + c$
Interpret real-life graphs

→

A5 Solving simultaneous
linear equations

→

A6 Drawing and
interpreting more
complex graphs

Rich task

You can draw squares in different orientations on
a square dotty grid.
Explain how to draw squares on the grid.

The adjacent sides of a square are always at right
angles. Investigate the relationship between the
gradients of adjacent sides of squares.
What gradients are possible for the sides of
squares drawn on a 6 × 6 dotty grid?

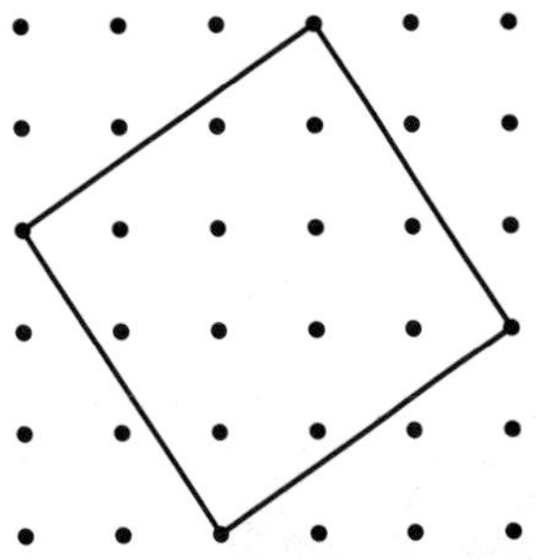

Straight line graphs

This spread will show you how to:

- Recognise that linear equations have straight-line graphs
- Plot straight-line graphs, given a linear equation

Keywords
Axes
Implicit
Linear
Plot

- The graphs of equations such as $y = 2x + 1$ (with no x^2 or higher powers) are straight lines. For example,

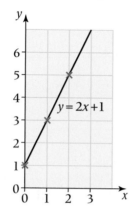

$Y = 2x + 1$ is a **linear** equation.

The graph of a linear equation is a straight line.

- You need three points to plot a straight line.

For example, to draw the graph of $y = 3x - 2$, first make a table of x and y values.

x	1	2	3
y	1	4	7

When $x = 1$, $y = (3 \times 1) - 2 = 1$

Draw x- and y–**axes** and **plot** the three points.

For an equation such as
'$2x + 3y = 12$' it is a little harder
to find the points.

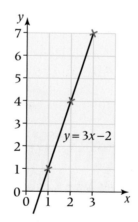

Two points are enough to fix a straight line but, as a check, work out a third point.

$2x + 3y = 12$ is an **implicit** equation – it does not give you the value of y directly.

Example

Plot the graph of $2x + 3y = 12$.

x	0	6	3
y	4	0	2

Draw the x and y axes and plot the points.

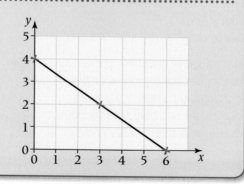

When the equation is in implicit form, as here, putting $x = 0$ and $y = 0$ gives you two points. For the third, choose an x-value between the first and second values.

1 Which of these equations have straight line graphs?

$y = 2x + 3$	$y = 7 - 3x$	$y = x^2 + 1$	$y = 5x$
$y = 7$	$y = 2x - x^3$	$2x + 7y = 8$	$x = -2$

2 a For each equation, copy and fill in the table of values.

i $y = 3x + 2$

x	0	1	2
y			8

ii $y = 2x - 4$

x	1	2	3
y	-2		

iii $2x + 5y = 10$

x	0		1
y		0	

b Draw x- and y-axes from −8 to 8 and plot the graphs.

Functional Maths — **A02**

3 I have a mobile phone. Each month, I pay £5 line rental. For every hour I then spend on the phone, I pay £7.
a Copy and complete this table of values to show my total phone bill for different lengths of time spent on the phone.

x (hours on phone)	1	2	3	4	5
y (total bill £)	12				

b Plot a graph to show hours against total bill.
c Use your graph to find the approximate cost if I spend 3 hours and 15 minutes on the phone one month.
d What is the equation of the graph you have drawn?

---

4 a The point (2, 5) lies on the graph $y = 2x + 1$. Does (3, 8) lie on this graph? Explain your answer.
b The point (2, 5) lies on the graph $y = 2x + 1$. Name another point that lies on this line.

5 a Plot the graphs $y = 3x - 1$ and $y = 3x + 2$ on the same axes.
b Explain why there is no point that lies on both of these graphs.
c Name the equation of another line that would have no points in common with these two.

Problem — **A03**

6 Would you prefer to get £3 per week pocket money, with 20p for every chore done (such as washing up) or £5 per week pocket money with 15p for every chore done? Use line graphs to report on your favoured option and to find how many chores you would need to do to receive the same amount under both options.

Further straight line graphs

This spread will show you how to:

- Recognise and understand the form of equations corresponding to horizontal, vertical and diagonal line graphs

Keywords

Diagonal
Horizontal
Intersect
Vertical

A straight line can be **diagonal**, **vertical** or **horizontal**.

The x-coordinate of every point on this vertical line is 2.

The y-coordinate can have any value.

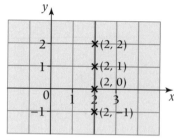

The equation of the line is $x = 2$.

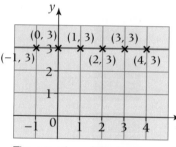

The y-coordinate of every point on this horizontal line is 3.

The x-coordinate can have any value.

The equation of the line is $y = 3$.

- Horizontal lines have equations of the form $x = c$.
- Vertical lines have equations of the form $y = c$.

c stands for a number.

Example

Give three points that would lie on each of these lines

a $y = 5$ **b** $x = -2$.

....

a $y = 5$

Since the y-coordinate is 5, possible points are $(1, \mathbf{5})$, $(2, \mathbf{5})$ and $(17, \mathbf{5})$

b $x = -2$

Since the x-coordinate is -2, possible points are $(\mathbf{-2}, 1)$, $(\mathbf{-2}, 2)$ and $(\mathbf{-2}, 11)$

Example

Where do the graphs $x = 4$ and $y = -1$ intersect?

•••

When lines **intersect** they cross.

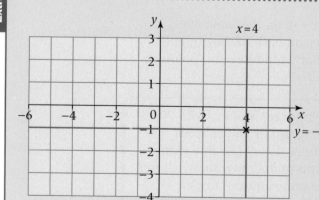

All points on the line $x = 4$ have x-coordinate 4.

All points on the line $y = -1$ have y-coordinate -1.

$y = -1$ The lines intersect at $(4, -1)$.

1 a Copy and complete the table by placing each of the fine equations in the appropriate column.

$$x = 9 \qquad y = 2x - 1 \qquad x = -0.5 \qquad y = 5 \qquad y = x^2 + x$$

Horizontal	Vertical	Diagonal	None of these

b Add an equation of your own to each column in the table.

2 Match each line with its equation.

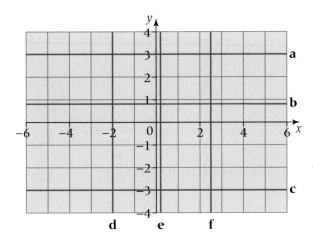

$$\boxed{y = -3} \quad \boxed{x = -2} \quad \boxed{y = 3} \quad \boxed{x = 2.5} \quad \boxed{x = \tfrac{1}{4}} \quad \boxed{y = \tfrac{3}{4}}$$

3 On one set of axes labelled from −6 to +6, plot these graphs.
a $x = 5$ **b** $y = 2$ **c** $x = 1.6$ **d** $y = -3$ **e** $y = 1$ **f** $x = -1\tfrac{1}{4}$

4 Where do these graphs intersect? Only plot the graphs if you need to.
a $x = 5$ and $y = 2$ **b** $x = 4$ and $y = -3$
c $x = -2$ and $y = 9$ **d** $y = -4$ and $x = -2$

5 a Give the equations of four lines which, when plotted, form the sides of a rectangle.
b Repeat part **a** for a square.
c Repeat part **a** for an isosceles right-angled triangle.

A03 Problem

6 a Which point with integer coordinates fits these clues?
Above $y = -1$, below $y = 3$, below $y = 2x + 1$,
above $y = 2 - x$ and left of $x = 2$.
b Write your own clues to describe the point (3, 4).

This spread will show you how to:

- Find the gradient and y-axis intercept of straight line graphs

Keywords
Coefficient
Constant
Gradient
Rise
Run
y-axis intercept

- The **gradient** of a straight line tells you how steep it is.

To work out the gradient find how many units the line **rises** for each unit it **runs** across the page.

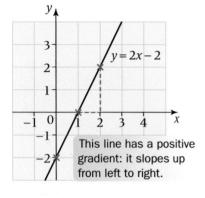

This line has a positive gradient: it slopes up from left to right.

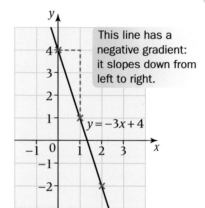

This line has a negative gradient: it slopes down from left to right.

For the line $y = 2x - 2$, gradient $= \dfrac{\text{rise}}{\text{run}} = \dfrac{2}{1} = 2$.　For the line $y = -3x + 4$, gradient $= \dfrac{\text{Rise}}{\text{run}} = \dfrac{3}{-1} = -3$.

- The gradient is the **coefficient** of x (the number of xs) in the equation of the line.

The **intercept** is the distance from the origin to where the line cuts the y-axis.

The line $y = 2x - 2$ cuts the y-axis at $(0, -2)$. The y-intercept is -2.
The line $y = -3x + 4$ cuts the y-axis at $(0, 4)$. The y-intercept is 4.

- The intercept is the **constant** term (the number) in the equation of the line.

Example

Find the gradient and intercept of the lines
a $y = 3x - 4$ **b** $y = \frac{1}{2}x + 5$

....

a gradient $= 3$, intercept $= -4$ **b** gradient $= \frac{1}{2}$, intercept $= 5$

If the equation is not in the form $y = ...$, rearrange it first, for example

$$3x + 2y = 12 \implies 2y = -3x + 12 \implies y = \tfrac{3}{2}x + 6$$

Now you can see that the gradient is $-\frac{3}{2}$ and the intercept is 6.

Example

Find the gradient and intercept of the lines
a $x + y = 4$ **b** $2x - 5y = 10$

....

a $x + y = 4$
      $y = 4 - x$
      $y = -x + 4$

b $2x - 5y = 10$
      $2x = 10 + 5y$
      $2x - 10 = 5y$ Divide by 5.
      $\frac{2}{5}x - 2 = y$

      gradient $= -1$, intercept $= 4$ gradient $\frac{2}{5}$, intercept $= -2$

1 Write the gradient and intercept of each line in the diagram.

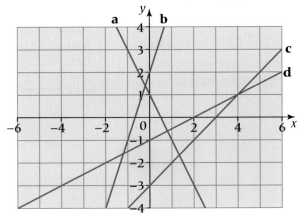

2 On a pair of axes labelled from -8 to $+8$, draw lines with these key characteristics.

a positive gradient

b y-intercept $= 3$

c gradient $= \frac{1}{4}$

d negative gradient and y-intercept $= -2$

e y-intercept $= 5$ and gradient $= 3$

f y-intercept $= 1$ and gradient $= \frac{2}{3}$

3 a Plot the line $y = 3x - 2$ on a pair of axes.
b Write its gradient and intercept.
c What do you notice about the connection between the gradient and intercept and the equation of the line?
d Repeat for the line $y = 2 - \frac{1}{2}x$.

A03 | Problem

4 The points $(2, 1)$ and $(5, 3)$ are shown.
a What is the gradient of the line joining these two points?
b Explain how you could have found the gradient without counting squares and only using the coordinates given.
c What is the gradient of a line joining $(2, 9)$ to $(31, 67)$?

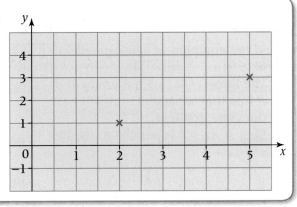

The equation $y = mx + c$

This spread will show you how to:

- Recognise that equations of the form $y = mx + c$ have straight-line graphs

You can find the **gradient** and **intercept** of a line from its equation.

- You can write the equation of any straight line in the form **$y = mx + c$**.

For example

$x + y = 5 \implies y = -x + 5$
$4x - y = 6 \implies 4x = 6 + y \implies y = 4x - 6$

- The gradient of the line $y = mx + c$ is m.

- The y-axis intercept of the line $y = mx + c$ is c.

m is the **coefficient** of x.

c is the **constant** (number).

$$\boxed{y = mx + c}$$

m is the gradient.
c is the y-axis intercept.

Example

What are the gradient and intercept of each of these lines?
a $y = 2x + 5$ **b** $y = 1 - 3x$

....

a Compare the equation to $y = mx + c$.
  $m = 2$ and $c = 5$
  Hence the gradient $= 2$ and the intercept $= 5$.
b In this equation $m = -3$ and $c = 1$.
  Hence the gradient $= -3$ and the intercept $= 1$.

- **Parallel** lines have the same gradient.

Example

Find the equation of a line parallel to $y = 4x + 5$.

....

The line $y = 4x + 5$ has gradient 4.
A line that is parallel to it will have the same gradient.
So, $y = 4x + 1$ is parallel to the line $y = 4x + 5$.

Many other equations are possible.

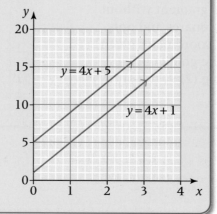

1 Match each line with its equation.

$y = 4x - 2$

$y = 3x + 1$

$y = x$

$y = 2 - 4x$

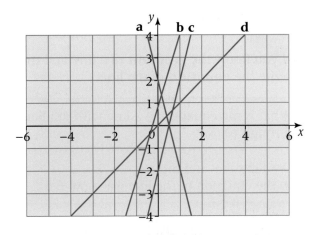

2 Copy and complete the table.

Equation	Gradient	Direction	Intercept
$y = 4x + 3$		positve	
$y = 3x + 4$			
$y = 9x - 2$			
$y = 4x - 5$			
$2y = 8x + 6$			

3 For each graph
  i write its gradient and intercept
  ii write the equation of the line.

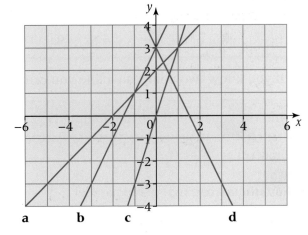

4 Write the equation of a straight line that is
  a parallel to $y = 3x + 3$
  b parallel to $y = 7 - 2x$ and cuts the y-axis at $(0, 3)$
  c a mirror image, in the y-axis, of $y = 3x + 1$.

5 a A straight line passes through $(0, 4)$ and $(2, 10)$.
    What is its equation?
  b A straight line is parallel to $2y = 6x - 9$ and goes through
    $(1, \ 7)$. What is its equation?

Finding the equation of a straight line graph

This spread will show you how to:

• Find the equation of a line joining several points

Keywords
Equation
Gradient
Intercept
$y = mx + c$

• If you know the **gradient** of a line and the y-axis **intercept** you can write the **equation** of the line.

Example

What is the equation of a line with gradient 9 passing through (0, 5)?

....

Gradient = 9 and intercept = 5,
so the equation of the line is $y = 9x + 5$.

Remember,
$y = mx + c$.

• If you know the gradient and a point on the line you can find the equation of the line.

Example

What is the equation of a line with gradient 8 that passes through the point (2, 7)?

....

Gradient = 8, so equation is $y = 8x + c$.
The line goes through (2, 7) so,

$7 = 8 \times 2 + c$ Put $x = 2$ and $y = 7$ in the equation $y = 8x + c$.
$7 = 16 + c$
$c = -9$

The equation of the line is $y = 8x - 9$.

• If you know two points on a line you can find the equation of the line.

Example

Find the equation of the line joining (1, 2) and (4, 3).

....

Gradient $= \dfrac{\text{rise}}{\text{run}} = \dfrac{1}{3}$

Equation of line is $y = \frac{1}{3}x + c$.

The line goes through (1, 2) so substitute 1 for x and 2 for y in $y = \frac{1}{3}x + c$.

$2 = \frac{1}{3} \times 1 + c$
$2 = \frac{1}{3} + c \implies c = \frac{5}{3}$

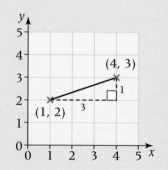

The equation is $y = \frac{1}{3}x + \frac{5}{3} \implies 3y = x + 5$

Check using the point (4, 3):
When $x = 4$, $3y = 4 + 5 = 9 \rightarrow y = 3$

1 Copy and complete the table.

Gradient	Intercept	Equation
3	5	
5	−2	
−2	7	
$\frac{1}{2}$	9	
$-\frac{1}{4}$	−3	
0	4	
1	0	

2 a Which of these lines will pass through the point (2, 8)?

$y = 4x$ $y = 2x + 3$ $y = 12 − 2x$ $y = 7x − 5$ $y = 5x − 1$

b Write the equations of two lines that pass through (1, 4).

3 Find the equations of the nine lines described in the table.

a Gradient of 7 and intercepts y-axis at (0, 5)	**b** Gradient of $\frac{1}{2}$ and passes through (0, 3)	**c** Parallel to a line with gradient 4 and passing through (3, 8)
d Gradient of 3 and passing through (4, 7)	**e** Gradient of −2 and cutting through (4, −3)	**f** Parallel to $y = \frac{1}{4}x − 1$ and passing through (0, −2)
g Passing through (0, 1) and (1, 5)	**h** Passing through (0, 2) and (5, 7)	**i** Passing through the midpoint of (1, 7) and (3, 13) with a gradient of 8

4 a What is the gradient of the line joining (0, 5) to (12, 41)?
b What is the equation of the line joining (0, 5) to (12, 41)?
c Repeat **a** and **b** for the points (3, 10) and (5, 6).

5 Where does the line $2y = 9x − 5$ cross
a the y-axis
b the x-axis
c the line $4y = x + 24$?

6 Find the equation of this line in the form $ax + by = c$, where a, b, c are values.

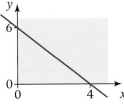

Linear graphs in real life

This spread will show you how to:

- Draw and interpret graphs modelling real-life situations
- Form linear functions, using the corresponding graphs to solve real-life problems

Keywords

Formula
Gradient
Graph
Intercept

- You can use a **graph** to represent a real-life situation.

Suppose you have a mobile phone. You pay £10 line rental each month, then 50p for every minute you spend making calls.

Work in
£50p = £0.5

Time on calls in minutes (x)	0	1	2	3	4
Price of calls in pounds	0	0.50	1.00	1.50	2.00
Line Rental (£)	10	10	10	10	10
Total cost in pounds (y)	10	10.50	11	11.50	12

To get the total cost, you add £10 (the line rental) to the call cost.
The call cost is the number of minutes on the phone multiplied by 0.50.
Hence,

Total cost = 0.50 × time on phone + 10
$$y = 0.5x + 10$$

This **formula** is linear and its graph is a straight line. The y-axis **intercept** is 10 and the **gradient** is 0.5.

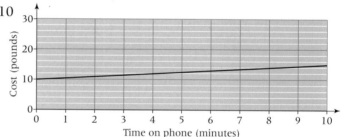

Example

Plot a graph to represent the total cost of hiring a party venue, if the owner charges £100 hire fee and £5 per guest.
Use the graph to estimate the number of guests if the total bill is £285.

If x is the number of guests and y is the cost (£s), then
$$y = 5x + 100$$

Draw a horizontal line from £285 to the graph.
Draw a vertical line from the graph to the horizontal axis.

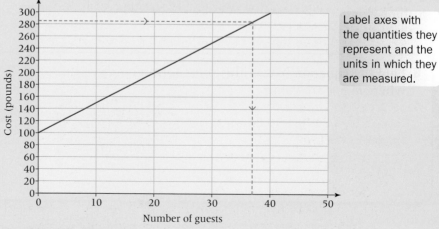

Label axes with the quantities they represent and the units in which they are measured.

For cost of £285, number of guests = 37.

1 a This graph is a conversion graph for miles to kilometres and vice versa.
Use the graph to convert
  i 20 miles into kilometres
  ii 60 kilometres into miles.

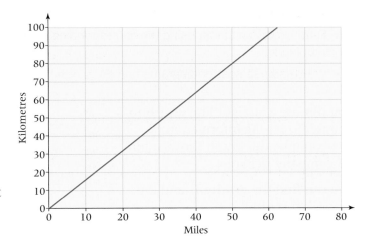

b If Dan ran 30 miles and Charlie ran 50 kilometres, who ran further?

c By finding the gradient of the line, give a formula to connect the number of miles (x) with the number of kilometres (y).

A02 Functional Maths

£1 = €1.45

2 Pauline and her family are going on holiday and exchange £400 spending money into euros (€) before they go. At the bank, the exchange rate is £1 = €1.15.

a Construct a graph that the family can take on holiday to convert any amount of their spending money from pounds to euros or vice versa.

b Use the graph to find
  i the cost, in euros, of a side trip which is advertised for £95
  ii the cost, in pounds, of a meal in a restaurant that comes to €85.

3 A campsite charges £15 per night per tent, plus an extra £3 per person.

a Construct a table of charges and, hence, a graph to show the cost of the campsite depending upon how many people stay in the tent. (The largest tent available is one that sleeps 15 people.)

b Use your graph to calculate how many people stayed in the tent if the total cost was £36.

c Explain why the total cost could never be £50.

d Suggest an equation for your graph, stating clearly the meaning of any letters you use.

e Use your equation to work out the cost of pitching a new Supertent that sleeps 27 people.

A03 Problem

4 Two competing electricity companies use these formulae to work out customers' bills.
The number of units of electricity used is x.
The price of the electricity is £y.

POWER UP!
$y = 3x + 5$

SPARKS ARE US!
$y = 2x + 15$

Using graphs, compare the pricing policies of the two companies and advise householders from which company they should buy their electricity.

Further linear graphs in real life

This spread will show you how to:

- Form linear functions, using the corresponding graphs to solve real-life problems
- Draw and use scatter diagrams and lines of best fit

Keywords
Gradient
Graph
Intercept

You can use a straight line **graph** to represent real-life information.
For example, this graph shows a hire car company's charges.

You can find the **gradient** and the y-axis **intercept**.

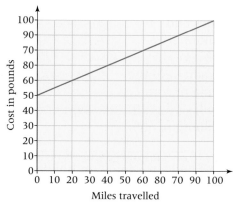

The line intercepts the y-axis at (0, 50).
For every 10 squares you move across, you travel 5 squares up:

$$\text{Gradient} = \frac{\text{rise}}{\text{run}} = \frac{5}{10} \text{ or } \frac{1}{2}$$

The equation of the graph is $y = \frac{1}{2}x + 50$.

- Graphs help to give you more information.
  - The intercept of 50 tells you that you are charged £50 for a car.
  - The gradient of $\frac{1}{2}$ tells you that for every 2 miles you travel, you are charged an extra £1.

p.144

Example

Interpret the line of best fit on this scatter diagram of 18 students' heights and weights.

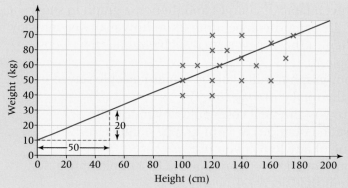

The gradient, $m = \frac{20}{50} = \frac{2}{5}$
The y-axis intercept, $c = 10$
The equation of the graph is $y = \frac{2}{5}x + 10$.

The gradient tells you that for every 5 cm you grow you gain 2 kg.
The intercept tells you that at 0 cm height, you weigh 10 kg – it does not always make sense to interpret the intercept on a straight line graph for a big age range.

1 Match each graph with an equation.

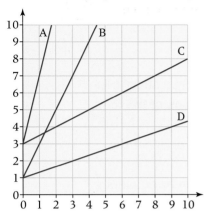

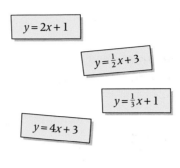

$y = 2x + 1$

$y = \frac{1}{2}x + 3$

$y = \frac{1}{3}x + 1$

$y = 4x + 3$

2 For each graph, find its equation in the form $y = mx + c$ and interpret the meaning of m and c, deciding if it is sensible to interpret c.

a

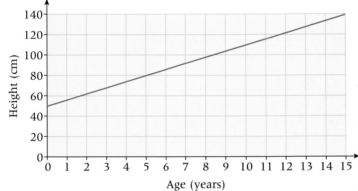

b

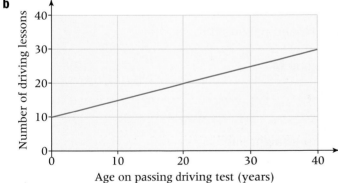

3 Interpret these equations representing real-life situations and discuss their limitations.

a $y = 0.3x + 2$ x is age in years

y is amount of pocket money in £.

b $y = 1\frac{4}{5}x + 32$ x is temperature in degrees Celsius (°C)

y is temperature in degrees Fahrenheit (°F)

Summary

Check out

You should now be able to:

- Recognise that an equation in the form $y = mx + c$ is a straight line
- Plot and draw graphs of straight lines with equations of the form $y = mx + c$
- Plot straight line graphs of functions in which y is or is not the subject
- Find the gradient and y-axis intercept of lines given by equations of the form $y = mx + c$
- Find the equation of the straight line by considering its gradient and y-axis intercept
- Understand the gradients of parallel lines

Worked exam question

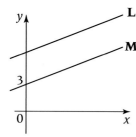

Diagram NOT accurately drawn.

The straight line **L** has equation $y = \frac{1}{2}x + 7$

The straight line **M** is parallel to **L** and passes through the point (0, 3).

Write down the equation for the line **M**.　　　　　　　　(2)

(Edexcel Limited 2007)

$y = mx + c$ is the equation of a straight line

The gradient of the line **L** is $\frac{1}{2}$

The gradient of the line **M** is $\frac{1}{2}$ and so $m = \frac{1}{2}$

$y = \frac{1}{2}x + c$ for line **M**

The intercept on the y-axis is 3 and so $c = 3$

$y = \frac{1}{2}x + 3$ for line **M**

> Write down the equation
> $y = \frac{1}{2}x + c$

> Substitute $x = 0$ and $y = 3$ into the equation
> $y = \frac{1}{2}x + c$ also gives $c = 3$

Exam questions

1 **a** Complete the table of values for $y = 4x - 3$.

x	−2	−1	0	1	2	3
y	−11		−3			9

(2)

b On a copy of the grid, draw the graph of $y = 4x - 3$, for values of x from −2 to 3

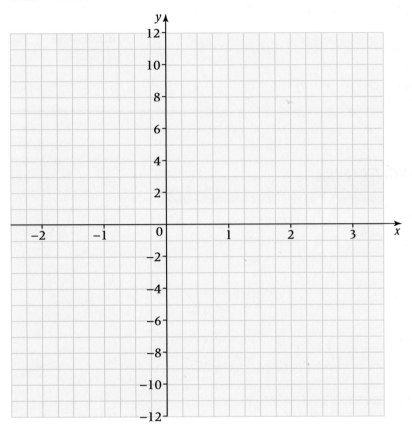

(2)

(Edexcel Limited 2009)

2 $y = 4x + 3$ is the equation of a straight line.

a What is the gradient of the straight line? (1)

b Find the coordinates of the point of intersection with the y-axis (2)

Functional Maths 1: Weather

Before creating a weather forecast, data is collected from all over the world to give information about the current conditions. A supercomputer and knowledge about the atmosphere, the Earth's surface and the oceans are then used to create the forecast.

Write down the temperature shown on each of these thermometers:

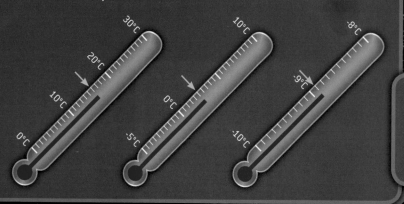

Each day, the Met Office receives and uses around half a million observations.

These tables show the national and UK weather records, last updated 24th November 2008:

HIGHEST DAILY TEMPERATURE RECORDS			
England	38.5 °C	10 August 2003	Faversham (Kent)
Wales	35.2 °C	2 August 1990	Hawarden Bridge (Flintshire)
Scotland	32.9 °C	9 August 2003	Greycrook (Scottish Borders)
Northern Ireland	30.8 °C	30 June 1976 12 July 1983	Knockarevan (County Fermanagh) Shaw's Bridge, Belfast (County Antrim)

LOWEST DAILY TEMPERATURE RECORDS			
Scotland	-27.2 °C	11 February 1895 10 January 1982 30 December 1995	Braemar (Aberdeenshire) Braemar (Aberdeenshire) Altnaharra (Highland)
England	-26.1°C	10 January 1982	Newport (Shropshire)
Wales	-23.3 °C	21 January 1940	Rhayader (Powys)
Northern Ireland	-17.5 °C	1 January 1979	Magherally (County Down)

Calculate the difference (in °C) between the maximum and minimum temperatures for each of the nations shown.

Use the internet to find out if these records have since been broken.

Wind direction is measured in tens of degrees relative to true North and is always given from where the wind is blowing. In the UK, wind speed is measured in knots, where 1 knot = 1.15mph, or in terms of the Beaufort Scale.

Look up the Beaufort Scale on the internet. Use it to describe the weather shown on this map.

Describe the wind speed (in knots) and direction (in tens of degrees and in words) shown by each of the arrows shown on this map:

▲
12
24
indicates a mean wind of 12 m.p.h., coming from the south, gusting 24 m.p.h.

Observed data can be used to make predictions, but there is always some level of uncertainty. This graph shows the range of uncertainty in temperature in Exeter with some indication of the most probable values:

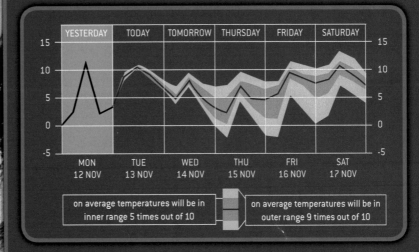

on average temperatures will be in inner range 5 times out of 10

on average temperatures will be in outer range 9 times out of 10

What predictions do you think a weather forecaster would have made about the temperature in Exeter during the week shown?

Justify your response by referring to the graph.

Fractions and decimals

The earliest fractions can be traced back to the Mesopotamians and ancient Egyptians. The Mesopotamians used base 60 for their fractions whilst the Egyptians preferred to write fractions as sums of unit fractions, for example $\frac{1}{2} + \frac{1}{3}$ to mean $\frac{5}{6}$. It was not until the sixteenth century that fractions as we know them today existed in Europe.

What's the point?

When written down a fraction can mean lots of different things: a number, a measure, an operator, a quotient and a ratio. Mathematicians are aware of these connections and this helps them use fractions more effectively, especially when they are changing between fractions, decimals and percentages.

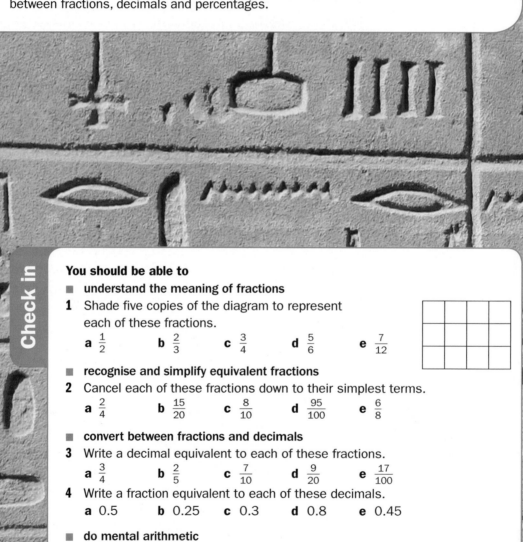

You should be able to

■ **understand the meaning of fractions**

1 Shade five copies of the diagram to represent each of these fractions.

    a $\frac{1}{2}$ **b** $\frac{2}{3}$ **c** $\frac{3}{4}$ **d** $\frac{5}{6}$ **e** $\frac{7}{12}$

■ **recognise and simplify equivalent fractions**

2 Cancel each of these fractions down to their simplest terms.

    a $\frac{2}{4}$ **b** $\frac{15}{20}$ **c** $\frac{8}{10}$ **d** $\frac{95}{100}$ **e** $\frac{6}{8}$

■ **convert between fractions and decimals**

3 Write a decimal equivalent to each of these fractions.

    a $\frac{3}{4}$ **b** $\frac{2}{5}$ **c** $\frac{7}{10}$ **d** $\frac{9}{20}$ **e** $\frac{17}{100}$

4 Write a fraction equivalent to each of these decimals.

    a 0.5   **b** 0.25   **c** 0.3   **d** 0.8   **e** 0.45

■ **do mental arithmetic**

5 Write notes to show how you would find a mental estimate for each of these calculations.

    a 19.2×28.9 **b** $355.72 \div 58.91$
    c $1206 - 816$ **d** $6987 + 6039$

Orientation

What I need to know	What I will learn	What this leads to
KS3 Understand and work with fractions Find the LCM	■ Do arithmetic with fractions ■ Convert between fractions, decimals and percentages	**N3** Calculating ratio and proportion **N6** Ordering fractions
N1 Arithmetic with decimals		**D3** Calculating with probability **D5** Combining probability

Rich task

In this diagram a square has been divided into different sized pieces. Calculate the fraction or percentage of the whole square that each piece represents.

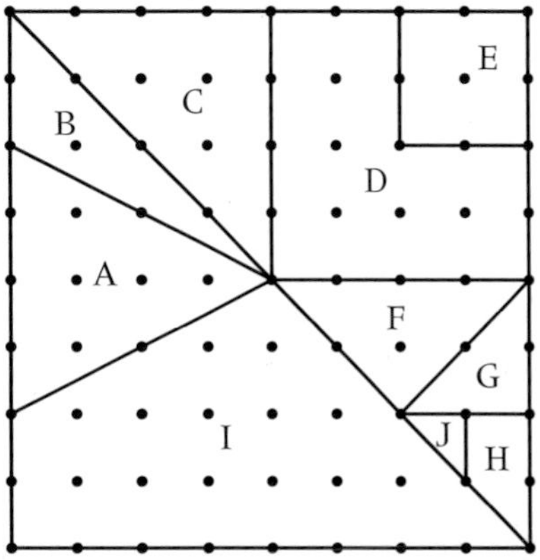

This spread will show you how to:

- Understand the terms equivalent and improper fractions and mixed numbers
- Add and subtract fractions

Keywords

Common denominator
Improper fraction
Mixed number

- You can only add and subtract fractions if they have common denominators.

$$\frac{2}{8} + \frac{5}{8} = \frac{7}{8}$$

- If the fractions have different denominators, change them to equivalent fractions with the same denominator, then add.

$$\frac{11}{12} - \frac{1}{3} = \frac{11}{12} - \frac{4}{12} = \frac{7}{12}$$

- If your answer is an improper fraction, change it to a mixed number.

$$\frac{3}{4} + \frac{2}{5} = \frac{15}{20} + \frac{8}{20} = \frac{23}{20} = 1\frac{3}{20}$$

- Cancel any common factors in the numerator and denominator.

$$\frac{4}{5} - \frac{3}{10} = \frac{8}{10} - \frac{3}{10} = \frac{5}{10} = \frac{1}{2}$$

$\frac{23}{20}$ is an **improper fraction** – its numerator is larger than its denominator.

$1\frac{3}{20}$ is a **mixed number** – a whole number and a fraction.

p.264

Example

Calculate   **a** $\frac{1}{2} + \frac{1}{3}$ **b** $\frac{3}{4} - \frac{1}{5}$ **c** $\frac{1}{4} + \frac{5}{6}$

a $\frac{1}{2} + \frac{1}{3} = \frac{3}{6} + \frac{2}{6} = \frac{3+2}{6} = \frac{5}{6}$ The lowest common denominator is 6.

b $\frac{3}{4} - \frac{1}{5} = \frac{15}{20} - \frac{4}{20} = \frac{15-4}{20} = \frac{11}{20}$ The lowest common denominator is 20.

c $\frac{1}{4} + \frac{5}{6} = \frac{3}{12} + \frac{10}{12} = \frac{3+10}{12} = \frac{23}{12} = 1\frac{1}{12}$ The lowest common denominator is 12.

A common mistake is to simply add or subtract the numerators and denominators.

Example

Both of these students calculated incorrectly.
Say what they did wrong and find the correct answer.

a $\frac{2}{3} + \frac{3}{4}$ **b** $\frac{7}{8} - \frac{2}{3}$

Jodi wrote   Abdul wrote

$\frac{2}{3} + \frac{3}{4} = \frac{2+3}{3+4} = \frac{5}{7}$ ✗ $\frac{7}{8} - \frac{2}{3} = \frac{7-2}{8-3} = \frac{5}{5} = 1$ ✗

a $\frac{2}{3} + \frac{3}{4} = \frac{8}{12} + \frac{9}{12}$ **b** $\frac{7}{8} - \frac{2}{3} = \frac{21}{24} - \frac{16}{24}$

$= \frac{17}{12}$ $= \frac{5}{24}$

$= 1\frac{5}{12}$

Jodie added the numerators and denominators instead of finding the common denominator first.

Abdul subtracted the numerators and denominators instead of finding the common denominator first.

a The answer $\frac{5}{7}$ cannot be correct. Both $\frac{2}{2}$ and $\frac{3}{4}$ are bigger than $\frac{1}{2}$, so the answer must be greater than 1.

b $\frac{7}{8}$ is smaller than 1, so when $\frac{2}{3}$ is taken away the answer must be less than 1.

1. The diagram shows that $\frac{1}{4} + \frac{1}{4} = \frac{1+1}{4} = \frac{2}{4} = \frac{1}{2}$

   Draw diagrams to show that

   a $\frac{3}{4} + \frac{2}{3} = \frac{2+1}{3} = \frac{3}{3} = 1$ **b** $\frac{1}{5} + \frac{3}{5} = \frac{1+3}{5} = \frac{4}{5}$

2. Calculate

   a $\frac{1}{5} + \frac{2}{5}$ **b** $\frac{1}{4} + \frac{3}{4}$ **c** $\frac{2}{7} + \frac{3}{7}$ **d** $\frac{3}{8} + \frac{5}{8}$

3. Draw diagrams (like the ones in question **1**) to illustrate your answers to question **2**.

4. Calculate   **a** $\frac{2}{3} - \frac{1}{3}$ **b** $\frac{4}{5} - \frac{1}{5}$ **c** $\frac{5}{6} - \frac{1}{6}$ **d** $\frac{9}{10} - \frac{3}{10}$

5. The diagram shows that $\frac{1}{3} + \frac{1}{4} = \frac{4}{12} + \frac{3}{12} = \frac{7}{12}$

   Draw diagrams to show that

   a $\frac{1}{3} + \frac{1}{2} = \frac{2}{6} + \frac{3}{6} = \frac{5}{6}$

   b $\frac{3}{5} + \frac{3}{10} = \frac{6}{10} + \frac{3}{10} = \frac{9}{10}$

6. Calculate   **a** $\frac{1}{5} + \frac{1}{10}$ **b** $\frac{2}{3} + \frac{1}{6}$ **c** $\frac{2}{5} + \frac{3}{20}$ **d** $\frac{1}{8} + \frac{1}{4}$

7. Draw diagrams (like the ones in question **5**) to illustrate your answers to question **6**.

8. Calculate   **a** $\frac{3}{8} - \frac{1}{4}$ **b** $\frac{5}{6} - \frac{1}{3}$ **c** $\frac{4}{5} - \frac{1}{20}$ **d** $\frac{3}{8} - \frac{1}{16}$

9. Calculate   **a** $\frac{4}{7} - \frac{1}{3}$ **b** $\frac{4}{5} - \frac{2}{3}$ **c** $\frac{8}{9} - \frac{5}{6}$ **d** $\frac{3}{5} - \frac{1}{4}$

10. The diagram shows what happens when you add fractions with a total that is greater than 1.

    In this example, $\frac{3}{4} + \frac{4}{5} = \frac{15}{20} + \frac{16}{20} = \frac{15+16}{20} = \frac{31}{20} = 1\frac{11}{20}$

    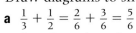

    Draw diagrams to find the answers to

    a $\frac{3}{5} + \frac{1}{2}$ **b** $\frac{5}{6} + \frac{3}{4}$

11. Calculate

    a $\frac{2}{3} + \frac{1}{2}$ **b** $\frac{4}{5} + \frac{1}{2}$ **c** $\frac{1}{3} + \frac{4}{5}$ **d** $\frac{4}{7} + \frac{1}{2}$

12. Change these improper fractions into mixed numbers.

    a $\frac{5}{4}$ **b** $\frac{9}{5}$ **c** $\frac{13}{8}$ **d** $\frac{17}{4}$

13. Change these mixed numbers to improper fractions.

    a $1\frac{3}{4}$ **b** $1\frac{7}{16}$ **c** $1\frac{5}{9}$ **d** $2\frac{4}{7}$

14. Calculate

    a $2\frac{3}{5} + 1\frac{1}{3}$ **b** $2\frac{3}{4} + 1\frac{5}{6}$ **c** $4\frac{3}{7} + 3\frac{1}{2}$ **d** $5\frac{4}{9} + 2\frac{3}{7}$

    e $3\frac{3}{5} - 2\frac{1}{4}$ **f** $2\frac{1}{2} - 1\frac{3}{4}$ **g** $3\frac{3}{4} - 1\frac{4}{5}$ **h** $7\frac{3}{7} - 2\frac{1}{2}$

This spread will show you how to:

- Calculate a fraction of a quantity
- Calculate with fractions effectively following simplification
- Use fractions to express an exact answer

- You find fractions of a quantity by multiplying.

  For example, two thirds of $5 = \frac{2}{3} \times 5 = \frac{2 \times 5}{3} = \frac{10}{3} = 3\frac{1}{3}$

p.112

- When the quantity and the denominator of the fraction have a **common factor**, **cancel** this factor before multiplying.

  For example, $\frac{2}{\cancel{3}_1} \times \cancel{24}^{8} = \frac{2}{1} \times 8 = 16$

Notice that you cannot give $3\frac{1}{3}$ as an **exact** answer in decimals because $\frac{1}{3}$ is a recurring decimal.

Example

Calculate   **a** $\frac{3}{4}$ of 28   **b** $\frac{5}{8}$ of 6   **c** $\frac{4}{9}$ of 12   **d** $\frac{5}{9}$ of 25

a $\frac{3}{\cancel{4}_1} \times \cancel{28}^{7} = 3 \times 7 = 21$

b $\frac{5}{\cancel{8}_4} \times \cancel{6}^{3} = \frac{5 \times 3}{4} = \frac{15}{4} = 3\frac{3}{4}$

c $\frac{4}{\cancel{9}_3} \times \cancel{12}^{4} = \frac{4 \times 4}{3} = \frac{16}{3} = 5\frac{1}{3}$

d $\frac{5}{9} \times 25 = \frac{125}{9}$
$= 13\frac{8}{9}$

Cancel by common factor 4 before multiplying.

You can also write $5\frac{1}{3}$ as $5.\dot{3}$, but it is simpler to leave it as a fraction.

In part **d**, you cannot cancel by 5. You can only cancel common factors where one is in a numerator (or a whole number) and the other is in a denominator.

You can extend this method to finding a fraction of a fraction of a quantity.

Example

Tom has £42. He spends $\frac{1}{3}$ of it on Monday. On Tuesday he spends $\frac{3}{4}$ of the remainder. How much does he spend on Tuesday?

Tom spends $\frac{1}{3} \times £42$ on Monday so he has $\frac{2}{3} \times £42$ on Tuesday. On Tuesday he spends

$$\frac{\cancel{3}^{1}}{\cancel{4}_{21}} \times \frac{\cancel{2}^{1}}{\cancel{3}_1} \times \cancel{42}^{21} = £21$$

1 Calculate these. Give your answers as fractions or mixed numbers.
 a $\frac{1}{2}$ of 7   **b** $\frac{1}{5}$ of 8   **c** $\frac{1}{3}$ of 10   **d** $\frac{1}{3}$ of 2   **e** $\frac{2}{5}$ of 6

2 Work out these. Show how common factors can be cancelled in each.
 a $\frac{1}{2}$ of 20   **b** $\frac{1}{4}$ of 84   **c** $\frac{1}{3}$ of 36   **d** $\frac{1}{5}$ of 65   **e** $\frac{2}{3}$ of 33

You should show all of your working for the questions **3**, **4** and **5**.
In particular, show how the calculations can be simplified by cancelling
common factors.

3 A reel holds 60 m of wire when new. $\frac{2}{5}$ of the wire has been used.
 a What length of wire has been used?
 b What length of wire is left on the reel?

4 Calculate the amount of liquid in these containers.
 a A 40 litre barrel that is $\frac{3}{8}$ full.   **b** A 240 cl jar that is $\frac{3}{4}$ full.
 c A 120 cl glass that is $\frac{2}{5}$ full.   **d** A 750 ml litre bottle that is $\frac{2}{3}$ empty.

5 Calculate
 a $\frac{5}{8}$ of 48 m   **b** $\frac{2}{9}$ of 36 km   **c** $\frac{4}{7}$ of 28 mm   **d** $\frac{3}{4}$ of 120 m

The answers to questions **6** and **7** are not be whole numbers. You
should show your working as before, and give your answers as fractions
or mixed numbers.

6 Calculate
 a $\frac{1}{6}$ of 10   **b** $\frac{1}{4}$ of 22   **c** $\frac{3}{10}$ of 15   **d** $\frac{1}{12}$ of 8   **e** $\frac{4}{9}$ of 21

7 Calculate these lengths.
 a $\frac{1}{9}$ of 24 miles **b** $\frac{5}{6}$ of 40 miles   **c** $\frac{5}{18}$ of 45 miles   **d** $\frac{3}{20}$ of 25 miles

8 Calculate these and convert your answers to (approximate)
decimal numbers.
 a $\frac{5}{12}$ of 16 m   **b** $\frac{4}{9}$ of 12 mm   **c** $\frac{3}{22}$ of 64 cm   **d** $\frac{3}{14}$ of 104 km

Problem

A03

9 An empty swimming pool is to be filled with water. It takes
12 hours to fill the pool and the full pool contains 98 m³ of
water. How much water will the pool contain after 5 hours?
Show your working.

10 Calculate these times using fractions, and then convert each of your
answers into hours and minutes.
 a $\frac{7}{12}$ of 9 hours **b** $\frac{5}{8}$ of 22 hours   **c** $\frac{7}{10}$ of 24 hours   **d** $\frac{7}{18}$ of 63 hours

11 Calculate
 a $\frac{2}{3}$ of £7   **b** $\frac{5}{12}$ of £40   **c** $\frac{3}{8}$ of £34   **d** $\frac{4}{9}$ of £66

Multiplying and dividing fractions

This spread will show you how to:

- Understand unit fractions and use them as multiplicative inverses
- Multiply and divide fractions

Keywords
Multiplicative
  inverse
Reciprocal
Unit fraction

The diagram shows the multiplication
$$\frac{2}{3} \times \frac{3}{4} = \frac{6}{12} = \frac{1}{2}$$
You get the same result if you multiply
the numerators together and multiply
the denominators together.

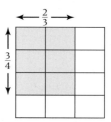

- To multiply fractions, multiply the numerators and then the denominators, and cancel any common factors.

Example

Find **a** $\frac{2}{3} \times \frac{4}{5}$ **b** $\frac{4}{9} \times \frac{3}{5}$ **c** $\frac{7}{4} \times \frac{5}{2}$

a $\frac{2}{3} \times \frac{4}{5} = \frac{2 \times 4}{3 \times 5}$ **b** $\frac{4}{9} \times \frac{3}{5} = \frac{4 \times 3}{9 \times 5} = \frac{4 \times \cancel{3}^{1}}{{}_{3}\cancel{9} \times 5}$ **c** $\frac{7}{4} \times \frac{5}{2} = \frac{7 \times 5}{4 \times 2}$

$= \frac{8}{15}$ $= \frac{12}{45}$ $= \frac{4}{15}$ $= \frac{35}{8} = 4\frac{3}{8}$

- Multiplying by a **unit fraction** is the same as dividing by its denominator. For example, multiplying by $\frac{1}{5}$ is the same as dividing by 5.
$$10 \times \frac{1}{5} = \frac{10}{5} = 2 \text{ and } 10 \div 5 = 2$$
- Dividing by a unit fraction is the same as multiplying by its denominator.
$$10 \div \frac{1}{2} = 10 \times 2 = 20$$

You combine these two ideas when you divide by a fraction.
For example

$$10 \div \frac{5}{2} = 10 \div 5 \div \frac{1}{2}$$ ÷5 is the same as $\times \frac{1}{5}$
$$= 10 \times \frac{1}{5} \times 2$$ $\div \frac{1}{2}$ is the same as ×2
$$= 10 \times \frac{2}{5}$$
$$= \frac{20}{5} = 4$$

A unit fraction has
a numerator of 1.

The **multiplicative
inverse** of an
integer is its
reciprocal.
For example, the
reciprocal of 5
is $\frac{1}{5}$.

- To divide by a fraction, multiply by its multiplicative inverse.

Example

Find **a** $\frac{3}{7} \div 5$ **b** $\frac{5}{12} \div \frac{3}{4}$ **c** $3\frac{1}{2} \div 2\frac{1}{5}$

a $\frac{3}{7} \div 5 = \frac{3}{7} \times \frac{1}{5}$ **b** $\frac{5}{12} \div \frac{3}{4} = \frac{5}{{}_{3}\cancel{12}} \times \frac{\cancel{4}^{1}}{3}$ **c** $3\frac{1}{2} \div 2\frac{1}{5} = \frac{7}{2} \div \frac{11}{5}$

$= \frac{3}{35}$ $= \frac{5}{9}$ $= \frac{7}{2} \times \frac{5}{11}$

$= \frac{35}{22} = 1\frac{13}{22}$

The **multiplicative
inverse** of a
fraction is the
original fraction
'turned upside
down'. The
inverse of $\frac{3}{5}$ is $\frac{5}{3}$.

Change mixed
numbers to
improper fractions
first.

1 Write the reciprocal (multiplicative inverse) of these integers.
  a 4 **b** 6 **c** 10 **d** 12

2 Write the multiplicative inverse of these fractions.
  a $\frac{1}{5}$ **b** $\frac{1}{9}$ **c** $\frac{1}{2}$ **d** $\frac{1}{3}$

3 Rewrite each of these divisions as multiplications.
  a $8 \div 5$ **b** $6 \div 4$ **c** $9 \div 5$ **d** $17 \div 3$

> For example, you can write $7 \div 5$ as $7 \times \frac{1}{5}$.

4 Copy and complete these sentences.
  a Dividing a number by 4 is the same as multiplying the number by ___.
  b Multiplying a number by $\frac{1}{2}$ is the same as dividing the number by ___.
  c Dividing a number by $\frac{1}{3}$ is the same as multiplying the number by ___.

5 Draw diagrams, like the one opposite, to show that
  a $\frac{1}{4} \times \frac{1}{2} = \frac{1}{8}$ **b** $\frac{3}{4} \times \frac{1}{3} = \frac{3}{12} = \frac{1}{4}$ **c** $\frac{2}{5} \times \frac{2}{3} = \frac{4}{15}$ **d** $\frac{1}{4} \times \frac{3}{5} = \frac{3}{20}$

6 Calculate, giving your answers in their simplest form.
  a $\frac{3}{4} \times \frac{1}{5}$ **b** $\frac{2}{3} \times \frac{2}{9}$ **c** $\frac{2}{7} \times \frac{1}{4}$ **d** $\frac{5}{16} \times \frac{4}{5}$
  e $\frac{5}{9} \times \frac{4}{7}$ **f** $\frac{7}{8} \times \frac{2}{21}$ **g** $\frac{4}{5} \times \frac{3}{13}$ **h** $\frac{8}{35} \times \frac{7}{24}$

7 Calculate, giving your answers as fractions in their simplest terms.
  a $3 \div 4$ **b** $6 \div 8$ **c** $4 \div 5$ **d** $8 \div 10$
  e $3 \div 7$ **f** $9 \div 5$ **g** $24 \div 7$ **h** $22 \div 8$

> For example, $13 \div 4 = 13 \times \frac{1}{4} = \frac{13}{4} = 3\frac{1}{4}$.

8 Calculate
  a $\frac{5}{8} \div 4$ **b** $\frac{3}{4} \div 6$ **c** $\frac{2}{3} \div 7$ **d** $\frac{3}{16} \div 9$
  e $4 \div \frac{1}{5}$ **f** $6 \div \frac{2}{3}$ **g** $5 \div \frac{2}{5}$ **h** $11 \div \frac{3}{7}$

9 Calculate
  a $\frac{1}{8} \div \frac{3}{8}$ **b** $\frac{1}{5} \div \frac{4}{5}$ **c** $\frac{1}{14} \div \frac{3}{7}$ **d** $\frac{1}{10} \div \frac{2}{5}$
  e $\frac{2}{3} \div \frac{3}{4}$ **f** $\frac{5}{8} \div \frac{7}{9}$ **g** $\frac{1}{12} \div \frac{3}{8}$ **h** $\frac{2}{7} \div \frac{3}{4}$

10 Calculate
  a $1\frac{1}{2} \times \frac{3}{4}$ **b** $2\frac{3}{4} \times \frac{2}{5}$ **c** $1\frac{1}{5} \times 2\frac{1}{2}$ **d** $1\frac{2}{3} \times 1\frac{4}{5}$
  e $2\frac{7}{8} \div \frac{3}{5}$ **f** $4\frac{1}{4} \div 3\frac{1}{2}$ **g** $5\frac{3}{8} \div 2\frac{3}{4}$ **h** $9\frac{1}{3} \div 2\frac{1}{4}$

This spread will show you how to:

- Distinguish between recurring and terminating decimals
- Convert fractions to decimals and percentages

Keywords
Denominator
Numerator
Recurring
Terminating

- To convert a fraction to a decimal divide the numerator by the denominator.

Example

Write these fractions as decimals. **a** $\frac{5}{8}$ **b** $\frac{5}{9}$ **c** $\frac{1}{7}$

a $\frac{5}{8} = 5 \div 8 = 8\overline{)5.000}^{\,0.625}$

b $\frac{5}{9} = 5 \div 9 = 9\overline{)5.000\ldots}^{\,0.555\ldots} = 0.\dot{5}$

c $\frac{1}{7} = 1 \div 7 = 7\overline{)1.000000000\ldots}^{\,0.142857142\ldots} = 0.\dot{1}4285\dot{7}$

0.625 is a **terminating** decimal.

0.555 ... $= 0.\dot{5}$ is a **recurring** decimal. The dot over the 5 shows the recurring digit.

$0.\dot{1}4285\dot{7}$ is a recurring decimal. The dots show the recurring group of digits.

To decide if a fraction will be a terminating or a recurring decimal, write it in its simplest form and look at the denominator.

- If the only factors of the denominator are 2 and/or 5 or combinations of 2 and 5 then the fraction will be a terminating decimal.
- If the denominator has any factors other than 2 and/or 5 then the fraction will be a recurring decimal.

Example

Say whether these fractions are terminating or recurring decimals.

a $\frac{9}{20}$ **b** $\frac{4}{30}$ **c** $\frac{7}{16}$ **d** $\frac{11}{13}$

a denominator $= 20 = 2 \times 2 \times 5 \rightarrow \frac{9}{20}$ is a terminating decimal

b $\frac{4}{30} = \frac{2}{15}$ denominator $= 15 = 3 \times 5 \rightarrow \frac{2}{15}$ is a recurring decimal

c denominator $= 16 = 2^4 \rightarrow \frac{7}{16}$ is a terminating decimal

d denominator $= 13 \rightarrow \frac{11}{13}$ is a recurring decimal

Simplify the fraction.

To convert a fraction to a percentage

- write it as a decimal
- multiply the decimal by 100%.

$100\% = \frac{100}{100} = 1$ so you are multiplying by 1 which does not change the value of the decimal.

Example

Write as percentages **a** $\frac{5}{8}$ **b** $\frac{5}{9}$ **c** $\frac{1}{7}$

a $\frac{5}{8} = 0.625 = 0.625 \times 100\% = 62.5\%$

b $\frac{5}{9} = 0.\dot{5} = 0.\dot{5} \times 100\% = 55.5\dot{5}\% = 55.6\%$ to 1 dp

c $\frac{1}{7} = 0.\dot{1}4285\dot{7} = 0.\dot{1}4285\dot{7} \times 100\% = 14.\dot{2}8571\dot{4}\% = 14.3\%$ to 1 dp

1 Write these fractions as decimals.

   a $\frac{1}{2}$ **b** $\frac{3}{4}$ **c** $\frac{2}{5}$

   d $\frac{1}{10}$ **e** $\frac{1}{5}$ **f** $\frac{1}{4}$

> You should be able to do all of these mentally.

2 Convert each of the decimals from question **1** to a percentage.

3 Convert each fraction to a percentage, using a written method. Show your working.

   a $\frac{5}{8}$ **b** $\frac{4}{5}$ **c** $\frac{7}{8}$

   d $\frac{3}{5}$ **e** $\frac{3}{8}$ **f** $\frac{1}{8}$

> You should practise using a written method here, even if you can do these mentally.

4 Use a calculator to check your answers to question **3**.

5 Use a calculator to convert these fractions to decimals.

   a $\frac{1}{16}$ **b** $\frac{7}{25}$ **c** $\frac{7}{125}$

   d $\frac{3}{40}$ **e** $\frac{7}{16}$ **f** $\frac{1}{32}$

6 Convert the decimal answers from question **5** to percentages.

7 Use a written method to convert each fraction to a decimal. Use the 'dot' notation to represent recurring decimals.

   a $\frac{1}{3}$ **b** $\frac{1}{6}$ **c** $\frac{2}{3}$

   d $\frac{1}{7}$ **e** $\frac{1}{9}$ **f** $\frac{5}{6}$

8 Convert your decimal answers from question **7** to percentages.

9 Use a calculator to check your answers to questions **7** and **8**.

10 Use an appropriate method to convert these fractions to decimals.

   a $\frac{3}{7}$ **b** $\frac{3}{16}$ **c** $\frac{17}{80}$

   d $\frac{5}{9}$ **e** $\frac{4}{25}$ **f** $\frac{5}{7}$

11 State whether each of these fractions will give a recurring decimal or a terminating decimal. Explain your answers.

   a $\frac{1}{25}$ **b** $\frac{3}{20}$ **c** $\frac{4}{11}$

   d $\frac{1}{126}$ **e** $\frac{1}{125}$ **f** $\frac{1}{128}$

A03 Problem

12 Shula says, 'I used my calculator to change $\frac{1}{13}$ to a decimal, and I got the answer 0.07692308. There is no repeating pattern, so the decimal does not recur.' Explain why Shula is wrong.

This spread will show you how to:

- Convert percentages and terminating and recurring decimals to fractions

Keywords
Denominator
Recurring
Terminating

- To convert a **terminating** decimal to a fraction:
  1 Write the decimal as a fraction with **denominator** 10, 100, 1000, ... according to the number of decimal places. For example, $0.45 = \frac{45}{100}$
  2 Simplify the fraction.
  $$\frac{45}{100} = \frac{9}{20}$$

2 decimal places so denominator is 100.

- To convert a percentage to a fraction, divide the percentage by 100.

  For example,
  $$43.7\% = \frac{43.7}{100} = 0.437 = \frac{437}{1000}$$

Example

Convert these to fractions   **a** 0.306   **b** 45%   **c** 32.5%   **d** 0.52

a $0.306 = \frac{306}{1000} = \frac{153}{500}$

b $45\% = \frac{45}{100} = \frac{9}{20}$

c $32.5\% = 0.325 = \frac{325}{1000} = \frac{65}{200} = \frac{13}{40}$

d $0.52 = \frac{52}{100} = \frac{26}{50} = \frac{13}{25}$

The next example shows how to convert **recurring** decimals to fractions. You multiply by a power of 10, then subtract one lot of the original decimal to produce a fraction.

Example

Write as fractions   **a** $0.\dot{4}$ **b** $0.\dot{5}\dot{6}$

a Let $x = 0.\dot{4}$ $10x = 4.\dot{4} = 4.444\ldots$
               $x = 0.\dot{4} = 0.444\ldots$
             $9x = 4$
              $x = \frac{4}{9}$

One recurring digit so multiply decimal by 10.

Subtract $10x - 1x$.

b Let $x = 0.\dot{5}\dot{6}$ $100x = 56.\dot{5}\dot{6} = 56.565656\ldots$
                  $x = 0.\dot{5}\dot{6} = 0.565656\ldots$
            $99x = 56$
             $x = \frac{56}{99}$

One recurring digit so multiply decimal by 100.

Subtract $100x - 1x$.

1 Convert these percentages to decimals.
  a 43%   **b** 86%   **c** 94%
  d 45.5%   **e** 3.75%   **f** 105%

2 Convert these decimals to fractions, using a mental method.
  a 0.5   **b** 0.25   **c** 0.2
  d 0.125   **e** 0.75   **f** 0.9

3 Convert these decimals to fractions.
  a 0.51   **b** 0.43   **c** 0.413
  d 0.719   **e** 0.91   **f** 0.871

4 Convert these percentages to fractions.
  a 49%   **b** 53%   **c** 73%
  d 81%   **e** 37%   **f** 19%

5 Convert these decimals to fractions. Give your answers in their simplest form.
  a 0.32   **b** 0.55   **c** 0.44
  d 0.155   **e** 0.64   **f** 0.265

6 Convert these percentages to fractions. Give your answers in their simplest form.
  a 55%   **b** 62%   **c** 84%
  d 65%   **e** 72%   **f** 18.5%

7 Which of these fractions will make recurring decimals? Explain your answer.
  $\frac{22}{25}$ $\frac{17}{20}$ $\frac{8}{11}$ $\frac{2}{5}$

8 Write each of these recurring decimals using 'dot' notation.
  a 0.111...   **b** 0.555...   **c** 0.75555...
  d 0.346346346...   **e** 0.7656565...

9 Convert these recurring decimals to fractions.
  a $0.\dot{2}$ **b** $0.\dot{6}$ **c** $0.\dot{2}\dot{5}$
  d $0.\dot{2}\dot{7}$ **e** $0.5\dot{4}\dot{5}$ **f** $0.\dot{6}0\dot{5}$

10 Convert these percentages to fractions.
  a $52.\dot{2}$%   **b** $5.\dot{5}$%   **c** $45.\dot{2}\dot{7}$%
  d $8.3\dot{5}$%   **e** $66.\dot{6}$%   **f** $8.\dot{2}0\dot{5}$%

---

11 Find a fraction equal to the recurring decimal $0.\dot{0}12345678\dot{9}$, giving your answer in its simplest form. Show your working.

A03 Problem

Summary

Check out

You should now be able to:

- Add, subtract, multiply and divide fractions
- Recognise the equivalence of fractions and decimals
- Convert between fractions, decimals and percentages
- Convert a recurring decimal to a fraction and vice versa

Worked exam question

a Work out the value of $\frac{2}{3} \times \frac{3}{4}$

Give your answer as a fraction in its simplest form. (2)

b Work out the value of $1\frac{2}{3} + 2\frac{3}{4}$

Give your answer as a fraction in its simplest form. (3)

(Edexcel Limited 2005)

a

$$\frac{{}^1\cancel{2}}{{}_1\cancel{3}} \times \frac{{}^1\cancel{3}}{{}_2\cancel{4}} = \frac{1}{2}$$

> Cancel the fractions before multiplying

OR

$$\frac{2}{3} \times \frac{3}{4} = \frac{6}{12}$$
$$= \frac{1}{2}$$

> Multiply the fractions and then cancel

b

$$1\frac{2}{3} + 2\frac{3}{4} = 1 + 2 + \frac{8}{12} + \frac{9}{12}$$
$$= 3 + \frac{17}{12}$$
$$= 4\frac{5}{12}$$

> Add the whole numbers then use a common denominator of 12

OR

$$1\frac{2}{3} + 2\frac{3}{4} = \frac{20}{12} + \frac{33}{12}$$
$$= \frac{53}{12}$$
$$= 4\frac{5}{12}$$

> Use a common denominator of 12 to add the fractions

Exam questions

1 **a** Write down the reciprocal of 4 (1)

   b Work out the value of $2\frac{4}{5} - 1\frac{3}{4}$

     Give your answer as a fraction in its simplest form. (3)

   c Sundas says that $4\frac{1}{3}$ is equal to 4.3

     Sundas is wrong.

     Explain why. (1)

(Edexcel Limited 2008)

2 Work out $2\frac{2}{3} \times 1\frac{1}{4}$

   Give your answer in its simplest form. (3)

(Edexcel Limited 2007)

3 **a** Work out $\frac{1}{3} + \frac{3}{5}$ (2)

   b Work out $2\frac{1}{4} \div \frac{3}{5}$ (3)

(Edexcel Limited 2006)

4 Work out $3\frac{3}{5} + 1\frac{2}{3}$

   Give your answer as a fraction in its simplest form. (3)

5 **a** Which of these fractions can be written as a recurring decimal?

     $\frac{1}{2}$ $\frac{2}{3}$ $\frac{3}{4}$ $\frac{5}{8}$

   b Explain your answer. (2)

Collecting and analysing data

In the run up to a general election, opinion polls are taken of which political party people are likely to vote for. The results of just a 1000 peoples' voting intentions are taken very seriously by the media and the politicians.

What's the point?

By selecting a representative sample, surveys allow statisticians to obtain reliable results from manageable amounts of data. The technique is used by polling organisations and throughout industry to monitor quality and productivity *etc*.

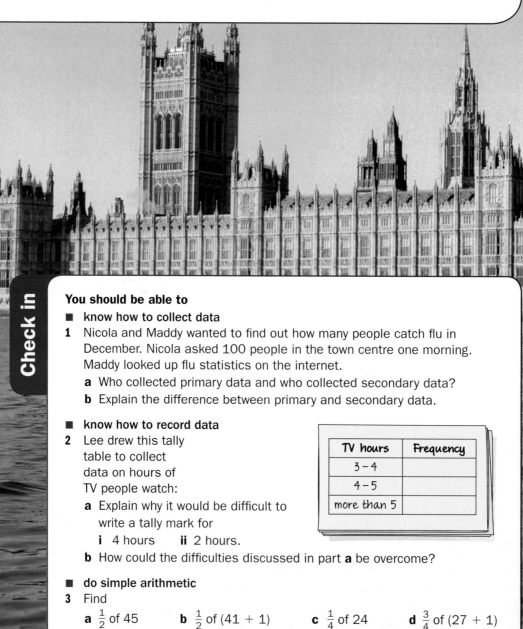

Check in

You should be able to

- **know how to collect data**
1 Nicola and Maddy wanted to find out how many people catch flu in December. Nicola asked 100 people in the town centre one morning. Maddy looked up flu statistics on the internet.
   - **a** Who collected primary data and who collected secondary data?
   - **b** Explain the difference between primary and secondary data.

- **know how to record data**
2 Lee drew this tally table to collect data on hours of TV people watch:

TV hours	Frequency
3 – 4	
4 – 5	
more than 5	

   - **a** Explain why it would be difficult to write a tally mark for
     - **i** 4 hours
     - **ii** 2 hours.
   - **b** How could the difficulties discussed in part **a** be overcome?

- **do simple arithmetic**
3 Find
   - **a** $\frac{1}{2}$ of 45
   - **b** $\frac{1}{2}$ of (41 + 1)
   - **c** $\frac{1}{4}$ of 24
   - **d** $\frac{3}{4}$ of (27 + 1)

What I need to know	What I will learn	What this leads to
KS3 Knowledge of collecting data Find averages and the range for data sets	■ Design a survey, select a sample and use a two-way table ■ Calculate measures of average and spread	**D4** Use statistics to describe data
N1 Order numbers Do basic arithmetic	■ Combine the mean of two data sets	Opinion polls, Journalism

Rich task

What is the average time taken to travel to school for pupils at your school?
Do boys take less time than the girls?
Do the pupils who live nearer the school take less time than the pupils who live further away?
Investigate and write a report on your results.

Designing a survey

This spread will show you how to:

- Design a questionnaire, recognising bias and taking steps to avoid it

Keywords
Biased
Data
Questionnaire
Survey

Organisations carry out statistical surveys to collect data that help them plan for the future.

You can use a **questionnaire** and collect **data** in a **survey**.

You have to choose suitable questions for a questionnaire.

Suitable questions,
- can be answered yes or no
- ask for facts.

Unsuitable questions,
- may be vague
- are leading or **biased**
- could be embarrassing.

Questions must have responses that,
- cover all possible answers
- do not overlap or have gaps.

Do you have an MP3 player?

How many CDs do you own?

none ☐ *1–10* ☐ *11–20* ☐ *over 20* ☐

Do you listen to a lot of music?

Do you agree that Coldplay is the best band in the world?

How cool are you?

Where do you buy your CDs?

internet ☐ *store* ☐ *other* ☐

How much do you spend on CDs each month?

£0 to £14.99 ☐   *£15 to £29.99* ☐

£30 and up ☐

Example

A recently restyled breakfast radio show conducted this survey.

1 What is your opinion of the new breakfast show?

    Fantastic ☐   Good ☐

2 How long do you listen to the show?

    10 min ☐   1 hour ☐   1–2 hours ☐

  a Comment on these questions.

  b Write two questions for the survey to find out if listeners like the new show and for how long they listen.

· ·

a Question **1** does not include all possible responses, for example if you do not like the new show or don't think it's an improvement. In question **2** there are gaps, between 10 min and 1 hour, and an overlap: **1** hour appears twice.

b 1 What is your opinion of the new breakfast show compared to the old one?

    Much better ☐   Better ☐

    Neither better nor worse ☐   Not as good ☐

  2 How much of the breakfast show do you listen to?

    all of it ☐   over an hour ☐

    half an hour to 1 hour ☐   less than half an hour ☐

1 Katy is doing a survey to find out how often people go to the cinema and how much they spend. She writes this question:

> How many times a month do you go to the cinema?

a What is wrong with this question?
b Write an improved question to find out how often people visit the cinema.
c Write a question to find out how much people spend when they go to the cinema.

2 Sally put this question in a questionnaire:

> Do you agree that tennis is the best sport?

a **i** What is wrong with this question?
  ii Write a better question to find out the favourite sport. Include some response boxes.

Sally also wants to find out how often people play sport.

b Design a question for Sally to use. Include some response boxes.

3 James wants to know which flavour crisps he should stock in the school tuck shop. He asks this question:

> Do you prefer plain or ketchup flavoured crisps?

a What is wrong with this question?
b Think about what crisps you and your friends like and design a better question for James to use. You should include some response boxes.
c James also put this question in his questionnaire:

> How many times have you visited the tuck shop?
> Once ☐   Lots of times ☐

  i Write two things that are wrong with this question.
  ii Design a better question for James to use. Include some response boxes.

4 Merlin wants to find out how far people would travel to see their favourite band perform. He writes this question:

> How far would you travel to see your favourite band?
> less than 1 mile ☐   5–10 miles ☐   any distance ☐

a **i** What is wrong with this question?
  ii Design a better question for Merlin to use. Include some response boxes.

Merlin also wants to find out how much people would pay for a ticket to see their favourite band.

b Design a question for Merlin to use. Include some response boxes.

In this exercise you can find out if any questions you write 'work' by testing them out on groups of people in your class. The larger the sample the more reliable the results.

Collecting data – choosing a sample

This spread will show you how to:

- Design a questionnaire, recognising bias and taking steps to avoid it
- Understand the concept of random sampling

Keywords
Bias
Population
Random
Sample

It may be time-consuming, too costly, too long or too impractical to collect data from everyone. In these cases you should ask a representative **sample**.

You must choose the sample so that it is not biased. For example, a survey of preferred music using a sample of friends is biased as friends are more likely to have similar opinions.

- A sample should represent a whole **population**.

One way of avoiding **bias** is to use a **random** sample.

- In a random sample each member of the population has the same chance of being included.

- Methods for choosing a random sample include:
  - taking names out of a hat
  - giving everyone a number and using a calculator or random number tables to pick numbers.

The population is the group of people or items being surveyed.

The larger the sample the more reliable the results.

A survey that includes everyone is called a census.

Example

James carries out a survey to find out if people in his town enjoy sport.
He stands outside a football ground and surveys people's opinions as they go in to watch a match.
Write two reasons why this is not a good sample to use.

....

People who watch football usually enjoy sport.
More men than women go to watch football so the survey could be gender biased.

Example

A train company carried out a survey about a local rail service.

They telephoned 100 people from a page of the telephone directory to answer a questionnaire on the rail service.

Write three reasons why this sample could be unrepresentative.

....

Only people who have a land-line telephone (and are not ex-directory) can be included in the sample.
Only people on one page of the telephone directory are included in the sample.
Some people may not be at home when they are phoned.

1 Katy is doing a survey to find out how often people go to the cinema and how much they spend. She stands outside a cinema and asks people as they go in.

Write a reason why this sample could be biased.

2 Sally wants to find out how often people play sport. Sally belongs to an athletics club. She asked members in her athletics club.

How could this sample be biased?

3 James wants to know which flavour crisps he should stock in the school tuck shop.
  a He asks his mum, dad, auntie and uncle.
    Explain why this is not a good sample to use.
  b He asks only Year 11 at his school.
    Explain why this sample could be biased.
  c Describe how James could take a sample of 50 people. (There are 1000 people in his school.)

4 Merlin wants to find out how far people would travel to see their favourite band perform.

  a He asks all his friends.

    Write a reason why this sample could be biased.

  b He goes into town one Saturday morning and asks anyone listening to music on a MP3 player.

    How could this sample be biased?

5 Jenny carries out a survey to find out the most popular band. She asks 10 of her friends – all girls.

How could this sample be biased?

6 Wayne carries out a survey to find out the most popular car colour. He stands on a street corner and notes the colour of the first 15 cars that pass by.

Write a reason why this sample could be biased.

7 Lisa wants to find out how people travel to work.

  a She asks people at a bus stop one morning.

    How could this sample be biased?

  b She opens the telephone directory at a random page and phones everyone on that page.

    How could this sample be biased?

Designing a data collection sheet – two-way tables

This spread will show you how to:

- Design and use data collection sheets and two-way tables

Keywords

Data
Two-way table

- You can use a data collection sheet to collect **data** from a questionnaire or experiment.

- You can use a **two-way table** to collate the two sets of results.

Example

Two questions on a questionnaire are:

'Are you male or female?'   and   'How old are you?'

Design a two-way table to collect this data.

	Under 10	10–19	20–29	30–40	40+	Total
Male						
Female						
Total						

- You can use data in a two-way table to find other results.

Example

The table gives information about Key Stage 4 students at a school.

	Boys	Girls
Year 10	68	117
Year 11	89	126

a Work out the percentage of Key Stage 4 students in Year 10 who are boys.
b Work out the percentage of Key Stage 4 students who are girls.

Find the totals in the table.

	Boys	Girls	Total
Year 10	68	117	**185**
Year 11	89	126	**215**
Total	**157**	**243**	**400**

There are 400 students at Key Stage 4 (68 + 117 + 89 + 126).

a There are 68 Year 10 boys: $\dfrac{68}{400} \times 100 = 17\%$

b There are 243 (117 + 126) girls altogether: $\dfrac{243}{400} \times 100 = 60.75\%$

A02 Functional Maths

1 Katy is doing a survey to find out how often people go to the cinema and how much they spend.

Design a suitable data collection sheet in the form of a two-way table that she could use.

2 Sally wants to find out how often people play sport.

She wants to divide the results into those from males and those from females.

Design a suitable data collection sheet in the form of a two-way table that she could use.

3 James wants to know which flavour crisps he should stock in the school tuck shop.
He also wants to know which year groups prefer which flavours.

Design a suitable data collection sheet in the form of a two-way table that he could use.

4 Merlin wants to find out how far people would travel to see their favourite band perform and how much they would spend on a ticket to watch them.

Design a suitable data collection sheet in the form of a two-way table that he could use.

5 Jenny carries out a survey to find out people's favourite band and how many of that band's CDs they own.

Design a suitable data collection sheet in the form of a two-way table that she could use.

6 Wayne carries out a survey to find the most popular car colour and the most popular make of car.

Design a suitable data collection sheet in the form of a two-way table that he could use.

7 Lisa wants to find out how people travel to work and how long it usually takes them.

Design a suitable data collection sheet in the form of a two-way table that she could use.

8 The table gives information about the number of students in Years 7–9 that attended a school disco.

	Year 7	Year 8	Year 9
Boys	42	58	96
Girls	78	104	122

  a How many students attended the disco?
  b Work out the percentage of students that were
    i Year 8 girls   **ii** in Year 7   **iii** boys.

Averages and spread

This spread will show you how to:
• Find an average and a measure of spread for a data set

Keywords
Average
Interquartile range
Lower quartile
Mean
Measure of spread
Median
Mode
Range
Upper quartile

You can summarise data using an **average** and a **measure of spread**.

p.146

• An average is a single value.
There are three types of average:
   – the **mode** is the value that occurs most often
   – the **median** is the middle value when the data are arranged in order
   – the **mean** is calculated by adding all the values and dividing by the number of values.

• Spread is a measure of how widely dispersed the data are.
Two measures of spread are:
   – the range
   – the **interquartile range** (IQR).

If there are one or more extreme values the IQR is a better measure of spread than the range.

> An extreme value is a value well outside the range of the rest of the data.

• Range = highest value − lowest value

• IQR = upper quartile − lower quartile

> **Lower quartile** $= \frac{1}{4}(n + 1)$th value.

> **Upper quartile** $= \frac{3}{4}(n + 1)$th value.

Example

Louise collected data on the number of times her friends went swimming in one month.

   4   7   22   1   6   2   1   5   6   6   4

Work out the: **a** range   **b** mode   **c** mean
                      d median   **e** interquartile range.

In order the data are: 1   1   2   4   4   5   6   6   6   7   22

a Range = 22 − 1 = 21

b Mode = 6

c Mean = (4 + 7 + 22 + 1 + 6 + 2 + 1 + 5 + 6 + 6 + 4) ÷ 11
  = 64 ÷ 11 = 5.8

> The mean does not have to be an integer even if all the data values are integers.

d There are 11 values.
  Median = $\frac{11 + 1}{2}$ = 6th value = 5

e Interquartile range = upper quartile − lower quartile

  Lower quartile = $\left(\frac{11 + 1}{4}\right)$th value   Upper quartile = $\left(\frac{3(11 + 1)}{4}\right)$th value
    = $\left(\frac{12}{4}\right)$th value   = 9th value
    = 3rd value   = 6
    = 2

  IQR = 6 − 2 = 4

> If there are n values
> Median value $= (\frac{n + 1}{2})$th value.

1 For these sets of numbers work out the
   i range   **ii** mode   **iii** mean
   iv median   **v** interquartile range.
   a 5, 9, 7, 8, 2, 3, 6, 6, 7, 6, 5
   b 45, 63, 72, 63, 63, 24, 54, 73, 99, 65, 63, 72, 39, 44, 63
   c 97, 95, 96, 98, 92, 95, 96, 97, 99, 91, 96
   d 13, 76, 22, 54, 37, 22, 21, 19, 59, 37, 84
   e 89, 87, 64, 88, 82, 88, 85, 83, 81, 89, 90
   f 53, 74, 29, 32, 67, 53, 99, 62, 34, 28, 27, 27, 27, 64, 27
   g 101, 106, 108, 102, 108, 105, 106, 109, 103, 105, 107, 104,
      104, 105, 105

2 For the set of numbers in question **1e**, explain why the interquartile range is a better measure of spread to use than the range.

3 For the set of numbers in question **1f**, explain why the mode is not the best average to use.

4 **a** Subtract 100 from each of the numbers in question **1g** and write down the set of numbers you get.
   b For your set of numbers in **a**, work out the
     i range   **ii** mode   **iii** mean
     iv median   **v** interquartile range
   c Compare your answers for the measures of spread in part **b i** and **v** and **1g i** and **v**.
     What do you notice?
   d Add 100 to your answers for the measures of average in part **b ii**, **iii** and **iv**.
     Compare these answers to the answers you got in question **1g**.
     What do you notice?
   e Give a reason for what you noticed in parts **c** and **d**.

A02 Functional Maths

5 A scientist takes two sets of measurements from her experiment.
Her results are:

Set A: 0 99 99 100 100 100 100 100 101 101 200
Set B: 0   0 99   99 100 100 100 101 101 200 200

   a For each set of measurements, work out the
     i range   **ii** mode   **iii** mean
     iv median   **v** interquartile range
   b Discuss what you notice about the measurements and your answers to part **a**.

A03 Problem

6 Repeat question **1d** with the addition of a twelfth number 90.

Mean of two combined data sets

This spread will show you how to:

- Find the mean of two combined data sets

- **Mean** $= \dfrac{\text{Sum of all values}}{\text{Number of values}}$

You can combine two data sets to form one larger data set by finding the mean of all the data.

Example

32 students, 12 boys and 20 girls, in class 8Z sat a maths test.
The boys' mean mark was 63%.
The girls' mean mark is 78%.
Work out the mean mark for class 8Z.

....

Boys: Total sum of marks $63 \times 12 = 756$
Girls: Total sum of marks $78 \times 20 = 1560$
Total sum of marks for boys and girls $756 + 1560 = 2316$
Mean mark of all students
    $2316 \div 32 = 72.375\% = 72\%$ to nearest whole mark.

Example

50 students answered a survey question about time spent on the internet one evening.
30 of the students were boys and 20 were girls.

The mean time spent on the internet by all 50 students was 18 minutes.
The mean time spent on the internet by the 30 boys was 24 minutes.

Work out the mean time spent on the internet by the 20 girls.

....

Total time spent on internet by all 50 students:
    $50 \times 18 = 900$ minutes

Total time spent on the internet by the 30 boys:
    $30 \times 24 = 720$ minutes

Total time spent on the internet by the 20 girls:
    $900 - 720 = 180$ minutes

Mean time spent on internet by the 20 girls:
    $180 \div 20 = 9$ minutes

Example

There are 13 boys and 16 girls in a class.

In a test, the mean mark for the boys was p.
In the same test, the mean mark for the girls was q.

Find an expression for the mean mark of all 29 students.

....

Mean $= \dfrac{13p + 16q}{29}$

1 An athletics club has 100 members, 60 boys and 40 girls.

The mean time the boys spent training one day was 86 minutes.
The mean time the girls spent training one day was 72 minutes.

Work out the mean time spent training on one day for all 100 members of the athletics club.

2 There are 120 students in Year 11 at St Edmunds school.
75 are girls and 45 are boys.

The mean time spent on homework each week for boys is 5.2 hours.
The mean time spent on homework each week for girls is 8.6 hours.

Work out the mean time spent on homework for all 120 students in Year 11 at St Edmunds school.

3 Of the students in Year 13 at St Edmunds school, 60 boys and 20 girls have passed the driving test.

The mean number of driving lessons that all 80 students had before passing the test was 19.75.
The mean number of driving lessons for the boys was 12.

Work out the mean number of driving lessons for the girls.

4 The mean mark in a statistics test for class 10Z was 84%.

There are 32 students in the class, 12 of whom are girls.
The mean mark in the test for these girls was 93%.

Work out the mean mark in the statistics test for the boys.

5 Thirty boys and girls were asked how many times they had visited the cinema in the past year.

The average number of times was 5.4.
The average number of times for the 12 boys that were asked was 2.5.

Work out the average number of times the girls in the group visited the cinema in the past year.

6 A fitness test was taken by 25 girls and 52 boys.

The average fitness score for the girls was 6.4, and the average fitness score for the boys was 9.2.

Work out the average fitness score for the whole group.

Large data sets – averages and range

This spread will show you how to:

- Use frequency tables to find the averages and range of a data set
- Use estimates of averages and range to summarise large data sets

Keywords

Frequency table
Mean
Median
Mode
Range

- You can put large amounts of data in a **frequency table**.
- You use the averages and the **range** to summarise the data.

Example

The table shows the length of the words in the answers to a crossword puzzle.

For these data, work out the
a mode
b median
c mean
d range.

Word length	Frequency
4	3
5	5
6	7
7	8
8	3
9	1

Word length	Frequency	Word length × frequency
4	3	4 × 3 = 12
5	5	5 × 5 = 25
6	7	6 × 7 = 42
7	8	7 × 8 = 56
8	3	8 × 3 = 24
9	1	9 × 1 = 9
Total	**27**	**168**

Total number of words.

Total number of letters.

a Mode = 7   Words with 7 letters have the highest frequency.

b $\frac{1}{2}(27 + 1) = 14$ so the 14th value is the median
The 14th value is in the 'Word length 6' group.
Median = 6

c Mean = $\dfrac{\text{Total number of letters}}{\text{Total number of words}}$

$= 168 \div 27 = 6.2$

p.280

d Range = 9 − 4 = 5   Longest – shortest word length.

1 The tables of data give information about the length of words in four different crosswords.
For each table, copy the table, add an extra working column and find the

i mode **ii** median **iii** mean **iv** range.

a

Word length	Frequency
4	2
5	5
6	4
7	2
8	2

b

Word length	Frequency
3	3
4	4
5	9
6	5
7	2

c

Word length	Frequency
4	6
5	3
6	5
7	4
8	2
9	5
10	2

d

Word length	Frequency
3	5
4	4
5	6
6	7
7	7
8	4
9	2

2 Jo had 16 boxes of matches.
She counted the number of matches in each box.
The table gives her results.

Number of matches	Frequency
41	2
42	7
43	4
44	3

Work out the mean number of matches in a box.

3 Brian played in 24 hockey matches one season.
The table gives information about the number of goals scored in these matches.

Number of goals scored	Frequency
1	6
2	9
3	4
4	2
5	3

Work out the mean number of goals scored.

Summary

Check out
You should now be able to:

- Design an experiment or survey
- Design a questionnaire and use data collection sheets and two-way tables
- Select and justify methods of sampling to investigate a population, including random sampling
- Identify possible sources of bias
- Calculate averages and range of data sets with discrete and continuous data

Worked exam question

Naomi wants to find out how often adults go to the cinema.
She uses this question on a questionnaire.

> "How many times do you go to the cinema?"
>
> ☐ ☐ ☐
>
> Not very often   Sometimes   A lot

a Write down two things wrong with this question. (2)

b Design a better question for her questionnaire to find out how often adults go to the cinema.
You should include some response boxes. (2)

(Edexcel Limited 2008)

a

1 'Not very often', 'Sometimes' and 'A lot' are vague and will be misunderstood.
2 There is not enough choice of responses. More response boxes are needed.

Write down two reasons.

*Another answer for **a** could be: No mention of time in the question.*

b

How many times did you go to the cinema last month?

0	1–2	3–4	5 or over
☐	☐	☐	☐

The question includes a time period.

The response boxes cover all possibilities.

Exam question

1 Ali found out the number of rooms in each of 40 houses in a town.
He used the information to complete the frequency table.

Number of rooms	Frequency
4	4
5	7
6	10
7	12
8	5
9	2

Ali said the mode is 9
Ali is wrong.

a Explain why. (1)
b Calculate the mean number of rooms. (3)
c Beccy found out the number of rooms in each of 80 houses
in the same town.
She used the information to complete the frequency table below.

Number of rooms	Frequency
4	10
5	12
6	15
7	18
8	17
9	8

Find the median number of rooms. (1)

d The median number of rooms in Ali's table is 6

Which of the two medians, Ali's or Beccy's, is more likely to
give the more reliable estimate for the median number of
rooms for a house in this town?
Give a reason for your answer. (1)

(Edexcel Limited 2007)

85

Angles and circles

In everyday life you usually rely on experience to decide if something is true.
In mathematics you start with something you already know is true and use steps
of logical reasoning to show how it follows that something else is true! This is called
a Mathematical proof.

What's the point?

In real life it's mainly pure mathematicians that prove the mathematics they use.
However, the skills gained in being able to set out a logical train of thinking lie at the
heart of every mathematician's work.

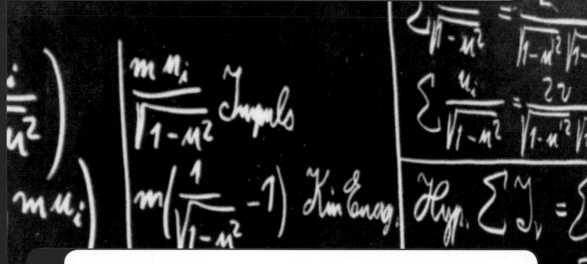

You should be able to

■ **use the properties of angles on a line and around a point**

1 The diagrams show angles on a straight line or at a point.
Work out the missing angles.

■ **know and use the sum of the angles in a triangle and quadrilateral**

2 Work out the missing angle in these shapes.

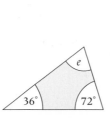

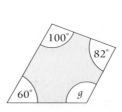
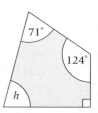

What I need to know

KS3 Experience of geometric arguments
Knowledge of the types of quadrilaterals

What I will learn

- Identify quadrilaterals and use their properties
- Prove and apply theorems about circles

What this leads to

Logical reasoning

Inside a museum is a circular room. The room is to be lit by spotlights. Each spotlight shines light in a beam 60° wide as shown in the diagram below.

How many spotlights will be needed to completely light the room and where should they be positioned?

Investigate using different spotlights with wider or narrower beams of light.

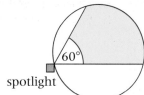

spotlight

Angles in straight lines

This spread will show you how to:
- Use parallel lines, alternate angles, corresponding angles and interior angles

- Angles are formed when two lines cross, as at this crossroads.

$$a = c \quad \text{and} \quad b = d$$

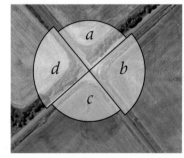

- **Vertically opposite** angles are equal.

Parallel lines are lines that never cross.

You need to remember these angle facts for parallel lines.

- **Alternate** angles are equal.

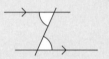

These are sometimes called Z angles.

- **Corresponding** angles are equal.

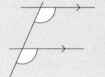

These are sometimes called F angles.

- **Interior** angles are **supplementary**. $a + b = 180°$

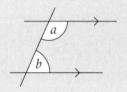

Supplementary angles add up to $180°$.

When you work out angles, you should always say which angle fact you are using.

Example

Find the missing angles. Give reasons for your answers.

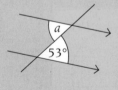

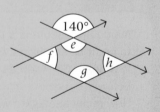

$a = 53°$ Alternate angles.

$b = 135°$ Vertically opposite.
$c = 135°$ Corresponding.
$d = 45°$ Interior angles.

$e = 140°$ Vertically opposite.
$f = 40°$ Interior angles.
$g = 140°$ Interior angles.
$h = 40°$ Interior angles.

1 Copy the diagrams.

  i Use colour to show alternate angles in each diagram.

  ii Use different colours to show corresponding angles in each diagram.

a    **b**    **c**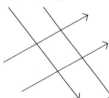

2 Find the missing angles in each diagram.
Write down which angle fact you are using each time.

a 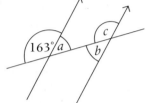   **b** 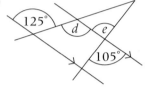   **c**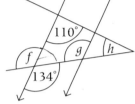

3 **a** Angles x, y and z are on a straight line.
Write down the value $x + y + z$.

  b Use alternate angles to work out the two
missing angles in the triangle.

  c Use your answers to **a** and **b** to show
that angles in a triangle add up to 180°.

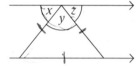

4 **a** Work out the missing angles in this diagram.

  b Describe the quadrilateral formed
between the pairs of parallel lines.

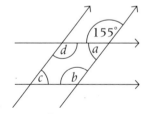

5 Work out the missing angles.

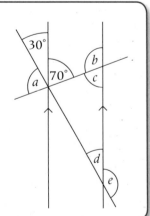

Angles in polygons

This spread will show you how to:
- Calculate and use the interior and exterior angles of polygons

Keywords
Exterior angle
Interior angle
Polygon
Vertex

A **polygon** is a closed shape with three or more straight sides.

- A regular polygon has all sides the same length and all interior angles equal.

A square
is a regular
quadrilateral.

The **interior angles** are inside the polygon.

The **exterior angles** are made by extending each side in the same direction.
Exterior angles are outside the polygon.

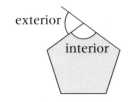

exterior

interior

- For any polygon the exterior angle sum = 360°.
- For a regular polygon with *n* sides, each exterior angle = 360° ÷ *n*.

A pentagon divides
into 3 triangles.
Angles in a triangle
add up to 180°.
So interior angle
sum of a
pentagon =
3 × 180 = 540°.

You can divide any polygon into triangles by drawing diagonals from a **vertex** (corner).

The number of triangles is always two less than the number of sides.

- For a polygon with *n* sides the interior angle sum = (*n* − 2) × 180°.

Example

In a regular octagon find
a an interior angle
b an exterior angle.

....

a An octagon has 8 sides and divides into 6 triangles.
Interior angle sum is 6 × 180° = 1080°
Each interior angle is 1080° ÷ 8 = 135°
b Each exterior angle is 360° ÷ 8 = 45°

Example

An irregular hexagon has angles 108°, 92°, 120°, 134°, 115° and *x*.
Find the size of angle *x*.
....

Any hexagon has 6 sides and divides into 4 triangles.
Sum of interior angles is 4 × 180° = 720°
Sum of the 5 given angles is
108° + 92° + 120° + 134° + 115° = 569°
So *x* = 720° − 569° = 151°

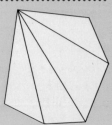

1 **a** Copy the table of regular polygons.
   b Draw each polygon and divide it into triangles by drawing diagonals from a vertex.
   c Complete the table.

Shape	△	▢	⬠	⬡	◯	◯
Number of sides						
Number of triangles the shape splits into						
Sum of the interior angles in the shape						
Size of one interior angle						
Size of one exterior angle						

2 Find the missing angles in these quadrilaterals.

a
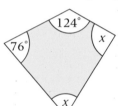

b
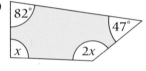

c

3 Find the missing angles in these irregular polygons.

a
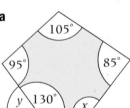

b
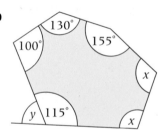

4 **a** Find the exterior angle marked x in each triangle.

i
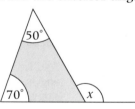

ii
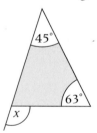

iii
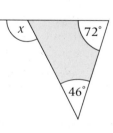

   b Use your answers to part **a** to help you copy and complete this statement.
   The exterior angle of a triangle is equal to_____

5 How many sides does a regular polygon have if
   a it has exterior angle 30°?
   b it has interior angle 135°?

This spread will show you how to:

- Understand and explain congruence and symmetry

Keywords
Congruent
Corresponding
Symmetry

- In **congruent** shapes, **corresponding** lengths are equal and corresponding angles are equal.

Example

Explain why these triangles are congruent.

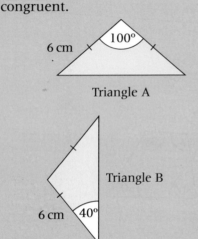

Triangle A

Triangle B

Both triangles are isosceles, with two equal sides of 6 cm.

Triangle A: angle 100° between equal sides, so other two angles must each be 40°.

Triangle B: base angles both 40°, so angle between two equal sides is 100°.

Side opposite 100° angle must be same length as side opposite 100° angle in Triangle A.

Three angles and three sides same in both triangles, so congruent.

Congruent shapes may be reflections, rotations or translations of each other.

Base angles of isosceles triangle are equal.
$180° - 100° = 80°$
$80° \div 2 = 40°$

$180 - (2 \times 40) = 100°$

The conditions for congruence are the same as required for the construction of a unique triangle

p.366

Example

Explain why these triangles are not congruent.

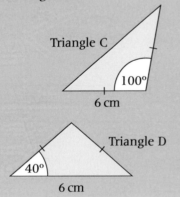

Triangle C

Triangle D

Both triangles are isosceles.

Both triangles have angles 40°, 40°, 100°.

Triangle D: lengths of the two equal sides not given. If they were 6 cm, as in Triangle C, then Triangle D would be equilateral.

Triangle D is not equilateral, so its equal sides are not 6 cm.

So the two triangles are not congruent.

Using base angles of isosceles triangle, as in the first example.

In an equilateral triangle, angles are 60°, 60°, 60°.

They have equal angles, but the side lengths are *not* equal.

A line of **symmetry** divides a shape into two congruent shapes.

A plane of symmetry divides a 3-D shape into two congruent 3-D shapes.

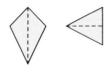

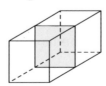

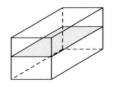

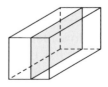

1 Explain whether or not these pairs of triangles are congruent.

a   b 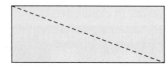   c

2 The dotted lines show two different ways of splitting a rectangle into two congruent shapes.

   a Draw two copies of the rectangle.
     Draw dotted lines to show two more ways of splitting the rectangle into congruent shapes.
   b For each, state whether or not the dotted line is also a line of symmetry.

3 Copy these shapes.
Draw dotted lines to split them into congruent shapes.

You may need more than one copy of each shape.

4 Copy these 3-D shapes.
Show how each 3-D shape can be split into two congruent 3-D shapes.

You may need more than one copy of each shape.

A03 Problem

5 How many planes of symmetry divide a cylinder into congruent shapes?

G2.4 Quadrilaterals

This spread will show you how to:
- Recall the definitions of special types of quadrilaterals
- Classify quadrilaterals by their geometric properties

Keywords
Kite
Parallelogram
Quadrilateral
Rectangle
Rhombus
Square
Trapezium

- A **quadrilateral** is a shape with four straight sides.
- Angles in a quadrilateral add up to 360°.

Properties of quadrilaterals

	Square	Rhombus	Rectangle	Parallelogram	Trapezium	Kite
1 pair opposite sides parallel	✓	✓	✓	✓	✓	
2 pairs opposite sides parallel	✓	✓	✓	✓		
Opposite sides equal	✓	✓	✓	✓		
All sides equal	✓	✓				
All angles equal	✓		✓			
Opposite angles equal	✓	✓	✓	✓		
Diagonals equal	✓		✓			
Diagonals perpendicular	✓	✓				✓
Diagonals bisect each other	✓	✓	✓	✓		
Diagonals bisect the angles	✓	✓		✓		
2 pairs of adjacent sides equal	✓	✓				✓

Example

Work out the area of this rhombus.

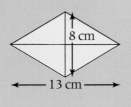

The diagonals bisect each other and are perpendicular.
Opposite angles are equal.
Diagonals bisect the angles.

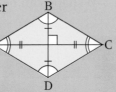

So the diagonal AC is a line of symmetry and the triangles ABC and ACD are congruent.

Each triangle has base 13 cm and height 4 cm.
Area of each triangle = $\frac{1}{2} \times 13 \times 4 = 26$ cm^2
Area of rhombus = $2 \times 26 = 52$ cm^2

Look at the table for the properties of a **rhombus**.

A line of symmetry divides a shape into two congruent shapes.

Area of triangle = $\frac{1}{2}bh$.

Example

Are these statements true or false? Give reasons for your answer.

a All squares are rhombuses

b Parallelograms are rectangles

a True, all properties of a rhombus are also properties of a square.

b False, a **parallelogram** does not have all its angles equal, nor are its diagonals perpendicular.

A **square** is a special type of rhombus with all angles equal.

A **rectangle** is a special type of parallelogram with all angles equal.

1 The lengths of the diagonals of a rhombus are 5 cm and 9 cm.
Find the area of the rhombus.

2 A square has diagonals 10 cm long.
  a Sketch the square.
  b Find the area of the square.

3 Jenny draws these three kites each with diagonals 6 cm and 14 cm.

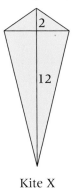
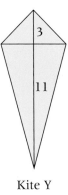
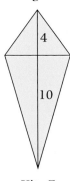

Kite X   Kite Y   Kite Z

DID YOU KNOW?

The kite is a rare bird of prey that hovers in the wind, which influenced the naming of the toy kite. This in turn influenced the naming of the kite shape.

  a Find the areas of kites X, Y and Z.
  b Describe how to work out the area of any kite.

4 The lengths of the diagonals of a kite are 8 cm and 15 cm.
Find the area of the kite.

Use your method from question **3**.

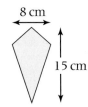

8 cm

15 cm

5 Here is a diagram of a parallelogram, P, and a rhombus, R.

P   R

A03 Problem

One diagonal been drawn inside each shape.
Has either shape been split into two congruent triangles?
Give a reason for your answer.

6 Are these statements true or false? Give reasons for your answers.
  a All squares are rectangles.
  b All kites are rhombuses.
  c All rhombuses are rectangles.

Circle theorems

This spread will show you how to:

- Recognise the parts of a circle, using correct vocabulary to describe them
- Calculate angles in a circle by using the circle theorems

Keywords
Arc
Chord
Diameter
Segment

A straight line joining two points on the circumference is a **chord**.

A chord divides the circle into two **segments**.

An **arc** is the part of the circumference that joins two points.

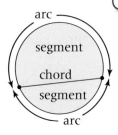

The **circumference** is the distance around the outside of a cicle.

The **diameter** is the longest chord.

- The angle at the centre of a circle is double the angle at the circumference from the same arc.

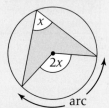

- Angles from the same arc in the same segment are equal.

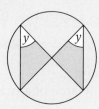

Example

Find the missing angles. Give a reason for each answer.

a

a $a = 70°$ Angle at the centre is double the angle at the circumference.

b

b $b = 360° - 104° = 256°$ Angles at a point.

c

c $c = 110°$ Angle at the centre is double the angle at the circumference.

The angle at the centre is a reflex angle.

d, e

d $d = 30°$ Angles on same arc are equal.
e $e = 65°$ Angles on same arc are equal.

f

f $f = 104°$ Angle at the centre is double the angle at the circumference.

1 Find the missing angles.
Give a reason for each answer.

a
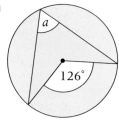

b
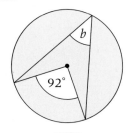

c
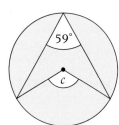

d
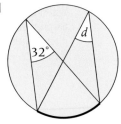

e
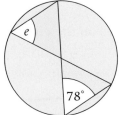

f
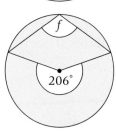

g
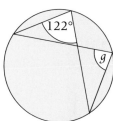

h
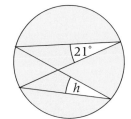

i
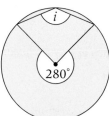

j
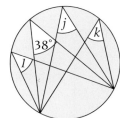

k
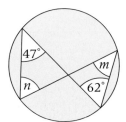

l
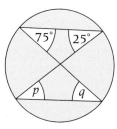

m
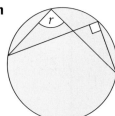

n
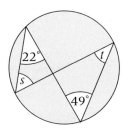

o
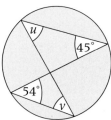

p
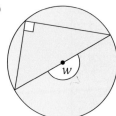

q
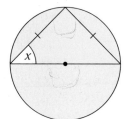

r
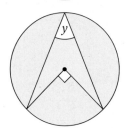

Further circle theorems

This spread will show you how to:
● Calculate angles in a circle by using the circle theorems

Keywords
Circumference
Cyclic
  quadrilateral
Semicircle

● An angle in a **semicircle** is a right-angle.

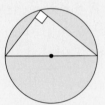

Angle at the centre is 180°. This is double the angle at the circumference. So the angle at the circumference is $\frac{1}{2}$ of 180° = 90°

A **cyclic quadrilateral** has all four vertices on the **circumference** of a circle.

● Opposite angles in a cyclic quadrilateral add up to 180°.

$a + b = 180°$
$x + y = 180°$

Example

Find the missing angles.
Give a reason for each answer.

You will need to use the angle facts from **G3.4**

a

a $a = 90°$

Angle in a semicircle is a right angle.

b

b $82° + b = 180$
$b = 98°$

Opposite angles in a cyclic quadrilateral add up to 180°.

c,d

c $c = 180° - 64° = 116°$
The obtuse angle at the centre is
$2 × 64° = 128°$

Opposite angles in cyclic quadrilateral add up to 180°.
Angle at the centre is double the angle at the circumference.

d $d = 360° - 128° = 232°$ Angles at a point.

e,f

e $e = 40°$
f $f = 180° - 40° = 140°$

Angles on same arc are equal.
Opposite angles in cyclic quadrilateral add up to 180°.

1 Find the missing angles.
Give a reason for each answer.

a
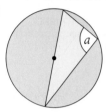

b
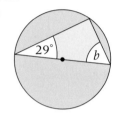

c
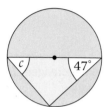

d
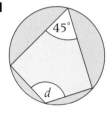

e
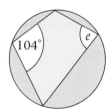

f
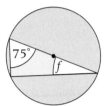

g
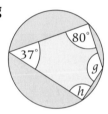

h
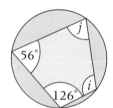

i
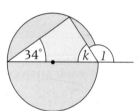

j
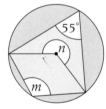

k
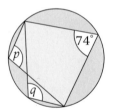

l
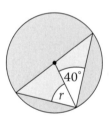

m
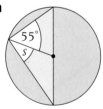

n
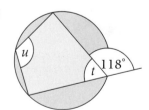

o
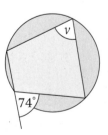

p
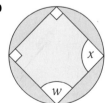

q
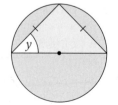

r
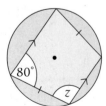

This spread will show you how to:

- Apply the tangent theorems to circle problems

- The angle between the **tangent** and the **radius** at a point is a **right angle**.

From a point outside a circle you can draw two tangents to the circle.

- Two tangents drawn from a point to a circle are equal.

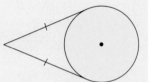

Example

Find the missing angles and the missing length.
Give a reason for each answer.

a

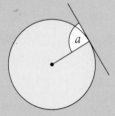

a $a = 90°$

Angle between tangent and radius at a point is a right angle.

b

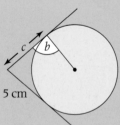

5 cm

b $b = 90°$

$c = 5\,cm$

Angle between tangent and radius at a point is a right angle.
Two tangents drawn from a point to a circle are equal.

c

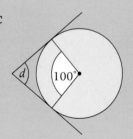

c The given angles are d and 100°. The two other angles are 90°
$d + 100° + 90° + 90° = 360°$
$d = 80°$

Angle between radius and tangent is a right angle.

Angles in a quadrilateral add up to 360°.

1 Find the missing angles and the missing lengths.
Give a reason for each answer.

a
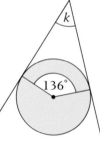

b
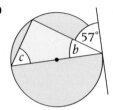

c
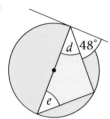

d
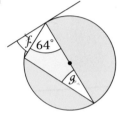

e
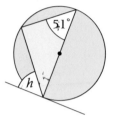

f
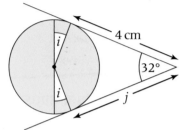

g

h
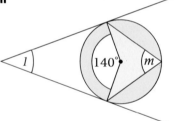

i
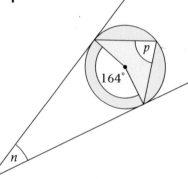

j
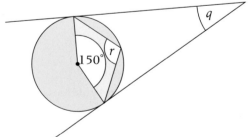

k
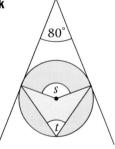

l
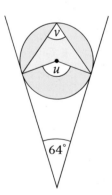

Summary

Check out
You should now be able to:

- Use properties of right-angled, equilateral isosceles and scalene triangles
- Understand and use the angle properties of parallel lines
- Calculate and use the interior and exterior angles of polygons
- Know the properties of special types of quadrilaterals
- Understand congruence and identify congruent shapes
- Use the correct vocabulary to describe parts of a circle
- Calculate angles in a circle using the circle theorems

Worked exam question

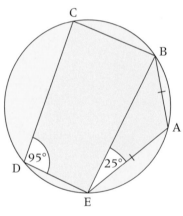

Diagram NOT accurately drawn.

A, B, C, D and E are five points on a circle.
Angle $BEA = 25°$ and angle $CDE = 95°$.
$AB = AE$.

a   **i** Work out the size of angle BAE.
    ii Give reasons for your answer.   (3)

b   Work out the size of angle CBE.   (1)

(Edexcel Limited 2007)

a

> **i** $25° \times 2 = 50°$
> $180° - 50° = 130°$
>
> **ii** There are 2 equal angles in the isosceles triangle.

Show the working out for the calculation.

You need to mention BOTH the two equal angles AND the isosceles triangle.

b

> $180° - 95° = 85°$

You do not need to give a reason.

Exam questions

1

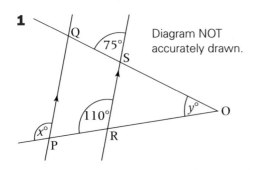

Diagram NOT accurately drawn.

PQ is parallel to RS.

OSQ and ORP are straight lines.

a **i** Write down the value of x.

   ii Give a reason for your answer. (2)

b Work out the value of y. (2)

(Edexcel Limited 2008)

2

Diagram NOT accurately drawn.

The diagram shows part of a regular 12-sided polygon.
Work out the size of the angle marked a. (3)

3

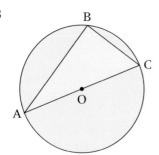

Diagram NOT accurately drawn.

A, B and C are points on the circumference of a circle, centre O.
AC is a diameter of the circle.

a **i** Write down the size of angle ABC.

   ii Give a reason for your answer. (2)

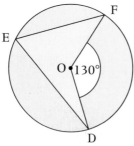

Diagram NOT accurately drawn.

D, E and F are points on the circumference of a circle, centre O.
Angle $DOF = 130°$.

b **i** Work out the size of angle DEF.

   ii Give a reason for your answer. (2)

(Edexcel Limited 2008)

Functional Maths 2: Sandwich shop

The manager of a catering company can use data about customer numbers in order to spot trends in customer behaviour and to plan for the future.

Simply Sandwiches

Simply Sandwiches

es, paninis, baguettes and salads

Simply Sandwiches

Customer numbers at 'Simply sandwiches' takeaway over a given two-week period were:

Day	Number of Customers	
	Week 1	Week 2
Monday	50	54
Tuesday	68	60
Wednesday	47	53
Thursday	58	57
Friday	52	56
Saturday	76	70
Total		

Simply Sandwiches

Work out an appropriate average number of customers for

a) each day of the week
b) the whole week in total.

How does your answer to b) differ if

c) you exclude Saturdays
d) a 24-person coach trip arrives on the second Wednesday?

Construct a pie chart to show what percentage of customers visited the sandwich shop on each day during this two-week period.

Comment on the spread of the data, referring to the data and your chart.

This frequency polygon shows customer numbers during each hour on the first Saturday, the busiest day during this two-week period:

What was the busiest/quietest time of day?

What can you say about the relationship between time of day and customer numbers? Would you expect every day of the week to have a similar pattern?

Justify your answers referring to the data.

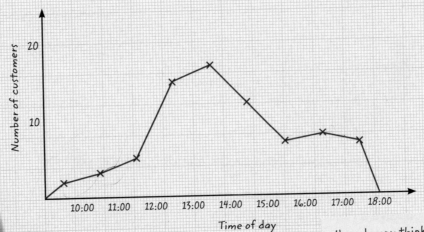

Check

CHECK NUMBER

143001

How do you think the manager could use such data about customer numbers?

A manager can use data about customer numbers to estimate how much stock to order each week. In reality, limitations due to space and the shelf life of products also apply.

In the second week of the two-week period at 'Simply sandwiches', total percentage sales of the different varieties of sandwiches were:

Variety	Ham	Cheese	Hummous	Tuna	Chicken
% sales	26	18	10	15	31

How many of each sandwich were sold during this week? (assuming every customer bought one sandwich)

The manager does a weekly stocktake every Sunday before placing the order for the following week.

The stocktake figures for this week were:

Product	Bread	Ham	Cheese	Hummous	Tuna	Chicken
Stock (packs)	6	2.5	3	2	1.5	1

Note that:

Each loaf of bread makes 20 sandwiches;

Each pack of ham, cheese and chicken contains 10 portions;

Each tub of hummous contains 8 portions;

Each can of tuna contains 14 portions.

Use the information given to estimate the amount of each product that the manager should order to last for the following week.

Record your estimations in a table.

The stock will be delivered on Wednesday.

A coach trip of 24 people arrives unexpectedly and places the following order on Tuesday, before the new stock arrives.

Sandwich	Ham	Cheese	Hummous	Tuna	Chicken
Quantity	4	7	5	3	5

Would 'Simply sandwiches' be able to cater for this order?

How would you advise the manager to prepare for such situations in the future?

Simply Sandwiches

sandwiches, paninis, baguettes and salads

Ratio and proportion

Lengths that are in the Golden ratio, $1 : \varphi$, where the Greek letter phi stands for the number 1.6180339887..., are thought to be the most pleasing to the eye. The ancient Egyptians used it in the construction of the pyramids and the Greeks used it in their architecture.

What's the point?

The concepts of ratio and proportion are central to the work of artists, designers and architects.

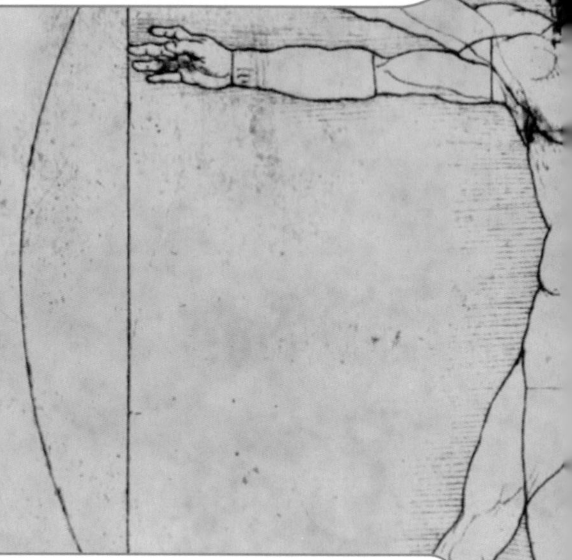

You should be able to

■ **calculate ratios**

1 Gill and Paul share £500 between them.
   Gill gets four times as much as Paul.
   How much does each person get?

■ **work with percentages**

2 In a sale, you can buy three items for the price of two.
   What percentage decrease on the original price does this represent?

■ **work with proportions**

3 Robyn mixes black and white paint to make two batches of grey paint.

   Batch A contains 3 litres of black paint and 7 litres of white paint.
   Batch B contains 5 litres of black paint and 5 litres of white paint.

   a Which batch is the darker colour?
      Explain your reasoning.
   b Robyn wants to make 30 litres of paint the same shade as Batch A.
      How much black paint and white paint will she need?
      Explain your answer.

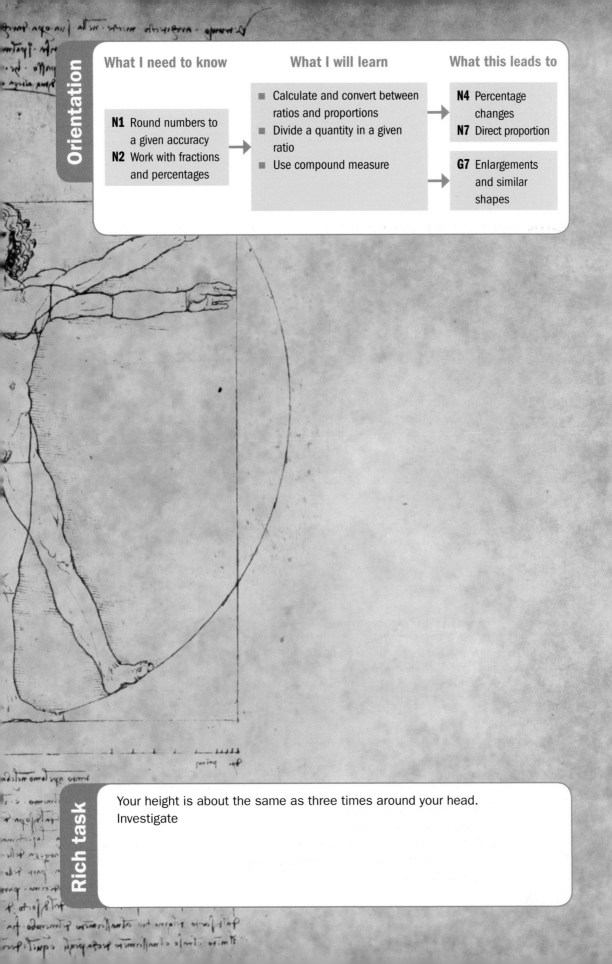

Orientation

What I need to know

N1 Round numbers to a given accuracy
N2 Work with fractions and percentages

What I will learn

■ Calculate and convert between ratios and proportions
■ Divide a quantity in a given ratio
■ Use compound measure

What this leads to

N4 Percentage changes
N7 Direct proportion

G7 Enlargements and similar shapes

Rich task

Your height is about the same as three times around your head.
Investigate

This spread will show you how to:

- Use ratio notation
- Simplify a ratio by cancelling common factors
- Divide a quantity in a given ratio

A **ratio** is used to compare quantities.

- You **simplify** a ratio by cancelling common factors.
  A class contains 16 girls and 12 boys. The ratio of boys to girls is 16 : 12, and you can simplify this ratio to 4 : 3.

- To divide a quantity in a given ratio (for example, dividing 20 m in the ratio 3 : 1)
  - first find the total number of parts: 3 + 1 = 4
  - now find the size of each part: 20 ÷ 4 = 5
  - multiply to find each share: 5 × 3 = 15, and 5 × 1 = 5

Example

In a group of 14 people, eight have brown eyes and six have green eyes. Write the ratio of eye colours in its simplest form.

brown : green = 8 : 6 = 4 : 3

Example

Simplify these ratios

a 1 m to 40 cm   **b** 6 hours to $2\frac{1}{2}$ days

a 1 m = 100 cm so the ratio of
1 m to 40 cm = 100 cm : 40
= 5 : 2

b $2\frac{1}{2}$ days = 60 hours so the ratio
6 hours to $2\frac{1}{2}$ days = 6 : 60
= 1 : 10

Both parts of the ratio must have the same units.

Example

Divide   **a** 120 cm in the ratio 2 : 3   **b** £108 in the ratio 3 : 5

a   2 + 3 = 5   Total number of parts = 5.
120 ÷ 5 = 24   One part = 24 cm.
2 × 24 = 48
3 × 24 = 72
The two lengths are 48 cm and 72 cm.
To check your answer add the two parts: 48 + 72 = 120, correct.

b   3 + 5 = 8
108 ÷ 8 = 13.5
3 × 13.5 = 40.5
5 × 13.5 = 67.5
The two amounts are £40.50 and £67.50.
Check your answer: 40.50 + 67.50 = 108, correct.

Remember to add £ signs and zeros.

1 Here are the numbers of boys and girls in some Year 11 classes.
   For each class, write down the ratio of boys to girls.
   a Class 11A has 17 boys and 13 girls
   b Class 11B has 11 boys and 19 girls
   c Class 11C has 14 boys and 15 girls

2 The table shows the number of
   students in three different classes
   who own pets. Write the ratio of pet
   owners to non-pet owners in each class.

Class	Pet owners	Non-pet owners
11A	7	23
11B	13	17
11C	16	13

3 Write the ratio of the number of vowels to
   the number of consonants in these words.
   Give your answers in their simplest form.
   a JAGUAR **b** PANTHER **c** TIGER **d** LEOPARD The vowels are
   e PAPERBACK **f** PERPETUATE **g** STAGGERS **h** MEIOSIS A, E, I, O and U.

4 Simplify these ratios
   a 6 : 4 **b** 12 : 3 **c** 5 : 10 **d** 2 : 8
   e 6 : 9 **f** 12 : 8 **g** 10 : 15 **h** 21 : 14

5 Divide each of these numbers in the ratio 3 : 2. Show your working.
   a 40 **b** 45 **c** 90 **d** 100
   e 250 **f** 2000 **g** 7500 **h** 9250

6 Divide each of these quantities in the ratio 2 : 1. Show your working.
   a 24 hours **b** 180° **c** 45 minutes **d** €360
   e 246 cm **f** 120 kg **g** 81 km **h** 54 miles

7 Divide £360 in these ratios
   a 1 : 1 **b** 1 : 2 **c** 2 : 1 **d** 3 : 5

8 Karla and Wayne share the tips they receive for working in a café.
   One Saturday, Karla's share was £12.50, and Wayne's was £7.50.
   a Write the ratio of Karla's tips to Wayne's tips in its simplest form.
   b The next week Karla and Wayne shared tips of £22 in the same
   ratio. Find their shares. Show all your working.

9 Mrs Jones wins £500 in a competition. She keeps 40% of the money.
   a How much money does Mrs Jones keep for herself?
   b Mrs Jones shares the rest between her children,
   Annie (8 years old) and Ben (12), in the ratio of their ages.
   Work out how much each child receives. Show your working.

A02 Functional Maths

.160

This spread will show you how to:

- Use ratio notation
- Simplify a ratio by cancelling common factors
- Divide a quantity in a given ratio

Keywords

Ratio
Simplify

You can use **ratio** to divide quantities into more than two amounts.

- To divide 55 m in the ratio 2 : 3 : 5
  - first add: $2 + 3 + 5 = 10$
  - then divide: $55\,m \div 10 = 5.5\,m$.
  - then multiply: $2 \times 5.5\,m = 11\,m$, $3 \times 5.5\,m = 16.5\,m$, and $5 \times 5.5\,m = 27.5\,m$

Check:
$11 + 16.5 + 27.5$
$= 55$

Example

Alan and Betty share a bingo prize of £48 in the ratio of their ages. Alan is 32, and Betty is 26. How much does each get?

Ratio of ages $= 32 : 26 = 16 : 13$ **Simplify** the ratio.

$16 + 13 = 29$, so one part $= £48 \div 29$ Add the parts.

Alan gets $\dfrac{£48}{29} \times 16 = £26.48$

Round to nearest penny.

Betty gets $\dfrac{£48}{29} \times 13 = £21.52$

Check:
£26.48 + £21.52
= £48

Example

Roni weighs 56 kg, Mike weighs 64 kg and Steffi weighs 72 kg. Write the ratio of their weights in its simplest form.

Roni's weight : Mike's weight : Steffi's weight $= 56 : 64 : 72 = 7 : 8 : 9$

Cancel by 8.

Example

The Smith, the Brown and the Jones families go on holiday to Cumbria. They stay in a farmhouse which they rent for £2000. They agree to share the rent according to the number of people in each family. There are 3 in the Smith family, 6 in the Brown family and 8 in the Jones family. How much does each family pay?

Ratio of family size $= 3 : 6 : 8$

$3 + 6 + 8 = 17$, so one part $= £2000 \div 17$

The Smiths pay $\dfrac{£2000}{17} \times 3 = £352.94$

The Browns pay $\dfrac{£2000}{17} \times 6 = £705.88$

The Joneses pay $\dfrac{£2000}{17} \times 8 = £941.18$

Check: £352.94 + £705.88 + £941.18 = £2000

1 Write these ratios in their simplest form.
   a 2 : 4   **b** 2 : 4 : 4   **c** 9 : 3 : 6
   d 10 : 20 : 8   **e** 8 : 20 : 12   **f** 15 : 35 : 10

2 Split £240 in these ratios.
   a 1 : 2   **b** 2 : 3   **c** 5 : 3
   d 1 : 2 : 3   **e** 5 : 1 : 2   **f** 7 : 2 : 3

3 Divide 250 m in these ratios, giving your answers correct to the
   nearest centimetre.
   a 1 : 4   **b** 2 : 3   **c** 2 : 5
   d 6 : 1   **e** 6 : 7   **f** 9 : 5

4 Peter, Bob and Yasmin share prize money of £7500 between them
   in the ratio 9 : 5 : 11. How much do they each receive?

5 Ann, Charles and Edward divide prize money of £850 between
   themselves in the ratio 4 : 3 : 2. Find the amount that each receives,
   giving your answers to the nearest penny.

6 Copy and complete the table to show
   how each quantity can be divided in
   the ratio given. Give your answers to
   a suitable degree of accuracy.

Quantity	Ratio	Share 1	Share 2	Share 3
200 km	2 : 3 : 5			
38 kg	1 : 2 : 3			
450 cm	2 : 3 : 8			
£720	4 : 5 : 10			
95 litres	1 : 6 : 7			

7 A fruit drink contains mango juice, pineapple juice and passion
   fruit juice, in the ratio 4 : 3 : 2 by volume. Calculate the volume
   of each type of juice in a 1-litre pack of the fruit drink.

8 Mrs Williams won a £500 Premium Bond prize, and decided to
   share it among her three children in the ratio of their ages.
   The children are aged 5, 7 and 9.
   a Calculate the amount that each child receives, giving your
   answers correct to the nearest penny.
   b Exactly one year later, Mrs Williams wins another £500 prize.
   Again, she decides to divide it among her children in the ratio
   of their ages. Calculate the amount that each child receives
   this time.

9 Robert and Kathleen are business partners. At start up, Robert
   invested £150 000 and Kathleen invested £200 000. Each year
   they share the profits in the ratio of their investments. In 2004
   the profit amounted to £29 540. Work out each partner's share.

This spread will show you how to:

- Describe and calculate proportions, using fractions, decimals or percentages
- Understand direct proportion and ratio

Proportions can be described and calculated using fractions, decimals or percentages.

There are 12 students in Class 3A.
$\frac{2}{3}$ of them are girls.
Number of girls = $\frac{2}{3} \times 12 = 8$

1000 ml of Quango contains 100 ml of fruit juice.
Proportion of fruit juice = $\frac{100}{1000}$
$= \frac{1}{10}$

Example

p.60

Find **a** $\frac{3}{5}$ of 85   **b** 28% of 360   **c** 120% of 45   **d** $\frac{7}{8}$ of 86

a $\frac{3}{5_1} \times 85^{17} = \frac{3}{1} \times 17 = 51$

> Cancel the 5s before multiplying.

b Estimate: 30% of 400 = 120
By long multiplication, $28 \times 36 = 1008$
So 28% of 360 = 100.8

> Work out 20% of 45 and add it.

c 120% of 45 = (100% of 45) + (20% of 45)
20% of 45 = 45 ÷ 5 = 9 ➡ 120% of 45 = 45 + 9 = 54

> Work out $\frac{1}{8}$ of 86 and then subtract.

d $\frac{1}{8}$ of 86 = 86 ÷ 8 = $10 + \frac{6}{8} = 10\frac{3}{4}$ ➡ $\frac{7}{8}$ of 86 = $86 - 10\frac{3}{4} = 76 - \frac{3}{4} = 75\frac{1}{4}$

Example

What proportion is   **a** 4 of 32   **b** 5 of 35   **c** 12 of 100?
Write your answers as percentages.

a $\frac{4}{32} = \frac{1}{8}$
so 4 is $\frac{1}{8}$ of 32
$\frac{1}{8} = \frac{1}{8} \times 100\%$
$= 12\frac{1}{2}\%$

b $\frac{5}{35} = \frac{1}{7}$
so 5 is $\frac{1}{7}$ of 35
$\frac{1}{7} = \frac{1}{7} \times 100\%$
$= 14.3\%$

c $\frac{12}{100} = \frac{3}{25}$
so 12 is $\frac{3}{25}$ of 100
$\frac{3}{25} = \frac{3}{25} \times 100\%$
$= 12\%$

Example

A 1 kg bag of 'Grow Up' fertiliser contains 45 grams of phosphate. A 500 gram packet of 'Top Crop' fertiliser contains 20 grams of phosphate.
What is the proportion of phosphate in each fertiliser?

Proportion of phosphate in 'Grow up' = $\frac{45}{1000} \times 100\%$
$= 4.5\%$

> Percentages are easy to compare.

Proportion of phosphate in 'Top Crop' = $\frac{20}{500} \times 100\%$
$= 4\%$

1 One tenth ($\frac{1}{10}$) of the weight of a soft drink is sugar. Find the amount of sugar in these weights of drink.

   a 750 g   **b** 45 g   **c** 1 kg   **d** 1250 g

You should be able to do all of these mentally.

2 Three fifths ($\frac{3}{5}$) of the volume of a fruit cocktail is orange juice. Find the amount of orange juice in these volumes of fruit cocktail. You should show all of your working.

   a 150 ml   **b** 380 ml   **c** 2 litres   **d** 280 cm^3

*See Example **1** part **a**.*

3 Calculate
   a 120% of 50 g   **b** 90% of 40 mm
   c 95% of 400 g   **d** 80% of 39 km

*Try these mentally. See Example **1** part **c**.*

4 What proportion is
   a 5 of 50   **b** 6 of 80   **c** 9 of 45   **d** 15 of 80?

5 Write your answers from question **4** as percentages.

6 Find these proportions, giving your answers as
   i fractions in their lowest terms   **ii** percentages.
   a 3 out of every 20   **b** 4 parts in a hundred
   c 8 out of 20   **d** 64 in every thousand

7 Find these.
   a 38.3% of 192 mm   **b** $\frac{3}{8}$ of £840

   c 19% of £52.00   **d** $\frac{7}{8}$ of 960 kg

A02 Functional Maths

8 A 250 ml glass of fruit drink contains 30 ml of pure orange juice. What proportion of the drink is pure orange juice? Give your answer as
   a a fraction in its lowest terms
   b a percentage.

9 Antifreeze contains 10% concentrated antifreeze; the rest is water.
   a How much concentrate is contained in 2 litres of antifreeze?
   b How much antifreeze can you make with 300 ml of concentrate?
   c How much water would you add to 200 ml of concentrate?

A03 Problem

10 Samantha wins £4500 in a competition. She gives $\frac{1}{3}$ to her mother and $\frac{1}{5}$ to her sister.
   a How much does she keep? Show your working.
   b What proportion of the prize money does she give away? Give your answer as a fraction and as a percentage.

This spread will show you how to:
- Solve problems involving ratio and proportion

Keywords
Fraction
Proportion
Ratio

- **Ratios** compare one number with another.
  The ratio of boys to girls is 3 : 2.

- **Proportions** tell you what **fraction** of the whole amount something is.
  $\frac{3}{5}$ of the students are boys.

- A ratio can be used to find a proportion.
  A drink is made by mixing squash and water in the ratio 1 : 4, so one part out of every five is squash.
  The proportion of squash in the drink is $\frac{1}{5}$.

Example

A concrete mixer contains 5 kg of cement and 20 kg of sand. Find

a the ratio of sand to cement
b the proportion of sand in the mixture.

a Ratio sand : cement = 20 : 5 = 4 : 1
b Proportion of sand = $\frac{20}{25} = \frac{4}{5}$

Write the weight of sand as a fraction of the total.

Example

Pink paint is made from one-third red paint and two-thirds white paint. Write the ratio of red paint to white paint.

Ratio red : white = $\frac{1}{3} : \frac{2}{3}$
                   = 1 : 2

Multiply by 3 to simplify.

Example

The ratio of hardback books to paperbacks on a bookshelf is 2 : 7. What proportion of the books are paperbacks?

Proportion of paperbacks = $\frac{7}{2 + 7}$
                         = $\frac{7}{9}$

Write the fraction $\frac{\text{paperbacks}}{\text{total}}$

Example

The ratio of boys to girls in a class is 2 : 3.
Alex is working out the proportion of the class that are boys.
He says it's $\frac{2}{3}$.
What is wrong with Alex's reasoning?

The answer can't be right – there are more girls than boys in the class.
The proportion of boys = $\frac{2}{2 + 3} = \frac{2}{5}$

Remember to add the parts when you find a proportion.

1 Orange paint is made from 2 parts red to 3 parts yellow paint.
   a Write the ratio of red paint to yellow paint in the mixture.
   b Find the proportion of red paint in the mixture, giving your
   answer as a fraction.

> The answer is
> *not* $\frac{2}{3}$!

2 A bowl contains 200 grams of flour and 100 grams of sugar.
   a Write the ratio of sugar to flour.
   b Find the proportion of sugar in the mixture.

3 A string of decorative lights has 7 red bulbs, 21 green bulbs and
   14 blue bulbs.
   a Write the ratio of the number of bulbs of each colour in its
   simplest form.
   b Calculate the proportion of
   i red bulbs   **ii** green bulbs   **iii** blue bulbs.

4 The table shows the ratio of blue paint to yellow paint in mixtures
   to make different shades of green paint. For each one, calculate the
   proportion of the mixture that is blue paint.

| **a** 1 : 1 | **b** 1 : 2 | **c** 2 : 3 | **d** 3 : 5 | **e** 4 : 3 | **f** 5 : 2 |

5 Write the proportion of yellow paint in each of the mixtures given in
   question **4**.

6 A fruit drink contains $\frac{2}{5}$ water and $\frac{3}{5}$ fruit juice. Find the ratio of
   water to fruit juice in the drink.

A02 Functional Maths

7 A swimming club had a total income of £2000. It spent 30%
   of this income on pool hire, and the rest on instructors' fees.
   a Calculate the amount spent on pool hire.
   b Work out the ratio of the amount spent on pool hire to the
   amount spent on fees.

8 Tom has £2200. He gives $\frac{1}{4}$ to his son and $\frac{2}{5}$ to his daughter.
   How much does Tom keep for himself? You must show all
   your working.

9 Andrew records how he spends his time over a 24-hour period.
   He spends $\frac{1}{3}$ of the time sleeping and $\frac{1}{6}$ of the time travelling.
   What proportion of his time does Andrew spend on activities
   other than sleeping and travelling? Show your working.

10 Mrs Smith inherits £16 000. She divides the money between
    her three children John, Sarah and Mark in the ratio 6 : 7 : 8,
    respectively. How much does Sarah receive? What proportion
    of the money does Mark receive?

This spread will show you how to:

- Convert between ratios, proportions, percentages, fractions and decimals

Keywords

Approximation
Percentage
Proportion
Ratio

You can describe **proportions** using **percentages**.

- Proportions expressed as percentages are easy to compare.

  'The proportion of sugar is $\frac{1}{2}$ in recipe A and $\frac{2}{5}$ in recipe B' is the same as

  'Recipe A contains 50% sugar and recipe B contains 40% sugar'.

Sometimes it is impossible to express a proportion exactly as a percentage. For example, $\frac{1}{7}$ is an exact proportion, but 14% (or even 14.29%) is only an **approximation** to it.

Example

The metal alloy, constantan, is made from nickel and copper in a **ratio** of 2 : 3. What percentage of the alloy consists of nickel?

$2 + 3 = 5$
Proportion of nickel $= \frac{2}{5}$
Percentage of nickel $= \frac{2}{5} = \frac{2 \times 20}{5 \times 20} = \frac{40}{100} = 40\%$

Add the parts of the ratio.

Example

Here is Emma's fruitcake recipe.

Fruitcake (1kg)

125 g of butter
150 g of flour
100 g of sugar
450 g of dried fruit
170 g of eggs
  5 g of spices

What percentage of the cake is butter?

Proportion of butter $= \frac{125}{1000} = \frac{1}{8}$

Percentage of butter $= \frac{125}{1000} \times 100\%$

$= 12.5\%$

Find the proportion of butter first.

Example

The ratio of the number of wins, draws and losses for a football team one season is 5 : 4 : 3. What percentage of the games are wins?

$5 + 4 + 3 = 12$
Proportion of wins $= \frac{5}{12}$

$= 42\%$ to nearest whole number.

1 A tub of fruit yogurt contains fresh yogurt and fruit in the ratio 9 : 1 by weight. Find the percentage of fruit in the contents of the tub.

2 **a** Write the ratio of the number of vowels to the number of consonants in the word CATERPILLAR.
   b Find the percentage of the letters in the word CATERPILLAR that are vowels. Show your working.

> Consonants are all those letters which are not vowels.

3 The compositions of three different alloys of copper are shown in this table. Find the percentage of copper in each alloy.

Alloy	Composition
Nickeline	4 parts copper, 1 part nickel
US nickel coinage	3 parts copper, 1 part nickel
Medal bronze	93 parts copper, 1 part tin

4 Scott and Maxine buy a present for their father. They share the cost in the ratio 3 : 2. What percentage of the cost does Scott pay?

5 The ratio of boys to girls in a class is 5 : 4. What percentage of the class is girls? Show your working.

6 The ABC mobile telephone company keeps records of calls made on its network. The ratio of the number of calls that are successfully connected to those that are missed for any reason is 4 : 1.
   a Find the percentage of calls on the network that are successfully connected.
   b The 123 mobile telephone company also keeps records of calls on their network. They say that 75% of calls are connected successfully.
   Find the ratio of successful to missed calls on the 123 network.

7 Dan and Phil share £3800 between them in the ratio 3 : 7, respectively.
   a Calculate the amount received by each person.
   b Find the percentage of the total amount of money that was received by Dan.

8 Leon and Frieda divide $500 in the ratio 61 : 33, respectively.
   a Calculate the amount received by each person.
   b Find the percentage of the total amount that Frieda receives.

9 German silver is made from copper, zinc and nickel in the ratio 16 : 5 : 3.
   a Calculate the mass of each metal in 500 g of German silver.
   b Work out the percentage of each metal in German silver.

This spread will show you how to:

- Round answers to an appropriate degree of accuracy
- Solve problems involving compound measures, including speed and density

Keywords
Compound
Density
Speed

Compound measures involve a combination of measurements and units.

p.32

p.378

- The **density** of a material is its mass divided by its volume.
- **Speed** is the distance travelled divided by the time taken.

The units for density can be grams per cubic centimetre (g/cm³), or kilograms per cubic metre (kg/m³).

- A formula triangle can be a useful way of remembering the relationships between the different parts of a compound measure.
  For example:
  speed = distance ÷ time
  distance = speed × time
  time = distance ÷ speed

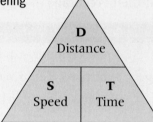

Speed can be measured in miles per hour (mph), kilometres per hour (kph) or metres per second (m/s).

p.30

Example

a A cube of side 3.5 cm has a mass of 600 g. Find the density of the cube in g/cm³, correct to 3 significant figures.
b A car travels 240 miles in 3 hours 45 minutes. Find the average speed of the car in miles per hour.
c Lubricating oil has a density of 0.58 g/cm³. Find
   i the mass of 2.5 litres of this oil
   ii the volume of 10 grams of the oil.

Volume of a cube = length³.

- -

a Volume = 3.5^3 cm³ = 42.875 cm³
   Density = 600 g ÷ 42.875 cm³ = 13.994 ... g/cm³ = 14.0 g/cm³ to 3 sf

Density = $\dfrac{\text{mass}}{\text{volume}}$

b Average speed = $\dfrac{\text{total distance}}{\text{total time}}$

   $= \dfrac{240}{3.75}$

   $= 64$ mph

Put the time in hours.

c i Mass = density × volume = 0.58 g/cm³ × 2500 cm³ = 1450 g
   ii Volume = mass ÷ density = 10 g ÷ 0.58 g/cm³ = 17.2 cm³

1 litre ≡ 1000 cm³

Example

A metal cuboid has a length of 2 cm, width 4 cm and height 6 cm.
a Find the volume of the cuboid.
b If the mass of the cuboid is 7.8 kg, find its density in g/cm³.

Volume of a cuboid = length × width × height.

- -

a Volume of cuboid = 2 × 4 × 6 = 48 cm³
b Density of cuboid = $\dfrac{\text{mass}}{\text{volume}} = \dfrac{7800}{48} = 163$ g/cm³ to 3 sf

Change the mass into grams.

A02 Functional Maths

1 Rod cycles 18 miles in 2 hours. Find his average speed, in miles per hour (mph).

2 If 4 metres of fabric costs £8.40, find the price of the fabric in pounds per metre.

3 A car travels 24 miles in 45 minutes. Find the average speed of the car in miles per hour (mph).

4 A train leaves Euston at 8:57 a.m. and arrives at Preston at 11:37 a.m. If the distance is 238 miles find the average speed of the train.

5 Copy and complete the table to show speeds, distances and times for five different journeys.

Speed (kph)	Distance (km)	Time
105		5 hours
48	106	
	84	2 hours 15 minutes
86		2 hours 30 minutes
	65	1 hours 45 minutes

6 A cube of side 2 cm weighs 40 grams.
   a Find the density of the material from which the cube is made, giving your answer in g/cm³.
   b A cube of side length 2.6 cm is made from the same material. Find the mass of this cube, in grams.

> Volume of cube = length³.

A02 Functional Maths

7 A box has a length and width of 22.50 mm, and a height of 3.15 mm. It has a mass of 9.50 g.
   a Find the density of the metal from which the box is made, giving your answer in g/cm³.
   b How many boxes can be made from 1 kg of the material?

> Volume of cuboid = length × width × height.

8 Emulsion paint has a density of 1.95 kg/litre. Find
   a the mass of 4.85 litres of the paint.
   b the number of litres of the paint that would have a mass of 12 kg.

A03 Problem

336

9 A steel cable weighs 2450 kg.
   The cable has a uniform circular cross-section of radius 0.85 cm.
   The steel from which the cable is made has a density of 7950 kg/m³.
   Find the length of the cable.

> Volume of a cylinder = πr^2 × length.

0.85 cm

l

Summary

Check out

You should now be able to:

- Use ratio notation and express a ratio in its simplest form
- Divide a quantity in a given ratio
- Convert between ratios, proportions, percentages, fractions and decimals
- Solve problems involving ratio and proportion
- Solve problems involving speed and density

Worked exam question

Here are the ingredients needed to make 8 pancakes.

> ## *Pancakes*
>
> **Ingredients to make 8 pancakes**
>
> *300 ml milk*
> *1 egg*
> *120 g flour*
> *5 g butter*

Jacob makes 24 pancakes.

a Work out how much milk he needs. (2)

Cathie makes 12 pancakes.

b Work out how much flour she needs. (2)

(Edexcel Limited 2008)

- -

a

$24 \div 8 = 3$
$3 \times 300 = 900$ ml

> Show these calculations for this method.

OR

a $300 + 300 + 300 = 900$ g

> Show this working for this method.

b

$12 \div 8 = 1\frac{1}{2}$
$1\frac{1}{2} \times 120 = 180$ g

> Show these calculations for this method.

OR

b 8 pancakes need 120 g of flour
  4 pancakes need 60 g of flour
 12 pancakes need 180 g of flour

> Show this working for this method.

Exam questions

1 Here are the ingredients for making cheese pie for 6 people.

> **Cheese pie for 6 people**
> 180 g flour
> 240 g cheese
> 80 g butter
> 4 eggs
> 160 ml milk

Bill makes a cheese pie for 3 people.
a Work out how much flour he needs. (2)

Jenny makes a cheese pie for 15 people.
b Work out how much milk she needs. (2)

(Edexcel Limited 2008)

2 There are some sweets in a bag.
18 of the sweets are toffees.
12 of the sweets are mints.
a Write down the ratio of the number of toffees to the number of mints.
Give your answers in its simplest form (2)

There are some oranges and apples in a box.
The total number of oranges and apples is 54
The ratio of the number of oranges to the number of apples is $1:5$
b Work out the number of apples in the box. (2)

(Edexcel Limited 2009)

A02

3 The density of concrete is 2.3 grams per cm^3.
a Work out the mass of concrete with a volume of 20 cm^3. (2)

480 grams of a cheese has a volume of 400 cm^3.
b Work out the density of the cheese. (2)

(Edexcel Limited 2006)

A03

4 A hopper can hold 80kg of wheat when full.
Wheat flow sinto an initially empty hopper at a rate of 125 grams per second.
How long does it take to fill the hopper? (3)

Expressions

Engineers and scientists use quadratic expressions to model and explain the behaviour of events and activities in real life. Without quadratic expressions there wouldn't be aircraft, mobile phones or satellite TV.

What's the point?

Expanding a pair brackets containing algebraic expression is just the generalisation of multiplying two numbers together. Similarly factorisation is the generalisation of the reverse process. These two processes however lie at the heart of further mathematical study – the quadratic expression.

Check in

You should be able to

■ **do basic arithmetic**

1 Work out these multiplications and divisions mentally.

   a 15×3 **b** 4×13 **c** $(-2) \times 13$ **d** 14×14

   e $56 \div 8$ **f** $91 \div 7$ **g** $150 \div (-3)$ **h** $1200 \div 40$

■ **know the order of operations**

2 Explain why the answer to each of these questions is 15.

   a $9 + 3 \times 2$ **b** $24 \div 3 + 7$ **c** $(3 + 2) \times 3$ **d** $3^3 - 4 \times 3$

■ **find common factors**

3 Find the highest common factor of these number pairs.

   a 6 and 9   **b** 8 and 12   **c** 20 and 30   **d** 12 and 18

   e 24 and 52   **f** 50 and 75   **g** 99 and 132   **h** 7 and 14

What I need to know

KS3 Work with and interpret algebraic expressions

N1 Work with negative numbers

What I will learn

- Collect like terms
- Use index notation
- Expand single and double brackets
- Factorise into single and double brackets

What this leads to

A4 Solve quadratic equations

N6 Index laws

A square grid is numbered from 1 to 100.

A 2 × 2 square is shaded in as shown on the grid.
The numbers in the opposite corners of the 2 × 2 square are multiplied together.

Investigate.

1	2	3	4	5	6	7	8	9	10
11	12	13	14	15	16	17	18	19	20
21	22	23	24	25	26	27	28	29	30
31	32	33	34	35	36	37	38	39	40
41	42	43	44	45	46	47	48	49	50
51	52	53	54	55	56	57	58	59	60
61	62	63	64	65	66	67	68	69	70
71	72	73	74	75	76	77	78	79	80
81	82	83	84	85	86	87	88	89	90
91	92	93	94	95	96	97	98	99	100

Writing and simplifying expressions in algebra

This spread will show you how to:
- Use the rules of algebra to write and manipulate algebraic expressions
- Simplify algebraic expressions by collecting like terms
- Use index notation

Keywords
Factorise
Index/indices
Like terms
Product
Simplify

There are conventions for writing expressions in algebra

- Do not include the multiplication sign $\qquad 3 \times p \rightarrow 3p$
- Write divisions as fractions $\qquad\qquad 3 \div p \rightarrow \frac{3}{p}$
- Write numbers first in products $\qquad\quad p \times 3 \rightarrow 3p$
- Write letters in products in alphabetical order $\quad 4 \times q \times r \times p \rightarrow 4pqr$

- To **simplify** expressions involving addition or subtraction, you collect **like terms** together.

$$\begin{array}{ll} \text{Like terms:} & \text{Unlike terms:} \\ 3z,\ 9z,\ -4z & 5p,\ 2p^2,\ 8q \end{array}$$

p.268

- To simplify expressions involving multiplication or division, you multiply the numbers then the letters.

You may need to use **indices**.

Example

a Write these using the rules of algebra.
  i $5 \times q \times 3 \times p$
  ii $y \times y \times y \times y \times y$
b Evaluate $3x + 2$ when $x = -4$.

....

a **i** $5 \times q \times 3 \times p = 15pq$
  ii $y \times y \times y \times y \times y = y^5$
b $3x + 2 = 3 \times (-4) + 2 = -12 + 2 = -10$

Numbers first, then letters in alphabetical order.

$3x$ means $3 \times x$

Example

Simplify these algebraic expressions.
a $3p + 9q - 2p + 7q$ $\quad$ **b** $5q - 7 + q^2$
c $7ab + 3ba$ $\qquad\qquad$ **d** $3t \times 4t \times 2t \times 2s$
e $\frac{15ab}{5b}$

....

a $3p + 9q - 2p + 7q = 3p - 2p + 9q + 7q$
$\qquad\qquad\qquad\qquad = p + 16q$
b $5q - 7 + q^2$ cannot be simplified as there are no like terms
c $7ab + 3ba = 7ab + 3ab = 10ab$
d $3t \times 4t \times 2t \times 2s = 48st^3$
e $\frac{15ab}{5b} = \frac{^3\cancel{15}a\cancel{b}}{\cancel{5}\cancel{b}} = 3a$

'$1p$' is simply 'p'.

'q' and 'q^2' are not like terms.

Cancelling – divide top and bottom by 5 and by b.

1 Write these expressions using the rules of algebra.

  a $5 \times w$ **b** $6 \div k$ **c** $y \times y$

  d $ab6$ **e** $k \times k \times 8 \times k$

2 Evaluate these expressions, given that $x = 6$.

  a $3x + 2$ **b** $10 - x$ **c** x^2

  d $\dfrac{10x - 16}{2}$ **e** $3x^2$

3 Simplify these expressions by collecting like terms.

  a $3a + 4b + 8a + 2b$ **b** $3t + 9 - t + 17$

  c $3x - 4y - 2x - 8y$ **d** $9p + p^2 + 5p$

  e $10xy + 10yx$ **f** $6ab + 2ba - ba$

4 Three students tried to simplify $3m + 5$. Which of them did it correctly?

      Sara   Paul   Abdul

     $3m + 5 = 8m$ $3m + 5 = 15m$ $3m + 5 = 3m + 5$

5 Simplify these expressions.

  a $4m \times 7n$ **b** $6m \times 2m$ **c** $\dfrac{20p}{2}$ **d** $\dfrac{14a}{7a}$

  e $2a \times 3b \times 4c$ **f** $k \times 2k \times 3k$ **g** $\dfrac{20ab}{5a}$ **h** $\dfrac{45c^2}{5c}$

6 Simplify the expressions in the grid and find the 'odd one out' for each row.

$3p + 2q + p + 5q$	$6p + 3q - 2p + 4q$	$5p - 3q - p + 5q$
$2m \times 3n$	$2 \times n \times m \times 5$	$6mn$
$\dfrac{24cd}{12c}$	$\dfrac{2d^2}{d^2}$	$\dfrac{2d^2}{d}$
$2n - 8$	$3m + 2n - m - 2m$	$3n - 2 - 6 - n$

7 a Write a simplified expression for the

    i perimeter   **ii** area

     of this rectangle

  b What are the measurements of a rectangle with perimeter $6x + 4y$ and area $6xy$?

$4p$

8

8 Copy this grid, replacing each expression with its simplified form (where possible).

$3a + 7b - 5a + 2b$	$3a \times 4a$	$\dfrac{20b}{5}$
$\dfrac{16ab^2}{8b}$	$2p + 7p^2 + 5p^3 + 8p$	$11abc + 2cab$
$5m - 4$	$3m \times 4m \times 5m$	$\dfrac{4a}{2a^3}$

This spread will show you how to:

- Multiply a single term over a bracket

Keywords
Bracket
Expand

A **bracket** in algebra means 'all multiplied by'.

$3(x + 5)$ means 'I have a number, add 5 then **multiply it all** by 3'

To write an expression without brackets, multiply all the terms inside the bracket by the term outside.
This is called **expanding** the bracket.

$3(x + 5)$ expanded is $3x + 15$

For negative terms, use the rules for multiplying by negative numbers:

p.8
- negative term $\times$ positive term $\rightarrow$ negative term

$$-3(x + 5) = -3 \times x + -3 \times 5$$
$$= -3x - 15$$

- negative term $\times$ negative term $\rightarrow$ positive term

$$-5(y - 8) = -5 \times y + -5 \times -8$$
$$= -5y + 40$$

Example

Expand these brackets.
a $5(m + 9)$ b $y(y - 7)$
c $3p(2p + 7 - q)$ d $-4m(m - 2)$

....

a $5(m + 9) = 5m + 45$
b $y(y - 7) = y^2 - 7y$
c $3p(2p + 7 - q) = 6p^2 + 21p - 3pq$
d $-4m(m - 2) = -4m^2 + 8m$

$y \times y$ is y^2

Be careful with negatives.

Example

Expand and simplify $3(t - 2) + 5(2 + t)$.

....

$$3(t - 2) + 5(2 + t) = 3t - 6 + 10 + 5t$$
$$= 3t + 5t - 6 + 10$$
$$= 8t + 4$$

Collect like terms.

Example

p.22

A rectangle of width x has length 1 cm more than the length.
Its area is 182 cm^2.
Show that $x^2 + x = 182$.

....

p.136

You are told that the length is 1 cm more than the width,
so the length is $(x + 1)$ cm.
Area of rectangle = length $\times$ width
$$182 = (x + 1) \times x$$
$$182 = x(x + 1)$$
$$182 = x^2 + x \text{ (as requested)}$$

Sketch a diagram to help:

1 Expand these brackets.

a $4(n + 5)$ **b** $6(b - 7)$ **c** $a(a + 3)$

d $a(b - c)$ **e** $4(2x + 3y - 4z)$ **f** $2h(h + 9)$

2 Expand these brackets.

a $-3(k + 9)$ **b** $-2(h - 5)$ **c** $-(w - 4)$

d $-(t - p)$ **e** $-k(k + 7)$ **f** $-9(2m - k + 4)$

g $-(x^2 - x - 8)$ **h** $-2(x^2 + 3)$ **i** $-3(1 - x)$

> Be careful with negatives.

3 Expand and simplify these expressions.

a $3(c + 2) + 7(c + 8)$ **b** $4(2x + 8) + 5(3x + 7)$

c $x(x + 8) + x(x + 2)$ **d** $5t(3t + 6) + 2t(t + 1)$

e $3(x - 7) + 4(x - 6)$ **f** $5(2 - x) + 7(x - 3)$

g $4(m - 6) - 2(m + 1)$ **h** $3(g - 3) - 7(2g - 6)$

i $2(p + 5) - (p - 4)$ **j** $(q - 4) - (3 - q)$

4 Expand and simplify $2x(x + 7) + x(9 - x) - 3x(2x - 7)$.

5 a Using brackets, write a formula for the area of this rectangle.

3

$2x - 1$

b Expand the brackets.

c The area of the rectangle is 15 cm².
Show that $6x - 18 = 0$.

6 An expression expands to give $24x + 16$.

a What could the expression have been if it involved one pair of brackets?

b What could the expression have been if it involved two single brackets in succession?

7 Write an expression involving brackets for the area of this trapezium.

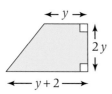

y

$2y$

$y + 2$

Expand the brackets and simplify your expression.

Expanding double brackets

This spread will show you how to:

● Expand double brackets

Keywords
Expand
FOIL
Product
Simplify

You can **expand** a double bracket in algebra by multiplying pairs of terms.

Each term in the first bracket multiplies each term in the second bracket:

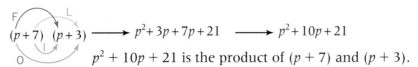

$$p^2 + 3p + 7p + 21 \longrightarrow p^2 + 10p + 21$$

$p^2 + 10p + 21$ is the product of $(p + 7)$ and $(p + 3)$.

F... **F**irsts
0 ... **O**uters
I ... **I**nners
L ... **L**asts

Example

Expand and simplify
$(x + 3)(x + 4)$

. . . .

$(x + 3)(x + 4) = x^2 + 4x + 3x + 12$
$\qquad\qquad\qquad = x^2 + 7x + 12$

F: $x \times x = x^2$
O: $x \times 4 = 4x$
I: $3 \times x = 3x$
L: $3 \times 4 = 12$

Example

Expand and simplify
a $(x + 3)(2x - 2)$ **b** $(3x + 2)^2$

. . . .

a $(x + 3)(2x - 2) = 2x^2 - 2x + 6x - 6$
$\qquad\qquad\qquad\quad = 2x^2 + 4x - 6$
b $(3x + 2)^2 = (3x + 2)(3x + 2)$
$\qquad\qquad\quad = 9x^2 + 6x + 6x + 4$
$\qquad\qquad\quad = 9x^2 + 12x + 4$

Use the rules for multiplying negative terms:
O: $x \times -2 = -2x$
L: $3 \times -2 = -6$

Example

A rectangle has length $x + 5$ and width $x - 2$.
Show that $A = x^2 + 3x - 10$, where A is the area.

. . . .

Sketch a diagram:

$x + 5$

$x - 2$

Area of rectangle = length × width
$\qquad\qquad A = (x + 5)(x - 2)$
$\qquad\qquad\quad = x^2 - 2x + 5x - 10$
$\qquad\qquad A = x^2 + 3x - 10$

1 Expand and simplify these expressions involving double brackets.

a $(x + 2)(x + 3)$ **b** $(p + 5)(p + 6)$

c $(w + 1)(w + 4)$ **d** $(c + 5)^2$

e $(x + 4)(x - 2)$ **f** $(y - 2)(y + 7)$

g $(t + 6)(t - 2)$ **h** $(x - 2)(x - 5)$

i $(y - 4)(y - 10)$ **j** $(w - 1)(w - 2)$

k $(p - 5)^2$ **l** $(q - 12)^2$

2 Expand and simplify

a $(2x + 1)(3x + 7)$ **b** $(5p + 2)(2p + 3)$

c $(3y + 4)(2y + 1)$ **d** $(2y + 6)^2$

e $(5t - 4)(2t + 4)$ **f** $(5w - 1)(3w + 9)$

g $(2x + 2y)(3x - 3y)$ **h** $(3m - 4)^2$

i $(2p + 5q)(3p - 8q)$ **j** $(2m - 3n)^2$

3 Write an expression for the areas of this rectangle and square.

a
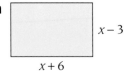
$x - 3$

$x + 6$

b

$2m - 3$

4 a Expand $(a + b)^2$.

   b Hence, or otherwise, calculate $1.32^2 + 2 \times 1.32 \times 2.68 + 2.68^2$

   c Write another calculation that you could work out using this expansion.

5 a The diagram shows a rectangle with an area of 75 cm^2.

Show that $6x^2 + 7x - 78 = 0$.

$2x + 3$

$3x - 1$

b This triangle also has an area of 75 cm^2. Show that $15x^2 = 14x + 158$.

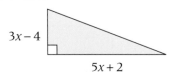
$3x - 4$

$5x + 2$

Factorising single brackets

This spread will show you how to:

- Factorise into single and double brackets

Keywords
Factorise
Highest common
factor (HCF)

The reverse of expanding a set of brackets is called **factorising**.
To factorise an expression, you put brackets in.

Expand

$12(x + 2)$ $12x + 24$

Factorise

You divide the terms by their **highest common factor (HCF)**.

HCF of $12x$ and 24 is 12

$12x + 24$ $12(x + 2)$

Check your answer
by expanding.

Example

Factorise fully
a $6x + 9$ **b** $12pq - 4pw$ **c** $5x + 10x^2 - 25xy$

· ·

a The HCF of $6x$ and 9 is 3.
  $6x + 9 = 3(2x + 3)$
b $12pq - 4pw$
  Deal with numbers first, then letters.
  HCF of 12 and 4 is 4.
  HCF of pq and pw is p.
  HCF of $12pq$ and $4pw$ is $4p$.
  So, $12pq - 4pw = 4p(3q - w)$
c The HCF of $5x$, $10x^2$ and $25xy$ is $5x$.
  $5x + 10x^2 - 25xy = 5x(1 + 2x - 5y)$

$5x \div 5x$ is 1

Example

a Factorise $y^2 - 3y$ **b** Factorise $(p + q)^2 - 2(p + q)$

· ·

a $y^2 - 3y = y(y - 3)$ **b** $(p + q)^2 - 2(p + q)$
                                            $= (p + q)((p + q) - 2)$
                                            $= (p + q)(p + q - 2)$

Each part in **b** has
$(p + q)$ in common

1 Factorise each of these fully by removing common factors.

a $2x + 4$ **b** $3y - 6$ **c** $12p + 36q$
d $25w - 5$ **e** $6xy + xw$ **f** $ab - 2bc$
g $pqr + qrt - qsw$ **h** $5xy - x$ **i** $2xy + 6x$
j $4ab - 6a^2$ **k** $25p^2 - 10p$ **l** $7x + 14xy$
m $2ac + 4a^2 - 8a$ **n** $15mn - 5m + 10m^3$ **o** $6p^4 - 12p$

2 All three students have completed their factorisations incorrectly.
Explain what they have done wrong.

Clare

$5x + 10xy$
$= 5x(0 + 2y)$

Ben

$6pq + 3p$
$= 3(2pq + 1)$

Vicky

$21p + 14pq$
$= 7p(14 + 7q)$

3 Factorise these expressions

a $10(x + y) + 13(x + y)$ **b** $(a - b)^2 + 5(a - b)$
c $6(q + r) - (q + r)^3$ **d** $(pt - w) + 6(pt - w)$

4 Factorise fully

a $ax + bx + ay + by$ **b** $cd + bd + cm + bm$
c $a^2 + ab + 2a + 2b$ **d** $cd + ce - me - md$

> First look at what
> each pair of terms
> has in common.

5 Write a factorised expression for
a The perimeter of this rectangle

4

$2x - 6$

b The perimeter of a square with sides $5b + 10$

6 Use factorisation to help you to evaluate these, without a calculator.

a $2 \times 1.86 + 2 \times 1.14$ **b** $3 \times 5.87 - 3 \times 0.37$
c $5.86^2 + 5.86 \times 4.14$ **d** $3.32 \times 6.68 + 3.32^2$

7 Show that the shaded area of this rectangle is $2(4x + 5)$.

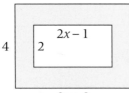

4

2

$2x - 1$

$3x + 2$

Factorising double brackets

This spread will show you how to:

- Factorise into single and double brackets

Keywords
Factor
Factorise
Quadratic

p.240

When you expand double brackets, you often get a **quadratic** expression with three terms. Therefore you can factorise a quadratic expression into double brackets.

A quadratic expression contains a squared term, such as x^2.

$$(x+2)(x+3) = x^2 + 3x + 2x + 6$$
$$= x^2 + 5x + 6$$
$$2 + 3 = 5 \qquad 2 \times 3 = 6$$

- The two numbers in the brackets **multiply** to give the number at the end and **add** to give the number of xs.

You can use this pattern to help you factorise a quadratic expression. You can check your answer by expanding the brackets.

Example

a Factorise $x^2 + 8x + 15$.
b Factorise $x^2 - 7x - 18$.

....

a Look for two numbers that multiply to give $+15$ and add to give $+8$. These are $+3$ and $+5$.
$x^2 + 8x + 15 = (x + 3)(x + 5)$
b Look for two numbers that multiply to give -18 and add to give -7.

It can help to write out all the factor pairs.
Consider the factor pairs of -18:
-1 and 18
-18 and 1
-3 and 6
-6 and 3
-2 and 9
-9 and 2

The two numbers are -9 and $+2$.
$x^2 - 7x - 18 = (x - 9)(x + 2)$

Consider the factor pairs of 15.
1 and 15
3 and 5

You can check by expanding.

Example

Factorise $y^2 - 3y - 28$.

....

$y^2 - 3y - 28 = (y - 7)(y + 4)$

Two numbers that multiply to give -28 and add to give -3.
$-7 \times 4 = -28$
$-7 + 4 = -3$

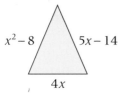

1 Factorise each of these into double brackets.
  a $x^2 + 6x + 8$ **b** $x^2 + 10x + 21$ **c** $x^2 + 11x + 28$
  d $x^2 + 11x + 24$ **e** $x^2 - 8x + 12$ **f** $x^2 - 9x + 18$
  g $x^2 - 13x + 36$ **h** $x^2 + x - 12$ **i** $x^2 - 2x - 35$
  j $x^2 + 6x - 27$ **k** $x^2 - 14x + 32$ **l** $x^2 + 18x - 40$

2 **a** Write an expression for the perimeter of this triangle.

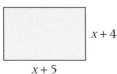

  b Factorise your expression.

3 Factorise each of these expressions.
  a $x^2 + 6x - 72$ **b** $x^2 - 10x - 24$ **c** $x^2 - 20x + 75$
  d $x^2 + 12x - 64$ **e** $x^2 - 64$ **f** $x^2 - 29x + 100$

4 Decide if each of these are single or double bracket factorisations.
  Factorise each fully.
  a $x^2 + 21x + 38$ **b** $5x^2 + 5x + xy$ **c** $x^2 + 22x + 121$
  d $x^2 + 7x - 18$ **e** $33 + p^2 + 14p$ **f** $2x^2 + 3xy$

5 Given that the area of this rectangle is 12 cm^2,
  show that $(x + 1)(x + 8) = 0$.

$x + 4$

$x + 5$

6 Use common factors and double brackets to factorise these
  expressions fully.
  a $2x^2 + 22x + 56$
  b $x^3 - 5x^2 - 24x$
  c $x^3 - 16x$

Example:
$x^3 + 5x^2 + 6x$
$= x(x^2 + 5x + 6)$
$= x(x + 2)(x + 3)$

7 Factorise $2.3^2 + 2 \times 2.3 \times 1.7 + 1.7^2$ and use this to show that
  the calculation results in 16.

This spread will show you how to:

- Understand the difference between identities, formulae and equations
- Use formulae from mathematics and other subjects

Keywords
Equation
Formula(e)
Identity
Substitute

- An **identity** is true for all values of x. For example
  $x(x + 1) \equiv x^2 + x$ Whatever value of x you try, this statement is always true.

$\equiv$ means 'is identical to'

- An **equation** is only true for a limited number of values of x.
  For example
  $2x + 1 = 5$ is only true when $x = 2$.
- A **formula** describes the relationship between two or more variables.
  For example, the formula for the area of a triangle is
  $A = \frac{1}{2}bh$
- You can **substitute** numbers into a formula to work out the value of a variable. For example

 p.22

6cm
← 8cm →

$A = \frac{1}{2}bh$

$= \frac{1}{2} \times 8 \times 6$

$= 24 \, \text{cm}^2$

Example

Decide if each of these statements is an identity, an equation or a formula.

a $x^3 - 2x = x(x^2 - 2)$ **b** $5x - 1 = 2x + 3$ **c** $A = bh$

....

a Expand the right-hand side: $x(x^2 - 2) = x^3 - 2x$
   $x^3 - 2x =$ left-hand side so the statement is an identity.
b $5x - 3 = 2x + 3$
   $3x - 3 = 3$ This statement is only true for one value of x.
      $3x = 6$
       $x = 2$
   $5x - 3 = 2x + 3$ is an equation.
c $A = bh$ is a formula showing the relationship between the length of a pair of parallel sides, the distance between them and the area of a parallelogram.

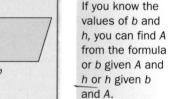
h
b

If you know the values of b and h, you can find A from the formula or b given A and h or h given b and A.

Example

Use the formula $A = 2(lw + wh + hl)$ to find A, the surface area of a cuboidal box, when $l = 50 \, \text{cm}$, $w = 30 \, \text{cm}$ and $h = 20 \, \text{cm}$

....

$A = 2(lw + wh + hl)$
  $= 2 \times (50 \times 30 + 30 \times 20 + 20 \times 50)$
  $= 2 \times (1500 + 600 + 1000)$
  $= 2 \times 3100$
  $= 6200 \, \text{cm}^2$

Write the formula.

Substitute the values of l, w and h.

1 Copy these statements and say whether they are identities, equations or formulae.

a $c = 2\pi r$	**b** $3x(x+1) = 3x^2+3x$	**c** $3x + 1 = 10$
d $y \times y = y^2$	**e** $2x + 5 = 3 - 7x$	**f** $A = \frac{1}{2}(a+b)h$
g $a^2 + b^2 = c^2$	**h** $20 - x = -(x - 20)$	**i** $2x^2 = 50$

2 A campsite charges for the pitching of a tent and the number of people, p, that stay in it. If C is the total cost in pounds, the formula used by the campsite is $C = 2p + 5$

  a Work out the cost of pitching a tent for 6 people.

  b If the cost is £23, how many people are sleeping in the tent?

  c Explain why the cost could never be £26.

3 Use these formulae to work out the required information.

  a $F = \frac{9}{5}C + 32$.   Find 12° Celsius in degrees Fahrenheit.

  b $P = \frac{1}{4}t - 8$.   Find P when $t = 32$.

  c $A = \dfrac{bh}{2}$.   Find h when $A = 10$ and $b = 5$.

  d $W = 3d^2$.   Find W when d is 6 and d when W is 75.

  e $C = 2a - b$.   Find C when a is -8 and b is -2.

  f $P + 2r = K$.   Find K when P is $\frac{3}{4}$ and r is $\frac{1}{8}$.

4 a The formula for the area of a trapezium is $A = \frac{1}{2}(a + b)h$, where a and b represent the lengths of the parallel sides and h is the perpendicular height.
  Find the area of this trapezium.

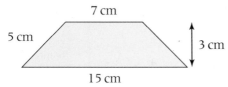

  b If another trapezium with the same perpendicular height has area 15 cm², suggest as many possibilities for the lengths of the parallel sides as you can.

Writing formulae

This spread will show you how to:
- Write a formula from given information

- You can **derive** a **formula** from information you are given.

Example

Derive a formula for the perimeter of this pentagon.

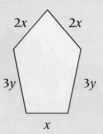

Let P represent the perimeter.
$P = 2x + 2x + 3y + x + 3y$
$P = 2x + 2x + 3y + 3y + x$
$P = 5x + 6y$

Perimeter is the distance all of the way around a shape.

Write formulae as simply as possible, using the rules of algebra.
For example

Write $S = \dfrac{D}{T}$, not $S = D \div T$ and $A = lw$, not $A = l \times w$

Example

p.48

a Write a formula to show your total amount of pocket money P (£s), if you receive £3 per month with an extra £2 for every job (j) you do at home.

b Write a formula for the total area of this shape. Let the total area be A.

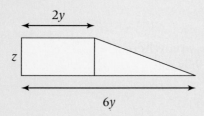

a If I don't do any jobs, I get £3.
If I do 4 jobs I get £3 plus $4 \times £2$, which is £11.
If I do 10 jobs I get £3 plus $10 \times £2$, which is £23.
So, if I do j jobs I get £3 plus $j \times £2$.
Hence
$P = 3 + 2j$

b A = area of a rectangle plus the area of a triangle
$= (z \times 2y) + \frac{1}{2}((6y - 2y) \times z)$
$= 2yz + \frac{1}{2}(4y \times z)$
$= 2yz + \frac{1}{2}(4yz)$
$= 2yz + 2yz$
$= 4yz$

Try the situation with various numbers, before putting it into algebra.

Write $2j$ not $j \times 2$.

The area of a rectangle is $A = lw$ and the area of a triangle is $A = \frac{1}{2}bh$.

1 For each shape, write your own formula to represent
  i the perimeter, P
  ii the area, A.

a

s

b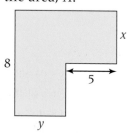

x

8

5

y

2 Write your own formula to represent these quantities.
  a The total cost, P, of buying d dollars if the exchange rate
     is $\$1 = £0.64$ and there is a fixed commision fee of £12.
  b The time in minutes, T, to complete my homework if I take 10
     minutes to get my books organised and then 35 minutes to do each
     piece, p.
  c The total phone bill, B, if you send n texts at 10p each, make
     m minutes of phone calls at 30p per minute and play a line rental
     of £5.60.

3 Megan is writing a formula to work out the volume of this prism.
  Her formula is
  $$V = \frac{abc}{2}$$

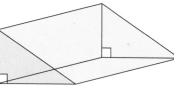

  Explain what a, b and c stand for.
  Explain why Megan's formula is correct.

p.340

Problem

A03

4 This cylinder has base radius r and height h.

  a Write a formula for V, the volume of the cylinder.
  b Nicholas has written this formula for the surface area, A, of the
     cylinder.

  $$A = 2\pi r^2 + 2\pi rh$$

  Explain what the $2\pi r^2$ part represents. Repeat for the $2\pi rh$ term.

Remember a
cylinder is a
prism. Volume
= area of cross
section × height.

A2

Summary

Check out
You should now be able to:

- Simplify algebraic expressions by collecting like terms
- Use index notation and simple laws of indices
- Multiply a single term over a bracket
- Expand the product of two linear expressions
- Factorise an algebraic expression using common factors
- Factorise quadratic expressions
- Distinguish in meaning between the words 'equation', 'formula', 'identity' and 'expression'
- Derive a formula
- Substitute numbers into a formula from mathematics and other subjects

Worked exam question

a   Expand and simplify $(x + 7)(x - 4)$ (2)

b   Expand $y(y^3 + 2y)$ (2)

c   Factorise $p^2 + 6p$ (2)

d   Factorise completely $6x^2 - 9xy$ (2)

(Edexcel Limited 2005)

a
$$(x + 7)(x - 4) = x^2 + 7x - 4x - 28$$
$$= x^2 + 3x - 28$$

> There should be 4 terms.

b
$$y(y^3 + 2y) \quad = y^4 + 2y^2$$

> These terms will not simplify any further.

c
$$p^2 + 6p \quad = p(p + 6)$$

d
$$6x^2 - 9xy \quad = 3(2x^2 - 3xy)$$
$$= 3x(2x - 3y)$$

> The common factor is 3 then x.

OR

d
$$6x^2 - 9xy \quad = x(6x - 9y)$$
$$= 3x(2x - 3y)$$

> The common factor is x then 3.

OR

d
$$6x^2 - 9xy \quad = 3x(2x - 3y)$$

> The common factor is 3x.

> For part **d**, factorise **completely** suggests there are at least 2 common factors.

Exam questions

1 **a** Simplify fully $\qquad 4a + 5b - 2a + b$ (2)
   b Factorise $\qquad x^2 - 6x$ (2)
   c Expand $\qquad x(3 - 2x^2)$ (2)
   d Factorise completely $\qquad 12xy + 4x^2$ (2)

2 **a** Simplify $\qquad m^2 \times m^3$ (1)
   b Simplify $\qquad \dfrac{n \times n^5}{n^2}$

(1)

3 **a** Simplify $\qquad a \times a \times a$ (1)
   b Expand $\qquad 5(3x - 2)$ (1)
   c Expand $\qquad 3y(y + 4)$ (2)
   d Expand and simplify $\qquad 2(x - 4) + 3(x + 2)$ (2)
   e Expand and simplify $\qquad (x + 4)(x - 3)$ (2)

(Edexcel Limited 2009)

4 $P = 4k - 10 \qquad\qquad P = 50$
   a Work out the value of k. (2)
      $y = 4n - 3d \qquad\qquad n = 2, d = 5$
   b Work out the value of y. (2)

(Edexcel Limited 2009)

5 The cost of hiring a car can be worked out using this rule.

> Cost = £90 + 50p per mile

The cost of hiring a car and driving m miles is C pounds.
   a Write a formula for C in terms of m. (2)

Zara hired a car.
The cost is £240
   b How many miles did Zara drive? (3)

(Edexcel Limited 2007)

Displaying and interpreting data

Statistics are vital in medicine were they are used to test the safety and performance of new drugs. Tests are performed on large groups of people and the analysis of the results is used to evaluate the safety and reliability of the new drug.

What's the point?

When data is analysed it is essential for that analysis to be correct. Statisticians use a range of techniques to analyse and compare large data sets. It is the use of their statistical techniques that ensures that a drug is safe to be on sale to the general public.

You should be able to

1   Draw a grid with both axes from −4 to +5.
    Draw these points on the grid.
    a (2, 4)   **b** (1, −3)   **c** (5, −3)   **d** (−2, 2)
    e (−4, −1)   **f** (0, 3)   **g** (4, 0)   **h** (0, −1)

2   If £1 = €1.40,
    a how many euros would you get for
        i £2   **ii** £4.50?
    b how many pounds would you get for
        i €4   **ii** €7?

3   Work out
    a 50% of 80   **b** 50% of 120   **c** 25% of 60   **d** 25% of 200
    e 25% of 120   **f** 75% of 200   **g** 75% of 160   **h** 75% of 840

Exam questions

1 a Simplify fully $4a + 5b - 2a + b$ (2)
  b Factorise $x^2 - 6x$ (2)
  c Expand $x(3 - 2x^2)$ (2)
  d Factorise completely $12xy + 4x^2$ (2)

(Edexcel Limited 2007)

2 a Simplify $m^2 \times m^3$ (1)
  b Simplify $\dfrac{n \times n^5}{n^2}$ (1)

3 a Simplify $a \times a \times a$ (1)
  b Expand $5(3x - 2)$ (1)
  c Expand $3y(y + 4)$ (2)
  d Expand and simplify $2(x - 4) + 3(x + 2)$ (2)
  e Expand and simplify $(x + 4)(x - 3)$ (2)

(Edexcel Limited 2009)

4 $P = 4k - 10$ $P = 50$
  a Work out the value of k. (2)
    $y = 4n - 3d$ $n = 2, d = 5$
  b Work out the value of y. (2)

(Edexcel Limited 2009)

A02

5 The cost of hiring a car can be worked out using this rule.

$$\boxed{\text{Cost} = £90 + 50\text{p per mile}}$$

The cost of hiring a car and driving m miles is C pounds.
  a Write a formula for C in terms of m. (2)

Zara hired a car.
The cost is £240
  b How many miles did Zara drive? (3)

(Edexcel Limited 2007)

Displaying and interpreting data

Statistics are vital in medicine were they are used to test the safety and performance of new drugs. Tests are performed on large groups of people and the analysis of the results is used to evaluate the safety and reliability of the new drug.

What's the point?

When data is analysed it is essential for that analysis to be correct. Statisticians use a range of techniques to analyse and compare large data sets. It is the use of their statistical techniques that ensures that a drug is safe to be on sale to the general public.

Check in

You should be able to

1 Draw a grid with both axes from -4 to $+5$.
   Draw these points on the grid.
   a $(2, 4)$ **b** $(1, -3)$ **c** $(5, -3)$ **d** $(-2, 2)$
   e $(-4, -1)$ **f** $(0, 3)$ **g** $(4, 0)$ **h** $(0, -1)$

2 If £1 = €1.40,
   a how many euros would you get for
      i £2   **ii** £4.50?
   b how many pounds would you get for
      i €4   **ii** €7?

3 Work out
   a 50% of 80   **b** 50% of 120   **c** 25% of 60   **d** 25% of 200
   e 25% of 120   **f** 75% of 200   **g** 75% of 160   **h** 75% of 840

What I need to know

KS3 Plot points on a graph

D1 Calculate measures of average and spread

What I will learn

■ Draw and interpret scatter diagrams, stem-and-leaf diagrams, box plots and line graphs

What this leads to

D4 Frequency diagrams Comparing data sets

Public relations

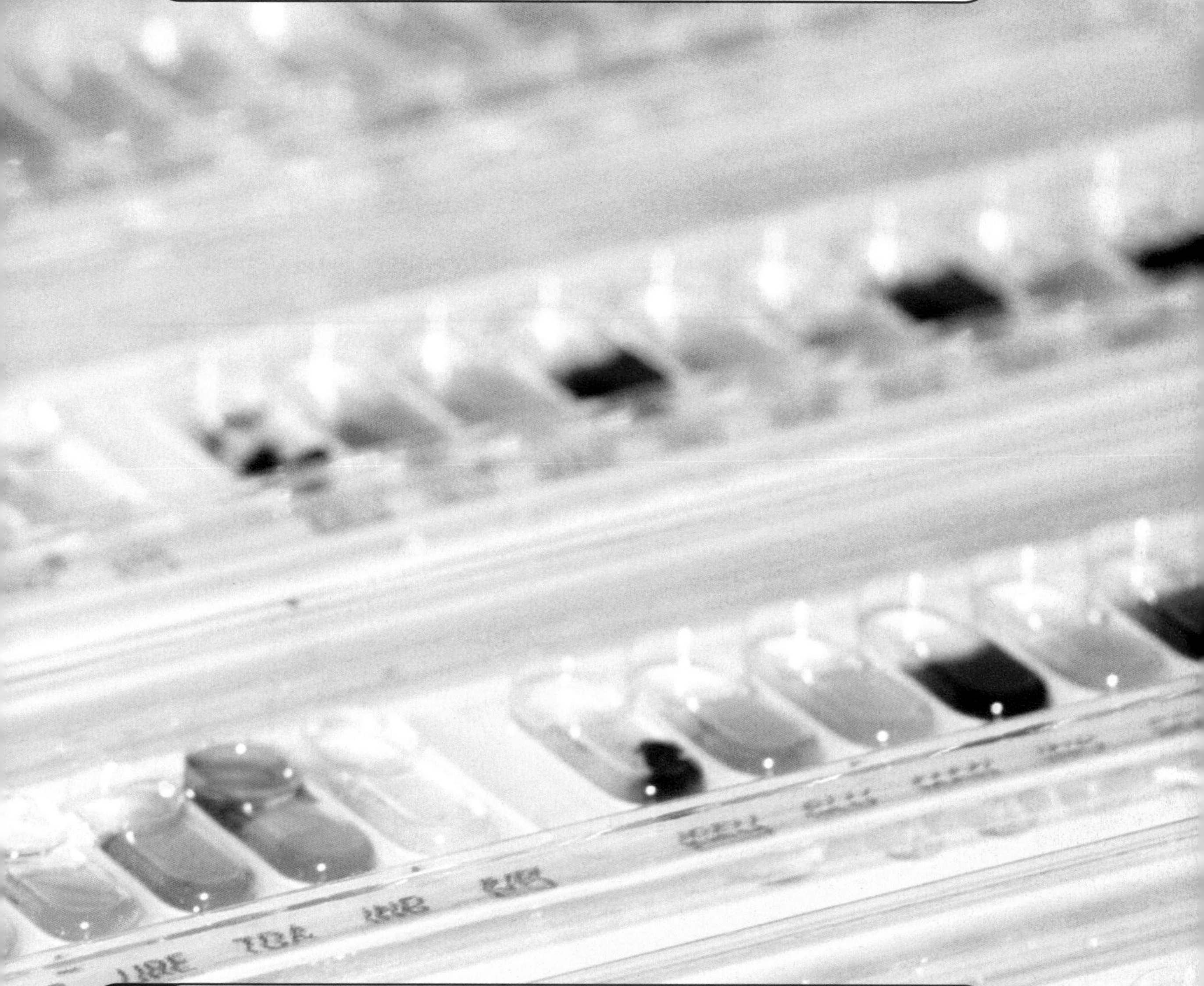

The taller you are the heavier you are.

Investigate if this statement is true by looking at different age groups.

Scatter diagrams

This spread will show you how to:

- Draw and use scatter diagrams
- Understand the concept of correlation

Keywords
Correlation
Scatter diagram

- A **scatter diagram** shows you the relationship between two numerical variables.

If there is a close relationship, or **correlation**, between the variables, the points will lie roughly on a straight line.

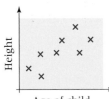

If the line slopes upwards there is **positive correlation**

If the line slopes downwards there is **negative correlation**

If the points are widely scattered, there is **no correlation**.

Example

Sally recorded the circumference and weights of eight pumpkins.

Circumference, cm	142	124	136	140	139	128	132	135
Weight, kg	28	21	25	26	26	23	23	24

a Draw a scatter diagram of these data.
b Describe the correlation shown.
c Describe the relationship between the circumference and weight of these pumpkins.

a

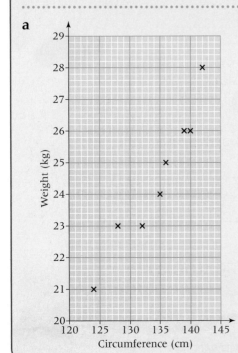

b There is positive correlation.
c The correlation shown suggests that larger pumpkins weigh more.

A02 **Functional Maths**

1 Kevin recorded the heights and weights of eight boys.

	Adam	Ben	Carl	Don	Evan	Fred	Gavin	Henry
Height, cm	157	136	142	140	138	152	156	160
Weight, kg	47	32	36	37	33	39	42	51

a Draw a scatter diagram of these data.
b Describe the correlation shown.
c Describe the relationship between the height and weight of these boys.

> You will need to use the graphs you draw in this exercise for exercise **D4.2**.

2 Iain recorded the percentages achieved in Statistics and Mathematics exams by 10 students.

Statistics %	78	82	74	75	93	70	66	62	77	89
Mathematics %	70	76	61	70	89	65	59	58	73	82

a Draw a scatter diagram of these data.
b Describe the correlation shown.
c Describe the relationship between the percentages achieved in Statistics and Mathematics for these students.

3 Louise recorded information about the average number of minutes per day spent playing computer games and the reaction times of nine students.

Minutes per day spent playing computer games	40	60	75	40	35	20	80	50	45
Reaction time, seconds	5.2	4.3	3.9	5.5	6.0	7.2	3.6	4.8	5.0

a Draw a scatter diagram of these data.
b Describe the correlation shown.
c Describe the relationship between the average number of minutes per day spent playing computer games and reaction time.

4 Bob caught nine fish during one angling session.
He recorded information about the weights and lengths of the fish he caught in this table.

Weight, g	500	560	750	625	610	680	600	650	580
Length, cm	30	32	50	44	39	48	40	45	36

a Draw a scatter diagram of these data.
b Describe the correlation shown.
c Describe the relationship between the weights and lengths of the fish.

Using scatter diagrams

This spread will show you how to:

• Draw and use lines of best fit

Keywords

Correlation
Line of best fit
Prediction
Scatter diagram

• You can use a **scatter diagram** that shows **correlation** to predict other results.

To predict results you first draw a **line of best fit**.

• A line of best fit should follow the trend of the plotted points

The line of best fit can slope downwards. It doesn't have to start at (0, 0).

Example

a Draw a line of best fit for the data Sally collected on the circumference and weight of eight pumpkins in D4.1.
b Use the line to predict
  i the weight of a pumpkin, circumference 137 cm
  ii the circumference of a pumpkin that weighs 22 kg.

p.50

a

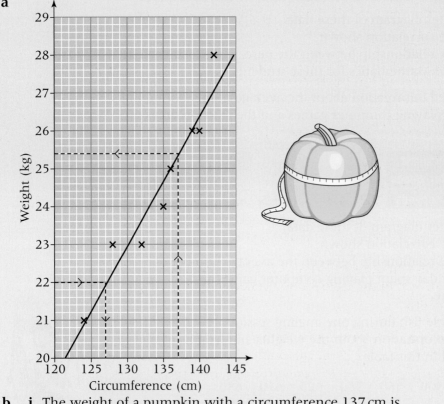

Draw a vertical line from 137 cm up to the graph, then across to the vertical axis.

b i The weight of a pumpkin with a circumference 137 cm is predicted to be 25 kg.

   ii The circumference of a pumpkin weighing 22 kg is predicted to be 127 cm.

It would not be sensible to use this graph to predict the weight of a pumpkin that is 120 cm around as this is outside the range of the collected data plotted in the graph.

1 Look at the graph you drew in Exercise **D4.1** question **1**.
  a Draw a line of best fit for the data Kevin collected on the height and weight of eight boys.
  b Use the line to predict
     i the weight of a boy 145 cm tall
     ii the height of a boy who weighs 45 kg.

2 Look at the graph you drew in Exercise **D4.1** question **2**.
  a Draw a line of best fit for the data Iain collected on Statistics and Mathematics results.
  b Use the line to predict
     i the percentage in Mathematics for Imogen who achieved 80% in Statistics
     ii the percentage in Statistics for Katherine who achieved 80% in Mathematics.
  c Give a reason why it would not be sensible to use this graph to predict the percentage achieved in the Mathematics exam for a person who achieves 46% in the Statistics exam.

3 Look at the graph you drew in Exercise **D4.1** question **3**.
  a Draw a line of best fit for the data Louise collected on time spent playing computer games and reaction time.
  b Use the line to predict
     i the reaction time of Jim, who spends on average 70 minutes per day playing computer games
     ii the time per day spent playing computer games for Ellis, who has a reaction time 6.4 seconds.
  c Give a reason why it would not be sensible to use this graph to predict the reaction time of Tom who spends on average 180 minutes per day playing computer games.

4 Look at the graph you drew in Exercise **D4.1** question **4**.
  a Draw a line of best fit for the data Bob collected on the weight and length of fish.
  b Use the line to predict
     i the length of a fish that weights 700 g
     ii the weight of a fish that is 42 cm long.

Stem-and-leaf-diagrams

This spread will show you how to:

- Draw stem-and-leaf diagrams, using them to find the median and range of data sets

Keywords
Interquartile range
Lower quartile
Median
Range
Stem-and-leaf
Upper quartile

You can represent small amounts of data on a **stem-and-leaf** diagram.

A stem-and-leaf diagram shows all the data and the overall shape.

- You can use a stem-and-leaf diagram to find statistics such as the **median**.

When you draw a stem-and-leaf diagram choose the stem according to the data.

For example, for heights given to nearest cm, data will be in the 100s:
165, 154, 178, ... so in the stem you use
$$15 \mid$$
$$16 \mid$$
$$17 \mid$$

Example

These are the percentages for a Statistics test taken by 23 students.

58	54	78	66	67	40	45	38	58	73	51	49
47	53	41	36	59	64	52	43	39	80	37	

a Draw a stem-and-leaf diagram for these data.
b For these data find the
  i range
  ii median
  iii lower quartile and upper quartile
  iv interquartile range.
c Describe the data.

a
```
3 | 8 6 9 7          3 | 6 7 8 9          Now put the leaves
4 | 0 5 9 7 1 3      4 | 0 1 3 5 7 9      in order.
5 | 8 4 8 1 3 9 2    5 | 1 2 3 4 8 8 9
6 | 6 7 4            6 | 4 6 7
7 | 8 3              7 | 3 8
8 | 0                8 | 0
```
Use the units digit as the leaves.

Use the tens digit as the stem.

Always give a key.

Key: 7 | 3 stands for 73%

b **i** Range = 44%
Highest value − lowest value: $80 - 36 = 44$.

ii Median = 52%
Data is in order so count up to find 12th value: $\frac{1}{2}(23 + 1) = 12$.

iii Lower quartile (LQ) 41%
Upper quartile (UQ) 64%
Lower quartile is the 6th value in the 1st quarter of data: $\frac{1}{4}(23 + 1) = 6$.

iv Interquartile range (IQR) 23%
Then count back 6 values to find upper quartile.

c Most students scored less than 60%.
IQR = UQ − LQ = $64 - 41 = 23$.

p.78

For these data sets

a Draw a stem-and-leaf diagram to show the information.
(Remember to include a key.)

b Find the
   i range   **ii** median   **iii** lower quartile and upper quartile
   iv interquartile range.

c Describe the data.

1 Percentage achieved in a Statistics test.

78 82 74 45 69 75 93 54 61 70 48 66
62 51 77 59 51 89 81 52 63 71 59

2 Minutes per day spent playing computer games.

40 26 75 84 33 39 28 66 67 71 80
37 52 47 63 49 41 44 58 69 43
73 55 59 43 61 38 29 30 77 60

3 Time taken, in minutes, to solve a crossword puzzle.

12 24 21 16   8   9   3 31 18 27 35
41 26 12 17   6   5 19 29 32 37 40
15 22 10 33 11   7 20 27 29 34

4 Weight in kg of Year 10 boys.

47 51 63 39 42 57 36 37 49 32 60 54
56 45 52 48 61 58 56 59 70 66 56

5 Height in cm of a group of Year 10 girls.

153 147 160 146 162 158 159 171 149 152 150 163
172 167 165 155 155 157 154 168 150 172 152

6 IQ scores of a Year 10 tutor group in a comprehensive school.

101 112 125 109   98 107 108 117 121 116   94
 91 105 106 114 118 126 131   92   88 129
 89 116 103 108 127 110 117 104 119 133

7 Time taken to the nearest minute to change a flat tyre.

10 17   3 19 22 27 16   9   6 30 23 21
12 21   9 25 23 18 33   8 32 15 11

Interpreting stem-and-leaf diagrams

This spread will show you how to:

● Draw and use back-to-back stem-and-leaf diagrams

Keywords

Average
Interquartile
range (IQR)
Median
Measure of
spread
Stem-and-leaf
diagram

You can represent two data sets of the same variable on a back-to-back **stem-and-leaf** diagram.
The data sets share a common stem.

● You can compare the two data sets using an **average** and a **measure of spread**.

The range measures the spread of all the data.
> Range = highest value – lowest value

The **interquartile range (IQR)** measures the spread of the middle half of the data.
> IQR = upper quartile – lower quartile

Example

Draw a back-to-back stem-and-leaf diagram to show the test results of a group of girls and a group of boys.

Girls   52   34   58   46   41   57   47   35   49   47   54

Boys   61   43   47   56   59   39   58   69   52   46   54

Compare the performances of the boys and the girls.

```
        Girls          Boys
        5  4 | 3 | 9
9 7 7 6 1 | 4 | 3  6  7
    8 7 4 2 | 5 | 2  4  6  8  9
            | 6 | 1  9        Key: 1 | 4 | 3 stands for 41% for girls, 43% for boys
```

There are 11 girls.

$\text{Median} = \frac{(11 + 1)}{2}\text{th result}$

= 6th result
= 47

Range = 58 − 34
= 24

$\text{Lower quartile} = \frac{(11 + 1)}{4}\text{th value}$

= 3rd result
= 41

Upper quartile = 54

IQR = 54 − 41
= 13

There are 11 boys.

$\text{Median} = \frac{(11 + 1)}{2}\text{th result}$

= 6th result
= 54

Range = 69 − 39
= 40

$\text{Lower quartile} = \frac{(11 + 1)}{4}\text{th value}$

= 3rd result
= 46

Upper quartile = 59

IQR = 54 − 41
= 13

1 The boys' average is 7 marks higher than the girls' average.
2 The IQR is the same for the boys and the girls.
3 The range of the boys' marks is greater than that of the girls.

For these data sets
a Draw a back-to-back stem-and-leaf diagram. (Remember to include a key.)
b For both data sets in each question work out the
   i median   **ii** interquartile range   **iii** range.
c Use your answers to part **b** to make comparisons between the two data sets.

1 Percentages achieved in two tests.
   Test A: 38   37   62   45   42   55   56   61   49   52
             47   58   43   51   44   56   41   44   53

   Test B: 65   72   57   79   66   48   53   54   41   75
             56   63   69   72   53   44   57   61   70

2 Height to the nearest cm of a sample of men and a sample of women.
   Men: 176   183   184   172   168   175   183   159   169   174
             160   180   178   167   182   188   171   178   158   165
             177   169   167

   Women: 157   148   151   167   174   165   169   158   153   155
             161   158   155   172   156   162   166   149   178   154
             152   150   162

3 IQ scores of two classes, X and Y.
   X: 105   123   131   117   118   104   98   96   103   112   110
         117   126   129   123   109   108   115   99   89   121   134
         106   105   122   124   116   110   118   115   130
   Y: 118   119   104   121   126   118   109   97   114   129   130
         107   116   87   93   128   121   118   113   103   102   114
         107   131   99   106   124   132   119   126   114

4 Reaction times to the nearest tenth of a second by a sample of boys and a sample of girls.
   Boys: 4.2   5.7   3.2   3.8   6.4   3.8   6.1   5.9   5.3   5.6   3.6   4.4
           5.2   3.2   3.8   5.8   4.7   4.5   6.2   6.8   7.1   6.6   7.2

   Girls: 4.4   4.6   5.2   4.3   6.7   7.2   8.0   4.0   8.2   7.7   6.3   7.6
           4.8   5.9   5.2   7.4   7.3   6.2   6.5   5.6   6.6   6.3   5.5

5 Times taken to the nearest minute for a sample of children to complete two jigsaw puzzles.
   Puzzle P: 13   17   19   10   8   22   31   11   24   27
                6   37   18   12   29   14   8   17   9
   Puzzle Z: 28   21   29   15   12   9   32   17   18   11
               19   16   8   33   24   14   25   17   23

Box plots

This spread will show you how to:
- Draw box plots

Keywords
Box plot
Box and whisker
Interquartile range
 (IQR)
Lower quartile
Median
Upper quartile

- You can represent data sets on a **box plot**.
- To draw a box plot you need the **median**, the **upper** and **lower quartiles** and the highest and lowest values.

Example

A group of 15 students took a science test.

These are their results.

52, 62, 71, 46, 41, 49, 36, 57, 60, 80, 41, 40, 79, 39, 64

a Find the median, the range and the upper and lower quartiles.
b Draw a box plot to represent these results.
c Find the interquartile range.

A box plot is sometimes called a **box and whisker diagram**.

a First arrange the data in order.

36, 39, 40, 41, 41, 46, 49, 52, 57, 60, 62, 64, 71, 79, 80

$$\text{Median} = \frac{(15 + 1)}{2}\text{th value} \qquad \text{Lower quartile} = \frac{(15 + 1)}{4}\text{th value}$$

$$= \text{8th value} \qquad\qquad\qquad = \text{4th value}$$

$$= 52 \qquad\qquad\qquad\qquad = 41$$

$$\text{Range} = 80 - 36 \qquad\qquad \text{Upper quartile} = \frac{3}{4}(15 + 1)\text{th value}$$

$$= 44 \qquad\qquad\qquad\qquad = \text{12th value}$$

$$\qquad\qquad\qquad\qquad\qquad = 64$$

b 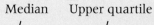 Lower quartile   Median   Upper quartile

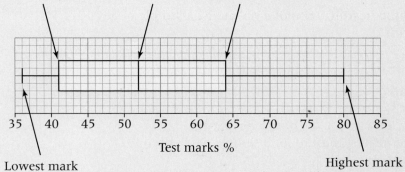

Use graph paper and remember to scale and label the axis.

Test marks %

Lowest mark   Highest mark

p.290

c Interquartile range = 64 − 41 = 23

The data sets for questions **1** to **7** were used in Exercise **D4.3** to draw stem-and-leaf diagrams.

Use the values you found for the median, lower quartile and upper quartile (or use the data to find them again) and draw a box plot to show the information for each of these data sets.

1 Percentage achieved in a Statistics test.

78 82 74 45 69 75 93 54 61 70 48 66
62 51 77 59 51 89 81 52 63 71 59

2 Minutes per day spent playing computer games.

40 26 75 84 33 39 28 66 67 71 80
37 52 47 63 49 41 44 58 69 43
73 55 59 43 61 38 29 30 77 60

3 Time taken, in minutes, to solve a crossword puzzle.

12 24 21 16   8   9   3 31 18 27 35
41 26 12 17   6   5 19 29 32 37 40
15 22 10 33 11   7 20 27 29 34

4 Weight in kg of Year 10 boys.

47 51 63 39 42 57 36 37 49 32 60 54
56 45 52 48 61 58 56 59 70 66 56

5 Height in cm of a group of Year 10 girls.

153 147 160 146 162 158 159 171 149
152 150 163 172 167 165 155 155 157
154 168 150 172 152

6 IQ scores of a Year 10 tutor group in a comprehensive school.

101 112 125 109   98 107 108 117 121
116   94   91 105 106 114 118 126 131
 92   88 129   89 116 103 108 127 110
117 104 119 133

7 Time taken to the nearest minute to change a flat tyre.

10 17   3 19 22 27 16   9   6 30 23 21
12 21   9 25 23 18 33   8 32 15 11

Line graphs

This spread will show you how to:

● Draw and interpret line graphs

Some types of data are collected over an extended period of time.
For example,

Electricity and gas bills are produced every quarter (three months).
Mobile phone bills are generated each month.
Unemployment reates arre published each month.

● Plotting the data on a line graph makes it easier to see any patterns.

Data often shows a short term, **seasonal variation**

For example, icecream sales are higher in Summer than in Winter.

and a longer term **trend**.

For example, inflation measures the annual increase or decrease in prices.

Example

Jenny's quarterly gas bills over a period of two years are shown in the table.

	Jan–March	April–June	July–Sept	Oct–Dec
2003	£65	£38	£24	£60
2004	£68	£42	£30	£68

Plot the data on a graph and comment on any pattern in the data.

Draw axes on graph paper with time on the horizontal axis.
Plot the coordinates as crosses on the grid.
Join them up with straight lines.

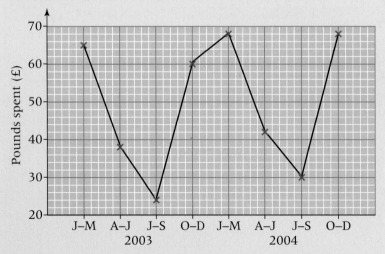

Plot time on the horizontal axis, J–M means Jan–March.

Gas bills are highest in the Winter months and lowest in the Summer months. This annual pattern appears to repeat itself.

There is a slight trend for the bills to rise from year-to-year.

For each of questions **1–6**
    a Plot the data on a graph
    b Comment on any patterns in the data.

1   The table shows Ken's monthly mobile phone bills.

Jan	Feb	Mar	April	May	June	July	Aug	Sept	Oct	Nov	Dec
£16	£12	£15	£18	£16	£18	£12	£10	£12	£15	£16	£20

2   The table shows Mary's quarterly electricity bills over a two-year period.

	Jan–March	April–June	July–Sept	Oct–Dec
2004	£45	£20	£15	£48
2005	£54	£24	£18	£50

3   The table shows monthly ice-cream sales at Angelo's shop during one year.

Jan	Feb	Mar	April	May	June	July	Aug	Sept	Oct	Nov	Dec
£16	£12	£15	£18	£38	£48	£52	£58	£18	£15	£16	£40

4   A town council carried out a survey over a number of years to find the percentage of local teenagers who used the town's library. The table shows the results.

year	1998	1999	2000	2001	2002	2003	2004	2005
%	14	18	24	28	25	20	18	22

5   Christabel kept a record of how much money she had earned from babysitting during three years.

	Jan–April	May–August	Sept–Dec
2001	£12	£18	£30
2002	£21	£33	£60
2003	£39	£42	£72

6   Steve kept a record of his quarterly expenses over a period of two years.

	Jan–March	April–June	July–Sept	Oct–Dec
2003	£35	£56	£27	£12
2004	£39	£68	£29	£18

Check out
You should now be able to:

- Draw and use scatter graphs
- Recognise correlation and draw and use lines of best fit
- Look at data to find patterns and exceptions
- Draw and interpret stem-and-leaf diagrams
- Draw and interpret box plots for small sets of data
- Draw and interpret line graphs

Worked exam question
Mrs Raja set work for the students in her class.
She recorded the time taken, in minutes, for each student to do the work.
She used her results to work out the information in the table.

	Minutes
Shortest time	4
Lower quartile	14
Median	26
Upper quartile	30
Longest time	57

On the grid, draw a box plot to show the information in the table.

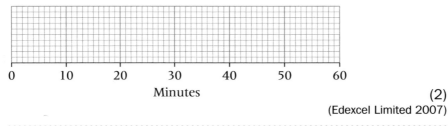

Minutes

(2)

(Edexcel Limited 2007)

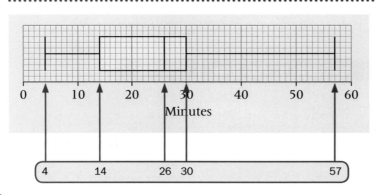

The vertical lines must be plotted accurately. The height of the box is not important.

Exam questions

1 Jason collected some information about the heights of 19 plants.
This information is shown in the stem and leaf diagram.

```
1 | 1  2  3  3
2 | 3  3  5  9  9
3 | 0  2  2  6  6  7
4 | 1  1  4  8
```

Key   4│8 means 48 mm

Find the median.

(2)

(Edexcel Limited 2008)

2 The scatter graph shows information about eight sheep.
It shows the height and length of each sheep.

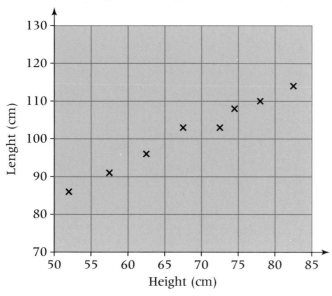

The table shows the height and the length of two more sheep.

Height (cm)	65	80
Length (cm)	100	110

a On a copy of the scatter graph, plot the information
from the table.

(1)

b Describe the relationship between the height and the
length of these sheep.

(1)

The height of a sheep is 76 cm.

c Estimate the length of this sheep.

(2)

(Edexcel Limited 2009)

Functional Maths 3: Recycling

The focus on protecting the environment from further damage is now stronger than ever. Recycling and reusing waste materials have become an important part of everyday life both for manufacturers and consumers.

This time-series chart shows the amounts of different materials recycled from households in England between 1997/98 and 2007/8.

What can you say about the different types of materials being recycled by households in England during this time? Do you notice any trends? Justify your response by referring to the data.

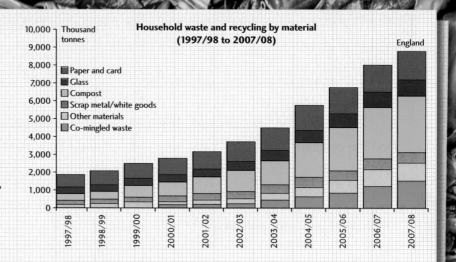

Copy and complete this table, using the data in the chart.

Year	1997/98	1998/99	1999/00	2000/01	2001/02	2002/03	2003/04	2004/05	2005/06	2006/07	2007/08
Co-mingled waste (amount)	0	0	0		221	268	469			1241	1563
Co-mingled waste (% of total)	0.00	0.00		7.33	6.94	7.16			12.65	15.39	
Other materials (amount)	230	257	355		235	269	385	516			989
Other materials (% of total)	12.31	12.27		5.94	7.38			8.92	10.74		
Scrap metal/white goods (amount)	231	253	265		369	419	465	577		601	598
Scrap metal/white goods (% of total)	12.36	12.08		11.02	11.58	11.20		9.97	7.83	7.45	
Compost (amount)	383		668		954	1189	1362	1960		2895	3189
Compost (% of total)	20.49	21.68		28.38	29.94	31.78		33.88	35.89	35.90	
Glass (amount)	335	347	383		426	470	568	670	760		902
Glass (% of total)	17.92	16.57		14.12	13.37	12.56		11.58	11.18	10.42	
Paper and card (amount)	690	783	842			1126	1272	1406		1535	1599
Paper and card (% of total)	36.92	37.39		33.21		30.10		24.30	21.70	19.04	
Total recycled (1000 tonnes)			2513	2812	3186					8063	

Compare the three largest components of recycled waste in 1997/98 and 2007/08. Can you think of any explanation for the difference? Justify your response, referring to the data.

A total of 25.3 million tonnes of household waste was collected in England in 2007/08, what percentage of this collected waste was re-used, recycled or composted?

The amount of household waste NOT re-used, recycled or composted was 7.0% lower in 2007/08 than in 2006/07. What was the total amount of household waste collected in tonnes in 2006/07?

Can you think of any reason for the trend shown by this data?

In 2007 the government set a target to reduce the amount of household waste in England not re-used, recycled or composted to 15.8 million tonnes. Do you think that this was a realistic target? Justify your response by referring to the data.

Manufacturers are responsible for designing packaging that is as environmentally friendly as possible while also protecting the product.

A company sells its own brand of baked beans in cans made of steel. The weight of these cans has been reduced by 13% every 10 years over the past 50 years. 50 years ago a can weighed 112g. What is the weight of a new can? By what proportion has the weight of a can changed over the last 50 years?

Glass milk bottles are 50% lighter than they were 50 years ago.

As well as reducing the consumption of raw materials, lighter packaging also saves money in other ways such as transport costs.

A supermarket sells tomatoes in packs of six. The packaging consists of a plastic tray with a lid as shown.

Given that on average this variety of tomato is spherical with a radius of 3cm, on average what percentage of the available volume of each package is empty?

Do you think that not having packaging would risk the quality of the tomatoes?

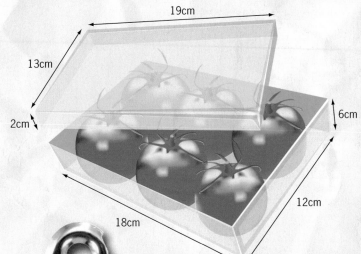

19cm

13cm

2cm

6cm

18cm

12cm

The product/pack ratio compares the weight of the packaging with the weight of the product it contains. Companies use this ratio to assess the suitability of the packaging used for each of their products. They often express it as a percentage to show how much of the overall weight is contributed by the packaging.

Look at some of the packaging you have at home. Could it be adapted to use less material without increasing the risk of damage to the product? If so, how?

How does the packaging used for perishable goods (e.g. food) differ from that used for non-perishable goods (e.g. electrical items)?

Research some well-known manufacturing companies on the Internet to find out about their packaging guidelines. Do they have different rules for different products (e.g. perishable/non-perishable goods)?

Percentages

In modern society people want to buy their own house, own a new car and go to university. To do this they need to borrow money from a bank or building society. These organisations lend the money but charge a fee, called 'interest', that is calculated as a percentage (fraction) of the amount borrowed.

What's the point?
Being able to solve problems involving percentages gives people greater control of their finances. It allows you to budget properly and be aware of the risks involved in borrowing too much money. In real life not being able to pay back enough money each month can lead to debt and bankruptcy

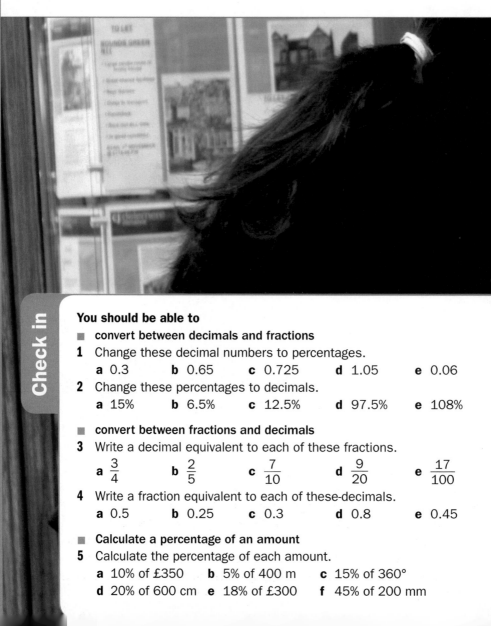

Check in

You should be able to

■ **convert between decimals and fractions**

1 Change these decimal numbers to percentages.

    a 0.3   **b** 0.65   **c** 0.725   **d** 1.05   **e** 0.06

2 Change these percentages to decimals.

    a 15%   **b** 6.5%   **c** 12.5%   **d** 97.5%   **e** 108%

■ **convert between fractions and decimals**

3 Write a decimal equivalent to each of these fractions.

    a $\frac{3}{4}$ **b** $\frac{2}{5}$ **c** $\frac{7}{10}$ **d** $\frac{9}{20}$ **e** $\frac{17}{100}$

4 Write a fraction equivalent to each of these-decimals.

    a 0.5   **b** 0.25   **c** 0.3   **d** 0.8   **e** 0.45

■ **Calculate a percentage of an amount**

5 Calculate the percentage of each amount.

    a 10% of £350   **b** 5% of 400 m   **c** 15% of 360°

    d 20% of 600 cm   **e** 18% of £300   **f** 45% of 200 mm

Orientation

What I need to know	What I will learn	What this leads to
N1 Multiply and divide by decimals **N2** Multiply and divide by fractions **N3** Understand proportional changes	■ Calculate percentage changes ■ Calculate (repeated) fractional changes ■ Calculate original amounts after a fractional change	**N7** Direct proportion Personal finance

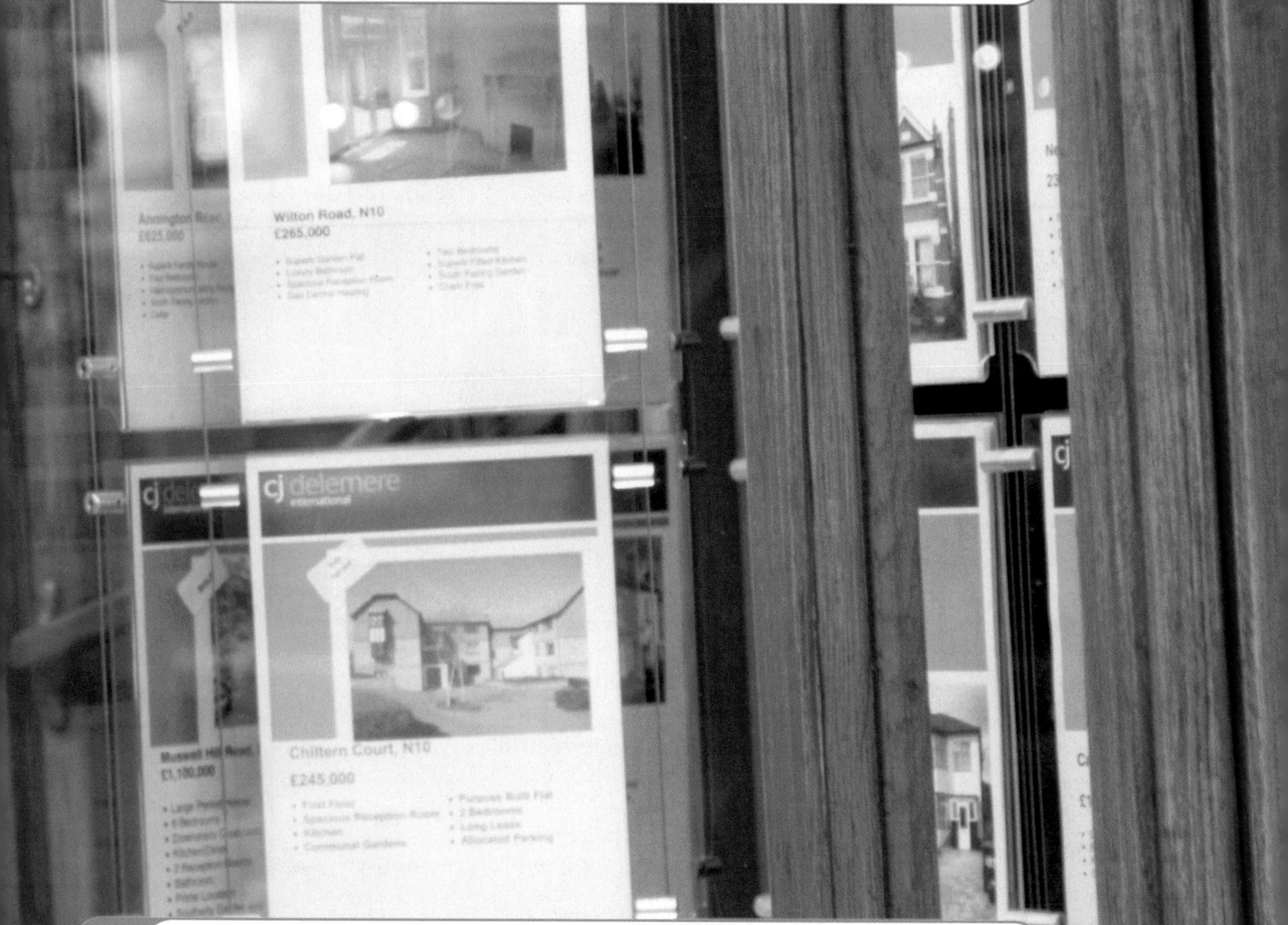

Rich task

Olivia buys a car for £15 000.
She takes out a loan which will last for 5 years.
The bank charges an interest rate of 8% per year.

a How much will Olivia have to pay back after 5 years?

b In real life people pay the money back in stages, say at the end of every year. Olivia pays some of the money back at the end of each year. She wants to pay exactly the same amount of money back each year. Calculate the size of Olivia's yearly payments, so that the loan is paid off in exactly 5 years.

This spread will show you how to:

- Calculate a percentage of a quantity

You often need to calculate a percentage of a quantity.

- Use mental methods to find simple percentages.

75% of £38: work out one quarter of £38 (which is £9.50), and multiply by 3.

- For more complicated examples, you can find 1% of the quantity, and then multiply by the required percentage.

27% of 48 m: first find 1% (which is 0.48 m), and multiply by 27.

- A quick method, especially when using a calculator, is to multiply by the appropriate decimal number.

To find 38% of a quantity, multiply by 0.38.

Example

Calculate　　**a** 45% of 60 cm　　**b** 34% of 85 kg　　**c** 16% of £25

a 50% of 60 = 30　　　　　　　　50% is a half.
  5% of 60 = 3
  45% of 60 cm = 30 cm − 3 cm　　= 27 cm

b $34\% = \frac{34}{100} = 0.34$　　　　0.34 is the **decimal equivalent** of 34%.

  34% of 85 kg = 0.34 × 85
  　　　　　　 = 28.9 kg　　　By calculator.
  　　　　　　 = 29 kg to 2 sf.

c 16% of £25 = $\frac{\overset{4}{\cancel{16}}}{\underset{\cancel{1}}{\cancel{100}}} \times \overset{1}{\cancel{25}}$

  　　　　　 = £4

Example

At Fitz High School, 95% of the students have never been absent.
There are 1180 students at the school.
How many of them have a perfect attendance record?

　　　　95% = 0.95
95% of 1180 = 0.95 × 1180
　　　　　　 = 1121

So 1121 students have a perfect attendance record.

Example

Tom is saving 6% of his salary in a pension fund.
His current salary is £25 000.
How much will he save this year?

10% of £25 000 = £2500
  5% of £25 000 = £1250
   1% of £2500 = £250
   6% of £2500 = £1250 + £250
   　　　　　　 = £1500

This calculation is easy to do mentally.

1 Find the percentages of these numbers mentally.
   a 25% of 42   **b** 90% of 140   **c** 20% of 1200
   d 60% of 500   **e** 30% of 440   **f** 11% of 900

2 Calculate these, using a mental method wherever possible.
   a 30% of £750   **b** 55% of 1800 m
   c 90% of 2800 kg   **d** 60% of €240

3 Use an appropriate method to work out these. Show all your
   working and do not use a calculator.
   a 9% of 1500   **b** 13% of 700   **c** 31% of 2400
   d 36% of 50   **e** 43% of 900   **f** 6% of 3200

4 Calculate these, using an appropriate mental or written method.
   Show all your working and do not use a calculator.
   a 23% of 4800 mm   **b** 61% of 3200 kg
   c 39% of €3700   **d** 17% of £2900

5 Write a decimal number equivalent to each percentage.
   a 50%   **b** 60%   **c** 25%   **d** 51%
   e 64%   **f** 22%   **g** 15%   **h** 70%
   i 7%   **j** 8.5%   **k** 0.15%   **l** 0.01%

6 Calculate these, using an appropriate method.
   Use a calculator where necessary.
   a 15% of 38   **b** 25% of 800   **c** 27% of 59
   d 96% of 104   **e** 41% of 41   **f** 80% of 25

7 Calculate these, rounding your answers to the nearest penny.
   a 16% of £24   **b** 63% of £85   **c** 93% of £15
   d 42% of £405   **e** 88% of £32   **f** 6% of £265

A02 Functional Maths

8 Mrs Jones has a conservatory built which costs £12 000.
   She pays an initial deposit of 15%. The remainder is to be
   paid in 24 equal monthly instalments. How much is each
   of these instalments?

9 A restaurant adds a 12% service charge to the bill. What will
   be the total cost for a meal that is £65.80 before the service
   charge is added?

10 A school has 1248 pupils, and 48% of them are girls. How
   many boys are there in the school? Show your working.

11 Julia earns a salary of £47 800 per year. She is awarded
   a 2.7% pay rise. Calculate her new salary.

This spread will show you how to:

- Solve problems, involving percentage increase and decrease

You often need to work out problems involving percentage **increase** and **decrease**.

p.320

- To work out the new amount after a percentage increase or decrease,
  - first find the increase or reduction

    Increase £25 000 by 5%
    5% of £25 000 = £1250

    then simply add or subtract
    £25 000 + £1250 = £26 250

  - Using a calculator, you can find the new amount by multiplying by the **decimal equivalent**.

    To find the result of a 12% decrease, multiply the original amount by 0.88.
    To find the result of a 23% increase, multiply the original by 1.23.

Example

Find the result when

a 45 cm is increased by 20%   **b** 44 kg is decreased by 17%.

....

a 20% of 45 cm = 9 cm,
new length 45 cm + 9 cm = 54 cm

b 100% − 17% = 83%
44 kg × 0.83 = 36.52 kg

New mass is
83% of original.

Example

This is how Jackie worked out 22% of 85 cm.

What is wrong here?

Ok – 22% off from 85 leaves 78%, so I'll work out 85 × 0.78 …

Examiner's tip
The words 'of' and 'off' can cause confusion – especially when you are in a hurry in an exam!

....

This question simply asks her to work out 22% **of** 85 cm – she is not being asked to work out a percentage reduction.

22% of 85 cm = 0.22 × 85
             = 18.7 cm

Example

Alan's car depreciates by 15% every year.
It is valued at £9750 now. What will be its value in one year's time?

....

Value after depreciation = 100% − 15% = 85% of original value
0.85 × 9750 = £8287.50 = £8300 to nearest £100

85% = 0.85

1 Calculate the results when these amounts are increased by the percentages given.

   a 72 by 50%   **b** 60 by 20%   **c** 45 by 10%
   d 600 by 3%   **e** 480 by 25%   **f** 500 by 40%

> You should be able to do questions **1** and **2** mentally.

2 Decrease each of these amounts by the percentages given.

   a 28 by 10%   **b** 45 by 20%   **c** 60 by 15%
   d 75 by 50%   **e** 380 by 40%   **f** 65 by 1%

3 Use a written method to calculate each of these percentage increases. Show your working and do not use a calculator.

   a 300 by 14%   **b** 200 by 32%   **c** 800 by 26%
   d 250 by 30%   **e** 750 by 83%   **f** 940 by 18%

4 Use a written method to find these percentage decreases.

   a 800 by 16%   **b** 700 by 24%   **c** 400 by 32%
   d 450 by 33%   **e** 350 by 61%   **f** 260 by 52%

5 Write the decimal number you must multiply by to find these percentage increases.

   a 20%   **b** 30%   **c** 45%   **d** 85%   **e** 6.5%

6 Write the decimal number you must multiply by to find these percentage decreases.

   a 40%   **b** 60%   **c** 35%   **d** 72%   **e** 18.5%

7 Use a calculator to find these percentage increases and decreases.

   a Increase 53 by 7%   **b** Decrease 42 by 4%
   c Increase 620 by 16%   **d** Decrease 300 by 18%

AO2 | Functional Maths

8 Calculate the new salaries after these pay increases.
   a £32 000 increased by 5%   **b** £18 450 increased by 4.7%
   c £26 500 increased by 3.2%   **d** £52 850 increased by 6%

9 Calculate the new prices after these price cuts.
   a £450 decreased by 22%   **b** £860 decreased by 35%
   c £1250 decreased by 42%   **d** £740 decreased by 3.5%

AO3 | Problem

10 Alan says, 'The cost of computer chips increased by 120% last year'.
   Billy says, 'That's not possible. They couldn't have gone up by more than 100%'.
   Explain why Billy is wrong.

11 Carina says, 'The cost of computer memory fell by 120% last year.'
   Deni says, 'That's not possible. Prices couldn't have fallen more than 100%'.
   Explain why Deni is correct.

Simple and compound interest

This spread will show you how to:

- Calculate simple and compound interest.

Keywords

Borrow
Compound interest
Interest
Invest
Principal
Rate
Simple interest

You earn **interest** when you **invest** in a savings account at a bank. However, you pay interest if you **borrow** money for a mortgage.

Interest is either
- **simple interest** – it is not added to the **principal**, or
- **compound interest** – added to the principal and will itself earn interest.

The original sum you invest is called the principal.

- To calculate simple interest, use the interest **rate** to work out the amount earned.

- If simple interest is paid for several years, the amount paid each time stays the same, because the interest is paid elsewhere and the principal stays the same.

- To calculate compound interest, work out the interest in the same way, but add the interest earned to the principal.

- If compound interest is paid for several years, the amount of interest earned each year increases, because the principal increases.

Example

Calculate the interest when £1000 is invested for 4 years at
a 5% simple interest (SI) **b** 5% compound interest (CI).

a SI for 1 year = 5% × £1000
= £50
SI for 4 years = 5% of £1000 × 4
= £50 × 4
= £200
Total SI = £200
Principal + SI = £1200

b 1st year's CI = 5% × £1000
= £50
new principal = £1000 + £50
= £1050
2nd year's CI = 5% × £1050
= £52.50
new principal = £1102.50
3rd year's CI = 5% × £1102.50
= £55.13
new principal = £1157.63
4th year's CI = 5% × £1157.63
= £57.88
new principal = £1215.51

Money invested at compound interest earns more than the same sum invested at simple interest.

- When you have to calculate compound interest over a number of years you can do it quickly using a calculator.

To increase by 6.5% multiply by decimal equivalent of 6.5% = 1.065

Example

£2000 is invested at 6.5% compound interest.
Find the principal after 15 years.

Principal at end of 1 year = £2000 × 1.065 = £2130
Principal at end of 2nd year = (£2000 × 1.065) × 1.065
= £2000 × 1.065^2 = £2268.45
After 15 years principal = £2000 × 1.065^{15} = £5143.68

Each year the principal increases by a factor of 1.065

1 Find the total interest earned when these amounts of money are invested at these annual rates of **simple interest**.

  a £100 at 5% for 1 year
  b £200 at 6% for 2 years
  c £1400 at 7.5% for 3 years
  d £650 at 3.5% for 10 years

2 Find the final value of each of these amounts invested at **compound interest**.

  a £250 at 5% for 1 year
  b £400 at 2% for 2 years
  c £1200 at 6% for 2 years
  d £1000 at 2% for 3 years

3 A sum of money is invested at 5% compound interest.

  • To find the amount after one year, multiply the principal by 1.05.
  • To find the amount after two years, multiply the principal by $1.05^2 = 1.1025$

  a What decimal number do you need to multiply the principal by to work out the amount after earning interest for one year at a rate of 6%?

  b What decimal number do you need to multiply the principal by to work out the amount after earning compound interest for two years at 6%?

4 Use your answers to question **3** to work out the final amount when a principal of £500 is invested.

  a for one year at an interest rate of 6%
  b for two years at 6% compound interest.

5 Find the decimal number you should multiply the principal by to find the final amount after earning compound interest at these rates.

  a 5% per year for 3 years
  b 6.5% per year for 5 years

6 Use your answers to question **5** to work out the final amount when

  a £5000 is invested for 3 years at 5% compound interest.
  b £800 is invested for 5 years at 6.5% compound interest.

Functional Maths **A02**

7 Which of these options earns the most interest when £5000 is invested for

  a 8 years at 6% simple interest
  b 6 years at 8% compound interest?
  Explain your answer.

Problem **A03**

8 **a** What decimal number do you multiply by, to find a 15% reduction?

  b Explain how you would then find a second 15% reduction.

Further percentage techniques

This spread will show you how to:

- Express a number as a percentage of another.
- Calculate the original amount before a percentage increase or decrease.

Keywords
Original amount

- To write one number as a percentage of another, start by writing the first number as a fraction of the other.

  - Write '23 as a percentage of 25' as a fraction as $\frac{23}{25}$.

- Now convert the fraction to a percentage.

  - $\frac{23}{25} = \frac{23 \times 4}{25 \times 4} = \frac{92}{100} = 92\%$.

- Sometimes you will need to use a calculator. For example, to find 9 as a percentage of 17: $\frac{9}{17} = (9 \div 17) \times 100\% = 52.9\%$ (to 1 decimal place).

Example

Find

a 13 as a percentage of 20   **b** 39 as a percentage of 75

c 8 as a percentage of 23

....

a $\frac{13}{20} = \frac{13 \times 5}{20 \times 5} = \frac{65}{100} = 65\%$

b $\frac{39}{75} = \frac{13}{25} = \frac{52}{100} = 52\%$

c $\frac{8}{23} = (8 \div 23) \times 100\% = 34.8\%$

- To find the **original amount** before a percentage change, use the final amount to work out 1% of the original amount, and then find the original amount by multiplying by 100.

Example

Find the original price of a denim jacket reduced by 15% to £32.30.

....

85% of original price = £32.30

1% of original price = £32.30 ÷ 85

100% of original price = £32.30 ÷ 85 × 100

= £38

$100\% - 15\% = 85\%$

Example

Following a 5% price increase, a car radio costs £168. How much did it cost before the increase?

....

105% of original price = £168

1% of original price = £168 ÷ 105 = £1.60

100% of original price = £1.60 × 100 = £160

1 For these pairs of numbers, find the smaller number as a percentage of the larger one. You should be able to do all of these mentally.
   a 16 and 20 **b** 7 and 25 **c** 6 and 50
   d 20 and 40 **e** 17 and 34

2 Use a written method to find
   a 12 as a percentage of 20 **b** 36 as a percentage of 75
   c 24 as a percentage of 40

3 Use a calculator to find
   a 19 as a percentage of 37 **b** 42 as a percentage of 147
   c 8 as a percentage of 209

4 Jason scored 53 out of 60 in a science test and 39 out of 45 in a maths test.
   a Convert each of his marks to a percentage.
   b Jason says, 'I did better in maths, because I only dropped 6 marks; I dropped 7 marks in science.' Explain why Jason is wrong.

5 A book costs £4 after a 20% price reduction. How much did it cost before the reduction? Show your working.

6 These numbers are the results when some amounts were increased by 10%. For each one, find the original number.
   a 55 **b** 44 **c** 88 **d** 121

7 Find the original cost of the following items.
   a A vase that costs £7.20 after a 20% price increase.
   b A table that costs £64 after a 20% decrease in price.

A02 Functional Maths

8 a Francesca earns £350 per week. She is awarded a pay rise of 3.75%. Frank earns £320 per week. He is awarded a pay rise of 4%. Who gets the bigger pay increase? Show all your working.
   b Bertha's pension was increased by 5.15% to £82.05. What was her pension before this increase?

A03 Problem

9 Carys is trying to find the original price of an item that is on sale. Its sale price is £56 and it was reduced by 20%. Carys' working is shown to the right. What is wrong with her working?

> 20% of 56 = 11.2
> 56 + 11.2 = 67.2
> The answer is £67.20

Reverse percentages

This spread will show you how to:

- Solve problems using reverse percentages

In a reverse **percentage** problem, you are given an amount after a percentage change, and you have to find the **original** amount.

Example

In a sale, a pair of shoes cost £38.25 after a 15% **decrease**.

Find the original price of the shoes.

Sale price is $(100 - 15)\% = 85\%$ of original price

$$85\% = 0.85$$

Original price $\times 0.85 =$ sale price

$$\text{Original price} = \frac{£38.25}{0.85} = £45$$

First work out the reverse percentage.

0.85 is the decimal equivalent of 85%.

Example

A table costs £88, including 17.5% VAT. Find the cost (to the nearest pound) before VAT was added.

£88 is 117.5% of cost before VAT.

$$117.5\% = 1.175$$

Cost before VAT $\times 1.175 = £88$

$$\text{Cost before VAT} = \frac{£88}{1.175} = £75 \text{ to nearest pound.}$$

Find the decimal equivalent.

Example

The price of unleaded petrol is increased by 4% to 93.7 p per litre. Find the price before the **increase**.

Increased price is 104% of old price.

$$100\% = 1.04$$

Old price $\times 1.04 =$ Increased price

$$\text{Old price} = \frac{93.7}{1.04} = 90.1 \text{ p per litre to near 0.1 p}$$

Example

A car costs £15 000, including VAT at 17.5%. Veejay is working out the cost without VAT. He thinks: 'I'll just work out 17.5% of £15 000 and take that away, to get £12 375'.

What is wrong with Veejays's reasoning?

Original price	+	VAT	=	£15 000
100%		17.5%	=	117.5%

$£15\,000 \div 1.175 = £12\,766$ (to nearest pound)

1   Use a mental method to find the result of these following percentage increases.
   a £100 is increased by 25%   **b** £50 is increased by 20%
   c 40 m is increased by 15%   **d** 36 cm is increased by 50%

2   Calculate mentally the result of these percentage decreases.
   a £60 is decreased by 25%   **b** 80 cm is decreased by 75%
   c 720 cm is decreased by 50%   **d** 120 mm is decreased by 15%

3   Write a decimal number that you could multiply a quantity by to find the results of these percentage changes.
   a An increase of 20%   **b** A decrease of 15%
   c An increase of 6%   **d** A decrease of 5%
   e A decrease of 6%   **f** A decrease of 17%

4   Calculate the result of these percentage increases and decreases, by multiplying by the correct decimal number. Show your method.
   You may use a calculator.
   a £64 decreased by 7%   **b** 45 kg increased by 14%
   c 10.4 seconds decreased by 3%   **d** 120 m increased by 65%
   e €240 increased by 8.5%   **f** $340 decreased by 11.5%

5   Calculate the original cost of these items, before the percentage changes shown.
   Show your method. You may use a calculator.
   a A hat that costs £46.50 after a 7% price cut.
   b A skirt that costs £32.80 after a price rise of 6%.

6   A computer costs £658, including VAT at 17.5%. Find the price, before VAT was added.

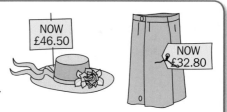

7   To decrease an amount by 8%, multiply it by 0.92.
   For a further decrease of 8%, multiply by 0.92 again, and so on.
   Use this idea to calculate
   a the final price of an item with an original price of £380, which is given two successive price cuts of 8%.
   b the final price of an item with an original price of £2400, which is given three successive price cuts of 10%.

Summary

Check out

You should now be able to:

- Calculate a fraction and a percentage of a quantity
- Calculate percentage increase and decrease
- Express a number as a percentage of another
- Calculate simple and compound interest
- Calculate the original amount after a percentage increase or decrease

Worked exam question

The price of all rail season tickets to London increased by 4%.

a Before this increase, the price of a rail season ticket from Reading to London was £2664
Work out the price after the increase. (3)

b The price of a rail season ticket from Cambridge to London increased by £121.60
Work out the price before this increase. (2)

c After the increase, the price of a rail season ticket from Brighton to London was £2828.80
Work out the price before this increase. (3)

(Edexcel Limited 2006)

a
$$4\% \text{ of } £2664 = 0.04 \times £2664$$
$$= £106.56$$
$$£2664 + £106.56 = £2770.56$$

Show both the multiplication and addition calculations.

OR

a
$$104\% \text{ of } £2664 = 1.04 \times 2664$$
$$= £2770.56$$

Show the multiplication calculation.

b
$$4\% = £121.60$$
$$1\% = £30.40$$
$$100\% = £3040$$

A different method could be:
$£121.60 \div 4 \times 100$

c
$$104\% = £2828.80$$
$$1\% = £27.20$$
$$100\% = £2720$$

A different method could be:
$£2828.80 \div 1.04$

You need to show your working whichever method you use.

Exam questions

A02

1 In April 2004, the population of the European Community was 376 million.
In April 2005, the population of the European Community was 451 million.

   a Work out the percentage increase in population.
     Give your answer correct to 1 decimal place.   (3)

   In April 2004, the area of the European Community was 3.2 million km^2.
   In April 2005, the area of the European Community increased by $\frac{3}{8}$

   b Work out the area of the European Community in April 2005.   (2)

(Edexcel Limited 2007)

2 Jack invests £3000 for 2 years at 4% per annum compound interest.
Work out the value of the investment at the end of 2 years.   (3)

(Edexcel Limited 2008)

3 In a sale, normal prices are reduced by 25%.
The sale price of a shirt is £16.50
Calculate the normal price of the shirt.   (3)

A03

4 This item appeared in a newspaper.

> **Cows produce 3% more milk**
>
> A farmer found that when his cow listened to classical music the milk it produced increased by 3%.
>
> This increase of 3% represented 0.72 litres of milk.

Calculate the amount of milk produced by the cow when it listened to classical music.   (3)

(Edexcel Limited 2006)

Sequences

There are lots of different types of sequences in mathematics. From the most basic arithmetic sequences such as the set of even numbers, to the curious Fibonacci sequence used to describe the growth of rabbit populations, through to the elegant and complex world of continued fractions.

What's the point?

Sequences are often infinite. Expressing any sequence as a formula (nth term) allows you to find any term in the sequence and begin to understand its properties.

You should be able to

■ **recognise patterns in sequences**

1 Complete the next two values in each pattern.

    a 2, 4, 6, 8, ...   **b** 100, 94, 88, 82, 76, ...   **c** 1, 2, 4, 7, 11, ...

    d 10, 7, 4, 1, ...   **e** 3, 6, 12, 24, ...   **f** $1, \frac{1}{2}, \frac{1}{3}, \frac{1}{4}, \frac{1}{5}, \ldots$

■ **use simple formulae**

2 Given that $n = 3$, put these expressions in ascending order.

    $2n + 7$ $4(n - 1)$ $2n^2$ $\frac{9}{n} + 15$ $15 - n$

■ **recognise special sequences**

3 This pattern has been shown in three different ways.

    It has a special name.

    What is it called?

    Describe how it got this name.

1	4	9	16
1×1	2×2	3×3	4×4
□			

What I need to know

KS3 Recognise patterns
in sequences of
numbers

A2 Write and manipulate
algebraic expressions
Write formulae
G2 Proof in geometry

What I will learn

Generate and
describe sequences
Find position-to-term
rules for sequences
Use proof and
counter examples

What this leads to

A-level
Maths

This L-shape, containing five
numbers, is drawn on a
10×10 grid numbered from
1 to 100.
We can call it L_{35} because
the largest number inside it is
35 – the L-number. The total of
the numbers inside L_{35} is 138.
Find a connection between the
L-number and the total of the
numbers inside the L shape.
What happens if you work on
a 9×9 grid, an 8×8 grid a
$m \times m$ grid?

1	2	3	4	5	6	7	8	9	10
11	12	13	14	15	16	17	18	19	20
21	22	23	24	25	26	27	28	29	30
31	32	33	34	35	36	37	38	39	40
41	42	43	44	45	46	47	48	49	50
51	52	53	54	55	56	57	58	59	60
61	62	63	64	65	66	67	68	69	70
71	72	73	74	75	76	77	78	79	80
81	82	83	84	85	86	87	88	89	90
91	92	93	94	95	96	97	98	99	100

This spread will show you how to:

- Generate common sequences and describe how number patterns are formed

Keywords
Sequence
Term

- A **sequence** is a set of numbers that often follow a pattern, for example
  5, 9, 13, 17, 21, ... are the first five **terms** of a sequence that goes up in 4s
  3, 6, 12, 24, 48, ... are the first five terms of a sequence that doubles
  1, 4, 9, 16, 25, ... is the sequence of square numbers
  ($1 \times 1, 2 \times 2, 3 \times 3$, etc.)
  1, 8, 27, 64, 125, ... are the cube numbers
  ($1 \times 1 \times 1, 2 \times 2 \times 2, 3 \times 3 \times 3$, etc.)

You can find sequences
in patterns, for example

Number of tiles is:
3, 5, 7, ...
The next diagram
would need 9 tiles.

Example

Write the next two terms in each sequence.

a 4, 5, 7, 10, 14, 19, ...   **b** 0.6, 0.7, 0.8, 0.9, ...

....

a 4, 5, 7, 10, 14, 19, ... The terms increase by 1, then 2, then 3.
The next two terms are $19 + 6 = 25$ and $25 + 7 = 32$.

b 0.6, 0.7, 0.8, 0.9, ... The terms increase by 0.1.
The next two terms are $0.9 + 0.1 = 1.0$ and $1.0 + 0.1 = 1.1$.

You may be
tempted to
continue 0.7,
0.8, 0.9, 0.10, ...
but 0.10 is less
than 0.9!

Example

Name each of these sequences.

a 1, 3, 5, 7, 9, ...   **b** 25, 36, 49, 64, ...

....

a These are the odd numbers.   **b** These are the square numbers,
starting at 5×5 (or 5^2).

Example

How many tiles would be in the tenth diagram in this pattern?

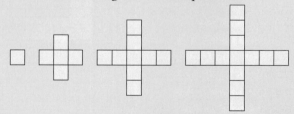

A table is a good
way of organising
results so that
you can see what
is happening.

....

Diagrams	1	2	3	4
Number of tiles	1	5	9	13

The number of tiles increases by 4 each time. If you continue this,
you get 1, 5, 9, 13, 17, 21, 25, 29, 33, **37**.
The tenth diagram would have 37 tiles.

1 Copy each sequence and add the next two terms.
  a 4, 9, 14, 19, 24, ___, ___   **b** 100, 93, 86, 79, 72, ___, ___
  c 1, 2, 4, 7, 11, ___, ___   **d** 9, 99, 999, 9999, 99 999, ___, ___
  e 1, 1, 2, 3, 5, 8, ___, ___   **f** 54, 27, 13.5, 6.75, ___, ___

2 Copy and fill in the missing numbers in each sequence.
  a 4, ___, 10, ___, 16, ...   **b** 4, ___, ___, 32, 64, ...
  c 95, ___, ___, ___, 87, ...   **d** 1, ___, 27, 64, ___, ...

3 Write the first five terms of each of these well-known number patterns.
  a Multiples of 3   **b** Powers of 2
  c Prime numbers   **d** Square numbers over 100

4 The triangular numbers form a sequence.

  a Copy the table and use the diagram to complete it.

Diagrams	1	2	3	4	5
Number of dots	1	3			

  b Hence, write the first 10 triangular numbers.
  c Why do you think the square numbers (1, 4, 9, 16, 25, ...) got their name?
  d How did the cube numbers get their name? Write the first five cube numbers.

A03   Problem

5 Find the tenth term in each of these number patterns.
  a $(1 \times 2), (2 \times 3), (3 \times 4), (4 \times 5), ...$

  b $\frac{1}{2}, \frac{2}{3}, \frac{3}{4}, \frac{4}{5}, ...$

  c $(5 \times 2), (5 \times 4), (5 \times 8), (5 \times 16), ...$
  d $(1 \times 1), (4 \times 8), (9 \times 27), (16 \times 64), ...$

6 Look at this number pattern.
    $7^2 = 49$
   $67^2 = 4489$
 $667^2 = 444\,889$
  a Write the next two lines in the pattern.
  b What is $66\,666\,666\,667^2$?
  c What is $\sqrt{4\,444\,444\,444\,888\,888\,889}$?

Generating sequences

This spread will show you how to:
- Use position-to-term rules to write sequences
- Find and describe the *n*th term of a sequence, using this to find other terms

Keywords

*n*th term
Position
Position-to-term rule
Sequence
Term

- You can use a **position-to-term** rule to write a **sequence**.

For example, a sequence is defined using the position-to-term rule, 'multiply by 5 and add 2'.
The first 5 terms are

Position	1	2	3	4	5
Term	$1 \times 5 + 2 = 7$	$2 \times 5 + 2 = 12$	$3 \times 5 + 2 = 17$	$4 \times 5 + 2 = 22$	$5 \times 5 + 2 = 27$

p.136

You can write a position-to-term rule in algebraic notation.
For example, T_n represents the **_n_th term** of the sequence
$$T_n = 3n - 2$$
To find T_1, the first term, put $n = 1$ in the rule.
Hence, $T_1 = 3 \times 1 - 2 = 1$
$T_2 = 3 \times 2 - 2 = 4$
$T_3 = 3 \times 3 - 2 = 7$
$T_4 = 3 \times 4 - 2 = 10$
The sequence $T_n = 3n - 2$ is 1, 4, 7, 10, ...

The rule connects the **position** of a **term** in a sequence to its value.

Example

Find the first five terms of the sequences with these position-to-term rules.

a $T_n = 5n + 7$
b $T_n = 3n^2$

....

a $T_1 = 5 \times 1 + 7 = 12$
$T_2 = 5 \times 2 + 7 = 17$
$T_3 = 5 \times 3 + 7 = 22$
$T_4 = 5 \times 4 + 7 = 27$
$T_5 = 5 \times 5 + 7 = 32$, ...
Hence, the sequence with $T_n = 5n + 7$ is 12, 17, 22, 27, 32, ...

b $T_1 = 3 \times 1^2 = 3$
$T_2 = 3 \times 2^2 = 12$
$T_3 = 3 \times 3^2 = 27$
$T_4 = 3 \times 4^2 = 48$
$T_5 = 3 \times 5^2 = 75$, ...
Hence, the sequence with $T_n = 3n^2$ is 3, 12, 27, 48, 75, ...

You may notice a pattern that allows you to generate the terms quickly, for example, this sequence goes up in 5s.

Remember 'BIDMAS', of the power comes before multiplication, so square before multiplying by 3.

1 Write the first five terms of each sequence.

   a $T_n = 8n + 2$ **b** $T_n = 5n - 4$ **c** $T_n = 7n$

   d $T_n = 10 - 2n$ **e** $T_n = n^2 - 3$ **f** $T_n = 2n^2$

2 Write the first five terms of each sequence.

   a $T_n = 9 - n$ **b** $T_n = (n + 1)(n + 2)$ **c** $T_n = \frac{1}{n}$

   d $T_n = (-n)^3$ **e** $T_n = (2n - 1)(n + 1)(n - 3)$ **f** $T_n = n^4$

3 Here are four position-to-term rules.

$$\boxed{T(n) = n^2} \qquad \boxed{T_n = 18 + 2n}$$

$$\boxed{T_n = n^3} \qquad \boxed{T_n = n(n + 5)}$$

   Which rules give

   a two sequences with an identical first term

   b two sequences whose third term is 24

   c two sequences whose tenth term is greater than 100.

4   **a** Generate all the terms and the formula for T_n, the nth term of the
     sequence described by this flow chart.

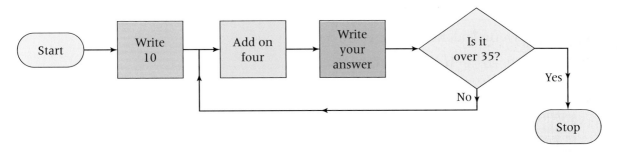

   b Design your own flow chart for the sequence 2, 5, 10, 17, 26, 37, 50.

A03 Problem

5 Write the algebraic position-to-term rule for

   a   **i** Two sequences with identical 1st terms

       ii Two sequences with identical 5th terms

   b If two sequences have identical 1st terms and identical 5th terms,
     do they have to be identical sequences? Discuss your answer.

This spread will show you how to:

- Find and describe the *n*th term of a sequence, using this to find other terms

Keywords
Formula
Linear
*n*th term

- In a **linear** sequence the terms go up or down by the same amount.

  12, 15, 18, 21, 24, ...   goes up in 3s, so it is linear

  100, 95, 90, 85, 80, ... goes down in 5s, so it is linear

You can find the ***n*th term** of a linear sequence by comparing the sequence to the times table to which it is related. For example the sequence 10, 17, 24, 31, 38, ... is connected to the $7\times$ table, hence:

Position	1	2	3	4	5
Multiples of 7	7	14	21	28	35
Sequence term	10	17	24	31	38

The terms go up in 7s.

By comparing the sequence to the $7\times$ table, you can see that you have to add 3 to the multiples of 7 to get the terms of the sequence. Hence *n*th term T_n = position $\times$ 7 + 3.

$$T_n = 7n + 3$$

You can use the *n*th term **formula** to find any term of the sequence quickly, for example, the 100th term of the above sequence is

$T_{100} = 7 \times 100 + 3$ so $T_{100} = 703$

Example

Find the *n*th term and, hence, the 100th term of

a 2, 8, 14, 20, 26, ... **b** 25, 20, 15, 10, 5, ...

....

a Compare the sequence to the multiples of 6.

Position	1	2	3	4	5
Multiples of 6	6	12	18	24	30
Term	2	8	14	20	26

This sequence goes up in 6s, so it must be connected to the 6 times table.

First term = 2 = 6 − 4   Second term = 8 = 12 − 4

Subtract 4 from the multiple of 6 to get each term.

$$T_n = 6n - 4$$

Hence, $T_{100} = 6 \times 100 - 4 = 596$

b Compare this sequence to the multiples of −5.

Position	1	2	3	4	5
Multiples of 5	−5	−10	−15	−20	−25
Term	25	20	15	10	5

If a sequence goes down use negative multiples.

First term = 25 = −5 + 30   Second term = 20 = −10 + 30

Add 30 to the multiple of −5 to get each term.

$$T_n = -5n + 30 \text{ or } T_n = 30 - 5n$$

Hence, $T_{100} = 30 - 5 \times 100 = -470$

On a number line, to get from −5 to 25, you need to add 30.

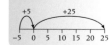

1 Find a formula for the nth term, T_n, of each sequence.

a 4, 9, 14, 19, 24, ...
b 1, 3, 5, 7, 9, ...
c 10, 12, 14, 16, 18, ...
d 1, 1.5, 2, 2.5, 3, ...
e −4, −2, 0, 2, 4, ...
f 1, 2, 3, 4, 5, ...
g The multiples of 13
h Counting up in 10s, starting from 4
i 10, 8, 6, 4, 2, ...
j 100, 95, 90, 85, 80, ...
k 50, $49\frac{3}{4}$, $49\frac{1}{2}$, $49\frac{1}{4}$, ...
l Counting down in multiples of 4 from 75

2 Write five linear sequences of your own that have a third term of 15. Find a formula for the nth term of each one, and use this to find the 100th term in each case.

3 Are these statements true or false?

a The 50th term of 2, 5, 8, 11, 14, ... is more than 150.
b The 50th term of 5, 9, 13, 17, 21, ... is even.
c The 100th term of 1000, 990, 980, 970, 960, ... is negative.

> Find the position-to-term rule for each sequence.

4 Here are some terms of a sequence. In each case find the formula for the nth term, T_n.

a The 5th term is 20, the 6th term is 28 and the 7th term is 36.
b The 100th term is 302, 101st term is 305 and the 102nd term is 308.
c The 153rd term is 260, the 154th term is 262 and the 155th term is 264.

A03 Problem

5 How could you find the nth term formula for these fractional sequences? See if you can work out what it would be in each case.

a $\frac{3}{7}$, $\frac{5}{10}$, $\frac{7}{13}$, $\frac{9}{16}$, $\frac{11}{18}$, ...
b $\frac{10}{30}$, $\frac{12}{27}$, $\frac{14}{24}$, $\frac{16}{21}$, $\frac{18}{18}$, ...
c $\frac{7}{1}$, $\frac{8}{4}$, $\frac{9}{9}$, $\frac{10}{16}$, $\frac{11}{25}$, ...
d $\frac{1}{11}$, $\frac{8}{9}$, $\frac{27}{7}$, $\frac{64}{5}$, $\frac{125}{3}$, ...

> Find separate formulae for the numerator and denominator.

6 a What is the nth term formula for $\frac{1}{1}$, $\frac{1}{2}$, $\frac{1}{3}$, $\frac{1}{4}$, $\frac{1}{5}$, ...?
b Plot this sequence on a graph, with the term number on the x-axis and the term on the y-axis.
c What happens if you continue this sequence indefinitely?
d Investigate sequences of your own that behave in a similar way.

This spread will show you how to:
- Explain how the formula for the nth term of a sequence works

Keywords
Formula
nth term
Pattern

- You can describe a **pattern** using a **formula**.
  This pattern is made of triangles.

   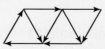

Number of triangles (n)	1	2	3	4
Number of arrows	3	5	7	9

+2   +2   +2

Number of arrows = 2 × number of triangles +1
so **nth term** of pattern = $2n + 1$
This works because each extra triangle needs two arrows and you need one arrow at the start.

This pattern is made from cubes.
The outside faces of the cubes in each block are painted.

How many faces are painted each time?

Block number (n)	1	2	3	4
Number of painted faces	6	10	14	18
4× table	4	8	12	16

In the nth block,
    number of painted faces = 4 × block number + 2
    so, nth term = $4n + 2$.

The terms go up in 4s.

Each cube in the pattern has 4 painted faces and the 2 end faces are painted.

Example

Find a formula connecting the number of edges and the number of hexagons and explain why it works.

Make a table.

Number of hexagons (n)	1	2	3	4
Number of edges	6	11	16	21
5× table	5	10	15	20

The terms go up in 5s.

Number of edges = 5 × number of hexagons + 1

$$E = 5n + 1$$

Each hexagon needs five edges to join it to the previous one, and there is one edge to start the whole pattern off.

 1 For each pattern, there is a formula. Explain why each formula works.

a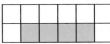

$E = 3s + 1$ where E = number of edges, s = number of squares

b

$W = B + 4$ where W = number of white tiles, B = number of coloured tiles

c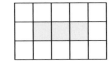

$M = L(L + 1)$ where M = number of nails and L = width of rectangle

2 Derive your own formulae for these patterns and explain why they work.

a Connect the number of coloured tiles (B) and the number of white tiles (W).

 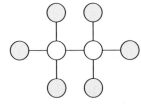 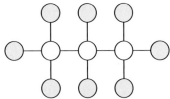

b i Connect the number of white circles (W) with the number of coloured circles (B).

ii Use the same pattern to connect the number of white circles (W) with the number of lines (L).

c Find a formula for

i the number of presents (P) swapped at a party with n people, if all the guests give one another a present

ii the number of handshakes (H) at a party with n people, if all the guests shake hands with each other.

Find the first 3 or 4 terms and make a table.

3 a A square jigsaw puzzle has n^2 pieces. It has C corner pieces, E edge pieces and M middle pieces. Write a formula for C, E and M in terms of n and show that your formula accounts for all the pieces.

b Extend to a rectangular puzzle with m pieces in its length and n pieces in its width.

A03 **Problem**

This spread will show you how to:

- Understand the difference between a practical demonstration and a proof
- Recognise the importance of assumptions when deducing results

Keywords
Counter-example
Demonstration
False
Proof

- You can show that a statement is **false**, by finding a **counter-example**.

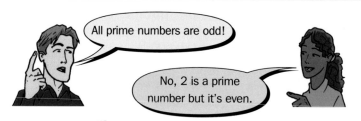

All prime numbers are odd!

No, 2 is a prime number but it's even.

2 is a counter-example, it doesn't fit the statement.

To prove that a statement is true, you can't just find an example.
There could still be an example that doesn't work! However, it is useful
to demonstrate the statement to yourself first with numbers. For example

> Show that the sum of two consecutive integers is always odd.
> $3 + 4 = 7 ...$ odd
> $36 + 37 = 73 ...$ odd

This is a **demonstration**.

- To prove a statement is true, you need to generalise it to all possible examples.

For example, show that the sum of two consecutive integers is always odd.
Let the consecutive integers be n and $n + 1$: Sum $= n + n + 1 = 2n + 1$.
$2n + 1$ is always odd since it is one more ($+1$) than a multiple of 2 ($2n$).

Example

Prove that the sum of two odd numbers is even.

·····································

This is a **proof** so use general expressions for the two odd numbers.
$2m + 1$ and $2n + 1$ are odd numbers. (They are each 1 more than a
multiple of 2.)
$(2m + 1) + (2n + 1) = 2m + 2n + 2$ or $2(m + n + 1)$ by factorisation.
$2(m + n + 1)$ is even because it is a number multiplied by 2.

First, it may help
to demonstrate it
to yourself:
$3 + 7 = 10$,
$9 + 11 = 20$,
$1 + 51 = 52$
etc.

p.130

Example

Prove that, in this triangle, $d = a + b$.

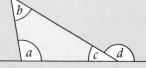

·····································

p.88

$a + b + c = 180°$ Angles in a triangle.
$c + d = 180°$ Angles on a straight line.
So $a + b + c = c + d$
$d = a + b$

1 Find a counter-example to show that each of these statements is untrue.
   a When you subtract 7 from a number, the answer is always odd.
   b When you square a number, the answer is always even.
   c When you treble a prime number, the answer is always odd.
   d When you find the product of two consecutive numbers, the answer is always odd.

2 The sum of any five consecutive integers is always a multiple of 5.
   a Demonstrate, with a few examples of your own, that this statement is true.
   b By letting the numbers be n, $n + 1$, $n + 2$, $n + 3$ and $n + 4$, prove the statement is true.

3 Repeat question **3** for these statements.
   a The sum of two even numbers is even.
   b Squaring an even number gives a number in the four times table.
   c The sum of an odd number and an even number is always odd.
   d If two consecutive numbers are multiplied, and the smaller number is subtracted from the result, you always get a square number.

A03 Problem

4 a Can you find a counter-example to this statement?

> Squaring a number will always give you a value greater than the number you started with.

   b What range of values does not support this statement?

5 Prove that these statements are true.
   a When taking three consecutive integers, the square of the middle integer is always one more than the product of the other two.
   b The answer to each equation in this pattern will always be 4.
   $$5 \times 8 - 4 \times 9 = 4$$
   $$6 \times 9 - 5 \times 10 = 4$$
   $$7 \times 10 - 6 \times 11 = 4 \dots$$
   c The difference between two square numbers will always be equal to the product of their sum and of their difference.

6 Prove that the angle in a semicircle is always a right angle.

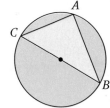

Join A to O. Let angle ACB be x and angle ABC be y.

p.96

Summary

Check out
You should now be able to:

- Generate terms of a sequence from number patterns and from diagrams
- Generate sequences of numbers using a position-to-term rule
- Find and use the nth term of an arithmetic sequence
- Justify the nth term of an arithmetic sequence by looking at the context
- Use a counter-example to show a statement is false

Worked exam question
Here are the first five terms of a number sequence.

    3   7   11   15   19

a Write down an expression, in terms of n, for the nth term of this sequence. (2)

Adeel says that 319 is a term in the number sequence.

b Is Adeel correct?
You must justify your answer. (2)

(Edexcel Limited 2006)

a

$n =$	1	2	3	4	5
	↓	↓	↓	↓	↓
	3	7	11	15	19

        +4   +4   +4   +4

nth term is $4 \times n + c$
Multiples of 4 are   4   8   12   16   20
nth term is $4n - 1$

> Write down the expression $4n + c$ for the nth term.

b

$4n - 1 = 319$
$4n = 319 + 1$
$4n = 320$
$n = 320 \div 4$
$n = 80$ and so Adeel is correct as 319 is the 80th term.

> Write down the equation $4n - 1 = 319$

Exam questions

1 Here are the first four terms of a number sequence.

    4   10   16   22

Write down an expression, in terms of n, for the nth term
of this sequence. (2)

2 Here are some patterns made from dots.

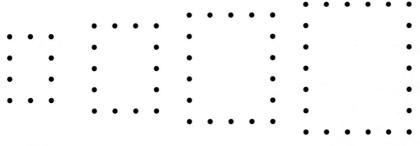

Pattern number 1   Pattern number 2   Pattern number 3   Pattern number 4

Write down a formula for the number of dots, d, in terms of the
Pattern number, n. (2)

(Edexcel Limited 2004)

3 The nth term of a sequence is $n^2 + 4$
Alex says
'The nth term of the sequence is always a prime number
when n is an odd number.'
Alex is wrong.
Give an example to show that Alex is wrong. (2)

(Edexcel Limited 2008)

4 Roberta says,
'The product of two prime numbers is always an odd number.'
She is wrong.
Explain why. (2)

Transformations

When you play a computer game your character is able to move around the screen because of mathematical transformations. Combinations of these transformations, often taking place in 3D worlds, allow the characters to move in many different ways.

What's the point?

Transformations move points and shapes from one place to another. Mathematicians use transformations not just to move shapes but also to move graphs and statistics. This helps them apply the power of mathematics to real life situations.

Check in

You should be able to

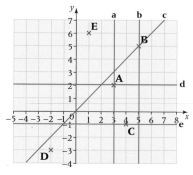

■ **plot points and give their coordinates**

1   Write the coordinates of the points A–E.

■ **draw lines and give their equations**

2   Write the equations of the lines a–e.

Exam questions

1 Here are the first four terms of a number sequence.

     4   10   16   22

Write down an expression, in terms of n, for the nth term
of this sequence.   (2)

2 Here are some patterns made from dots.

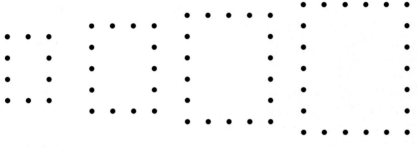

Pattern   Pattern   Pattern   Pattern
number 1   number 2   number 3   number 4

Write down a formula for the number of dots, d, in terms of the
Pattern number, n.   (2)

(Edexcel Limited 2004)

3 The nth term of a sequence is $n^2 + 4$
Alex says
'The nth term of the sequence is always a prime number
when n is an odd number.'
Alex is wrong.
Give an example to show that Alex is wrong.   (2)

(Edexcel Limited 2008)

4 Roberta says,
'The product of two prime numbers is always an odd number.'
She is wrong.
Explain why.   (2)

Transformations

When you play a computer game your character is able to move around the screen because of mathematical transformations. Combinations of these transformations, often taking place in 3D worlds, allow the characters to move in many different ways.

What's the point?

Transformations move points and shapes from one place to another. Mathematicians use transformations not just to move shapes but also to move graphs and statistics. This helps them apply the power of mathematics to real life situations.

Check in

You should be able to

■ **plot points and give their coordinates**

1 Write the coordinates of the points A–E.

■ **draw lines and give their equations**

2 Write the equations of the lines a–e.

What I need to know	What I will learn	What this leads to
KS3 Carry out simple transformations	■ Reflect, rotate and translate shapes ■ Describe and combine transformations	**G4** Enlarging shapes
A1 Plot points and simple lines		Computer science, Design

Rich task

A shape (object) is reflected in the line $y = x$. To find the new coordinates of each vertex of the shape (image), simply swap the x and y coordinates around. Investigate rules for finding the image coordinates for this and other transformations.

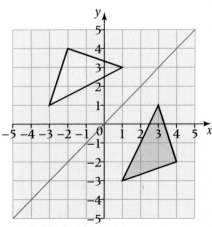

Reflection

This spread will show you how to:

- Identify properties preserved under reflection
- Understand congruence
- Describe reflections, using mirror lines

Keywords
Congruent
Mirror line
Perpendicular
Reflection

Just as you can see your reflection in a mirror, you can reflect a shape in a mirror line.

Corresponding points on the object and image are the same distance from the **mirror line**.

p.92

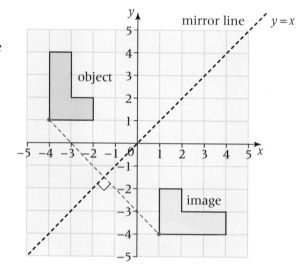

Corresponding angles and lengths are the same in the image and the object.

The object and the image are **congruent**.

The line joining a point and its image is **perpendicular** to the mirror line.

Reflection flips the shape over.

p.40

Example

Reflect the triangle with vertices at (1, 2), (2, 7) and (4, 7) in the line $x = 1$.

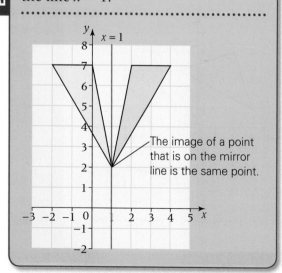

The image of a point that is on the mirror line is the same point.

Example

Reflect the pink triangle in the line $y = 3$.

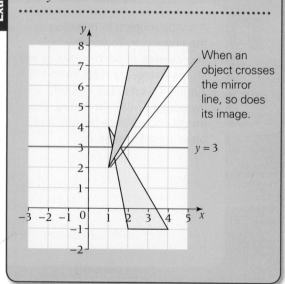

When an object crosses the mirror line, so does its image.

1 Copy this diagram.
   a Reflect the triangle T in the line $x = 2$.
     Label the image U.
   b Reflect the triangle T in the line $y = 1$.
     Label the image V.

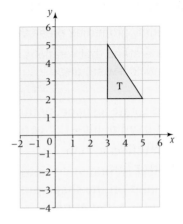

2 Copy this diagram.
   a Reflect the kite K in the line $x = -1$.
     Label the image L.
   b Reflect the kite K in the line $y = 1$.
     Label the image M.

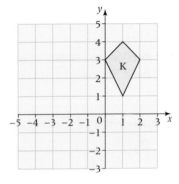

3 Copy this diagram and extend the y-axis to -8.
   a Reflect the quadrilateral Q in the x-axis.
     Label the image R.
   b Reflect the quadrilateral Q in the line $y = x$.
     Label the image S.

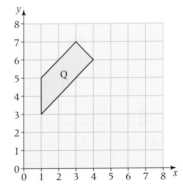

4 Copy this diagram.
   a Reflect triangle A in the y-axis.
     Label the image B.
   b Reflect triangle B in the y-axis.
     Label the image C.
   c What do you notice? Does this always happen?
     Check with some other reflections.

A03 Problem

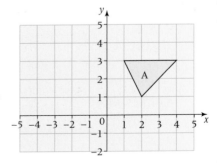

This spread will show you how to:

- Identify properties preserved under rotation
- Understand that rotations are specified by a centre and an (anticlockwise) angle

Keywords
Angle
Centre of rotation
Congruent
Rotation

You can rotate a shape by turning it about a fixed point – the **centre of rotation**.

This object is rotated 90° anticlockwise. Every point on the shape moves through the same **angle**.
The centre of rotation is (1, 0).

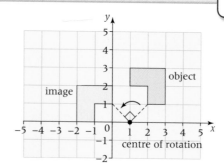

- In a **rotation**, corresponding angles and lengths are the same in the image and the object. The object and the image are **congruent**.

Corresponding points on the object and image are the same distance from the centre of rotation.

Example

Rotate the triangle through −90° about centre (1, 2).

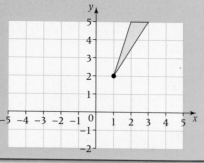

Rotate each vertex through −90° Use tracing paper to help.

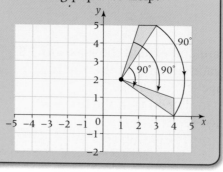

You measure angles anti-clockwise, so −90° means 90° clockwise.

A vertex at the centre of rotation does not move.

Rotation through 180° is a half turn, the same clockwise or anticlockwise.

Example

Rotate the pink triangle through 180° about centre O.

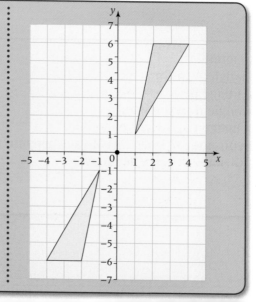

O is the origin, (0, 0).

1 Copy this diagram.
   a Rotate the triangle T through 180° about (0, 0). Label the image U.
   b Rotate the triangle T through 180° about (2, 2). Label the image V.

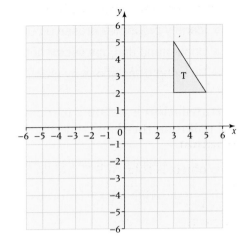

2 Copy this diagram.
   a Rotate the kite K through 180° about (1, 1). Label the image L.
   b Rotate the kite K through −90° about (2, 0). Label the image M.

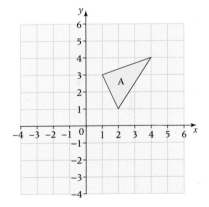

3 Copy this diagram.
   a Rotate the triangle A through 90° clockwise about (1, 0). Label the image B.
   b Rotate the triangle A through 90° anticlockwise about (0, 1). Label the image C.

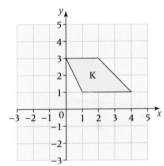

4 Copy this diagram.
   a Rotate shape P through −90° about (2, 1). Label the image Q.
   b Rotate image Q through −90° about (2, 1). Label the image R.
   c Rotate shape P through 180° about (2, 1). What do you notice? Explain why this happens.

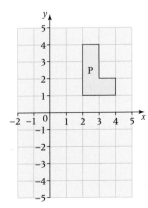

Translation

This spread will show you how to:

- Identify properties preserved under translation
- Understand and use vector notation
- Describe translations by giving a distance and direction-(or vector)

Keywords
Translation
Vector

A **translation** is a sliding movement.
All points on the shape slide the same distance in the same direction.

- In a translation the object and the image are congruent.

You can describe a translation using a **vector**.

- Vector $\binom{a}{b}$ means moving a units in the
x-direction and b units in the y-direction.

The L-shape is translated by the vector $\binom{3}{4}$.

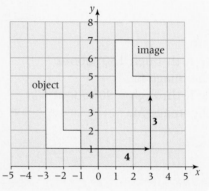

Example

Translate triangle A by the vector $\binom{-3}{4}$.

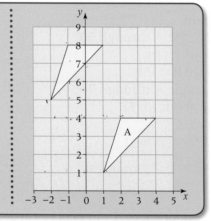

$\binom{-3}{4}$ means move
−3 in the
x-direction, so 3
squares to the left.

Move 4 in the
y-direction, so 4
squares up.

Example

Translate triangle T by the vector $\binom{0}{-3}$.

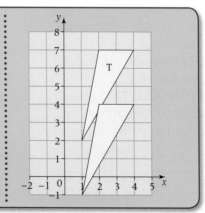

$\binom{0}{-3}$ means move
0 in the x-direction,
so the shape
only moves in the
y-direction.

Move −3 in the
y-direction, so 3
squares down.

1 Copy this diagram.

　a Translate the triangle T by the vector $\begin{pmatrix} 3 \\ 4 \end{pmatrix}$.

　　Label the image U.

　b Translate the triangle T by the vector $\begin{pmatrix} 5 \\ -2 \end{pmatrix}$.

　　Label the image V.

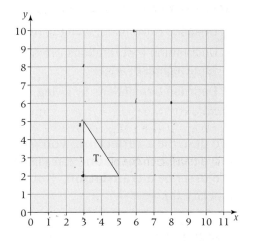

2 Copy this diagram.

　a Translate the kite K by the vector $\begin{pmatrix} -4 \\ 3 \end{pmatrix}$.

　　Label the image L.

　b Translate the kite K by the vector $\begin{pmatrix} 2 \\ -4 \end{pmatrix}$.

　　Label the image M.

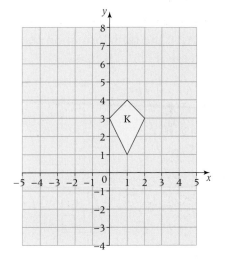

3 Copy this diagram.

　a Translate shape A by the vector $\begin{pmatrix} 6 \\ -4 \end{pmatrix}$.

　　Label the image B.

　b Translate the image B by the vector $\begin{pmatrix} -6 \\ 4 \end{pmatrix}$.

　　What do you notice? Does this always happen?
　　If so, why?

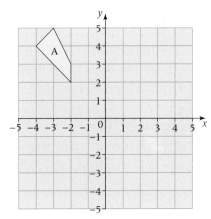

Describing transformations

This spread will show you how to:

- Describe reflections, using mirror lines
- Understand that rotations are specified by a centre and an (anticlockwise) angle
- Describe translations by giving a distance and direction (or vector)

Keywords

Maps
Reflection
Rotation
Transformation
Translation

- To describe a **reflection**, you give the equation of the line.
- To describe a **rotation**, you give the centre and the angle of rotation.
- To describe a **translation**, you give the distance and direction or you specify the vector.

Reflections, rotations and translations are all **transformations**.

Example

Describe the transformation that maps shape A on to

a shape B   **b** shape C   **c** shape D.

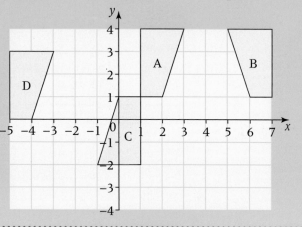

Maps means changes.

....

a In a reflection the mirror line bisects the line joining corresponding points on the object and image.

Shape B is a reflection of shape A in the line $x = 4$.

b The vertex $(1, 1)$ does not move during the rotation, so it must be the centre of rotation.

Shape C is a rotation of shape A through $180°$.

c Shape D is a translation of shape A by the vector $\begin{pmatrix} -5 \\ -1 \end{pmatrix}$.

1 Describe fully the
transformation that maps
  a shape A onto shape B
  b shape A onto shape C
  c shape A onto shape D
  d shape B onto shape D
  e shape C onto shape D
  f shape D onto shape C.

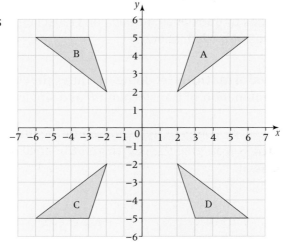

2 Describe fully the
transformation that maps
  a shape J onto shape K
  b shape L onto shape K
  c shape M onto shape K
  d shape L onto shape M
  e shape J onto shape M
  f shape M onto shape J.

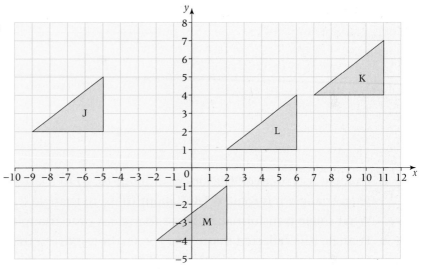

3 Describe fully the
transformation that maps
  a shape W onto shape X
  b shape W onto shape Y
  c shape W onto shape Z
  d shape Z onto shape W
  e shape X onto shape Y
  f shape Z onto shape X.

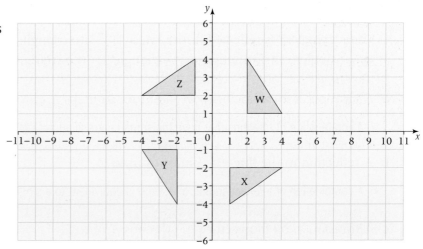

Combining transformations

This spread will show you how to:

- Transform 2-D shapes by translation, rotation and reflection and combinations of these transformations

Keywords
Reflection
Rotation
Transformations
Translation

You can combine **transformations** by doing one after the other. You can describe a combination of transformation as a single transformation.

Example

In this diagram, triangle A undergoes three pairs of transformations.

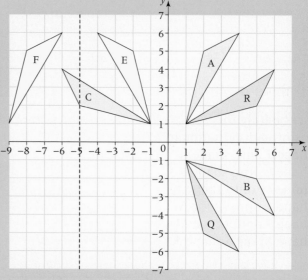

a Triangle A is rotated 90° clockwise about (0, 0) to triangle B.
Then triangle B is rotated 180° about (0, 0) to triangle C.
What single transformation maps triangle A onto triangle C?

b Triangle A is reflected in the line $y = 0$ (the x-axis) to triangle Q.
Then triangle Q is rotated through 90° anticlockwise about (0, 0) to triangle R.
What single transformation maps triangle A onto triangle R.

c Triangle A is reflected in the line $x = 0$ (the y-axis) to triangle E.
Then triangle E is reflected in the line $x = -5$ to triangle F.
What single transformation maps triangle A onto triangle F?

....

a A rotation of 90° anticlockwise about (0, 0) maps A onto C.

- A combination of rotations that have the same centre is equivalent to a single **rotation**.

b A reflection in the line $y = x$ maps A onto R.

- A combination of a reflection and a rotation is equivalent to a single **reflection**.

c A translation by the vector $\begin{pmatrix} -10 \\ 0 \end{pmatrix}$ maps A onto F.

- A combination of reflections is a **translation** when the mirror lines are parallel otherwise it is a rotation.

1 Copy this diagram.
  a Rotate triangle A 90° anticlockwise about centre (0, 0). Label the image B.
  b Rotate triangle B 90° anticlockwise about centre (0, 0). Label the image C.
  c Describe fully the single transformation that takes triangle A to triangle C.
  d Rotate triangle B 90° clockwise about centre (2, 1). Label the image D.
  e Describe fully the single transformation that takes triangle A to triangle D.

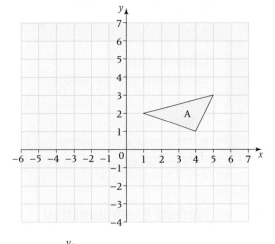

2 Copy this diagram.
  a Reflect triangle E in the *y*-axis. Label the image F.
  b Reflect triangle F in the *x*-axis. Label the image G.
  c Describe fully the single transformation that takes triangle G to triangle E.
  d Reflect triangle F in the line $x = 2$. Label the image H.
  e Describe fully the single transformation that takes triangle E to triangle H.

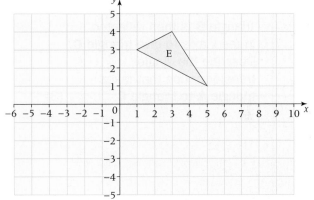

3 Copy this diagram.
  a Reflect triangle J in the x-axis. Label it K.
  b Rotate triangle K 180° about centre (0, 0). Label it L.
  c Describe fully the single transformation that takes triangle L to triangle J.

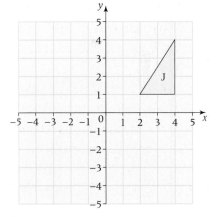

4 Copy this diagram.
  a Translate trapezium R by the vector $\begin{pmatrix} 5 \\ -3 \end{pmatrix}$. Label-the image S.
  b Translate trapezium S by the vector $\begin{pmatrix} -1 \\ 2 \end{pmatrix}$. Label-the image T.
  c Describe fully the single transformation that takes trapezium T to-trapezium R.

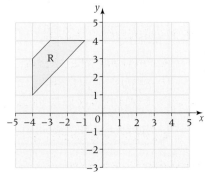

197

Summary

Check out

You should now be able to:

- Identify properties preserved under reflection, rotation and translation
- Describe and transform 2-D shapes using reflections, rotations, translations and a combination of these transformations

Worked exam question

Triangle **A** is reflected in the y axis to give triangle **B**. Triangle **B** is then reflected in the x axis to give triangle **C**. Describe the single transformation that takes triangle **A** to triangle **C**.   (3)

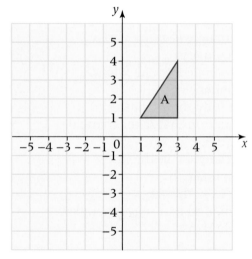

(Edexcel Limited 2006)

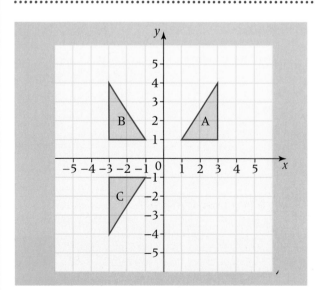

Reflect triangle **A** in the y axis to give triangle **B**.

Reflect triangle **B** in the x axis to give triangle **C**.

An alternative answer is an enlargement, scale factor −1 about (0, 0).

Rotation of 180° about (0, 0)

Exam question

1

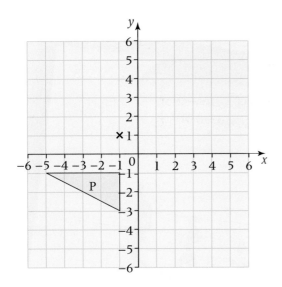

a On an accurate copy of the diagram, rotate triangle **P** 180°
about the point $(-1, 1)$.
Label the new triangle **A**. (2)

b Translate triangle **P** by the vector $\begin{pmatrix} 6 \\ -1 \end{pmatrix}$.
Label the new triangle **B**. (1)

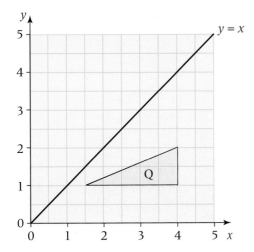

c On an accurate copy of the diagram, reflect triangle **Q** in
the line $y = x$.
Label the new triangle **C**. (2)

(Edexcel Limited 2008)

Mathematics is used widely in sport, particularly when taking measurements and recording results.

Here are the results and wind speeds for the fastest thirteen all-time 100m Men's sprinters as of 20th September 2009:

Rank	Time (s)	Wind speed (m/s)	Athlete	Nation	Date
			Bolt	JAM	16/08/2009
1	9.58	+0.9	Gay	USA	20/09/2009
2	9.69	+2.0	Powell	JAM	02/09/2008
3	9.72	+0.2	Greene	USA	16/06/1999
4	9.79	+0.1	Bailey	CAN	27/07/1996
		+0.7	Surin	CAN	22/08/1999
5	9.84	+0.2	Burrell	USA	06/07/1994
		+1.2	Gatlin	USA	22/08/2004
7	9.85	+0.6	Fasuba	NIG	12/05/2006
		+1.7	Lewis	USA	25/08/1991
		+1.2	Fredericks	NAM	03/07/1996
		−0.4	Boldon	TRI	19/04/1998
10	9.86	+1.8	Obikwelu	POR	22/08/2004
		+0.6			

What level of accuracy is reported for
a) the result times   b) the wind speeds?

Plot a time-series graph for these results. Can you see any trend? When (if ever) do you think the 9.50s barrier will be broken? Explain your answer referring to the data.

What is the   a) fastest   b) slowest
actual time that each of the athletes could have run to give their reported result?

FASTEST TEN ALL-TIME 100m WOMEN'S SPRINTERS
AS OF 20th SEPTEMBER 2009:

File   Edit   View   History   Bookmarks   Window   Help

http://berlin.iaaf.org/results/racedate=08-16-2009/sex=M/discCode=100/combCode=hash/roundCode=f/results.html#detM_100_hash_f

Fastest ten all-time 100m Women's sprinters as of 20th September 2009

Google   Google Maps   OUP   Wikipedia

CHOOSE YOUR COUNTRY!
Select a country

NEWS   PHOTOS   VIDEO   AUDIO

By Date   By Event   Entry List   Medal Table   Placing Table   Entry Standards

Here are the results and wind speeds for the fastest ten all-time 100m Women's sprinters as of 20th September 2009:

Rank	Time (s)	Wind speed (m/s)	Athlete	Nation	Date
1	10.49	0.0	Griffith-Joyner	USA	16/07/1988
2	10.64	+1.2	Jeter	USA	20/09/2009
3	10.65	+1.1	Jones	USA	12/09/1998
4	10.73	+0.1	Fraser	JAM	17/08/2009
		+2.0	Arron	FRA	17/08/1998
6	10.74	+1.3	Ottey	JAM	07/09/1996
7	10.75	+0.4	Stewart	JAM	10/07/2009
8	10.76	+1.7	Ashford	USA	22/08/1984
9	10.77	+0.9	Privalova	RUS	06/07/1994
		+0.7	Lalova	BUL	19/06/2004

Florence Griffith-Joyner's World Record is quoted as being 'probably strongly wind assisted' because it is suspected that the wind speed measurer was faulty.

Comment on this, referring to the data and using your diagrams and statistics.

Use time-series diagrams and statistics to compare the Women's and Men's results. Refer to any trends you notice.

A maximum tail wind of +2.0m/s is allowed for 'wind legal' results.
Head winds are not taken into account.

Tail winds follow the athlete and are recorded as +ve.

Head winds act against the athlete and are recorded as -ve.

Draw a scatter diagram of result time against wind speed for

a) the Men's results   b) the Women's results.

Do you think there is any correlation between wind speed and time?

Justify your response by referring to the data.

Probability

There are many things in life which are uncertain. Will it be sunny tomorrow? Will my football team win the Premier League? Will I be able to afford a house in the future? The mathematics used to deal with uncertainty is called probability.

What's the point?

Probability is a way of quantifying the uncertainty of an event, which is usually expressed as a fraction, decimal or percentage. When the Met Office gives a weather forecast, they use a complex mathematical model of the earth's climate with many variables, to predict the probability of sunshine in a particular region which they often express as a percentage.

Check in

You should be able to

- add and subtract fractions

1 Work out

   a $1 - \frac{2}{5}$ **b** $1 - \frac{4}{7}$ **c** $1 - \frac{3}{8}$ **d** $1 - \frac{4}{11}$

   e $\frac{2}{3} + \frac{1}{6}$ **f** $\frac{1}{5} + \frac{1}{4}$ **g** $\frac{1}{3} + \frac{5}{8}$ **h** $\frac{1}{4} + \frac{3}{8}$

- multiply a fraction by an integer

2 **a** $\frac{2}{5} \times 100$ **b** $\frac{1}{9} \times 360$ **c** $\frac{3}{8} \times 56$ **d** $\frac{7}{30} \times 240$

- convert between fractions and decimals

3 Change these fractions to decimals.

   a $\frac{7}{10}$ **b** $\frac{3}{4}$ **c** $\frac{3}{8}$ **d** $\frac{2}{5}$

   e $\frac{1}{3}$ **f** $\frac{1}{16}$ **g** $\frac{1}{9}$ **h** $\frac{5}{8}$

Orientation

What I need to know	What I will learn	What this leads to
KS3 The language of probability	■ Calculate theoretical and experimental probabilities ■ Use two-way tables ■ Calculate expectations	**D5** Combining probabilities
N2 Add fractions and decimals Multiply an integer by a fraction.		

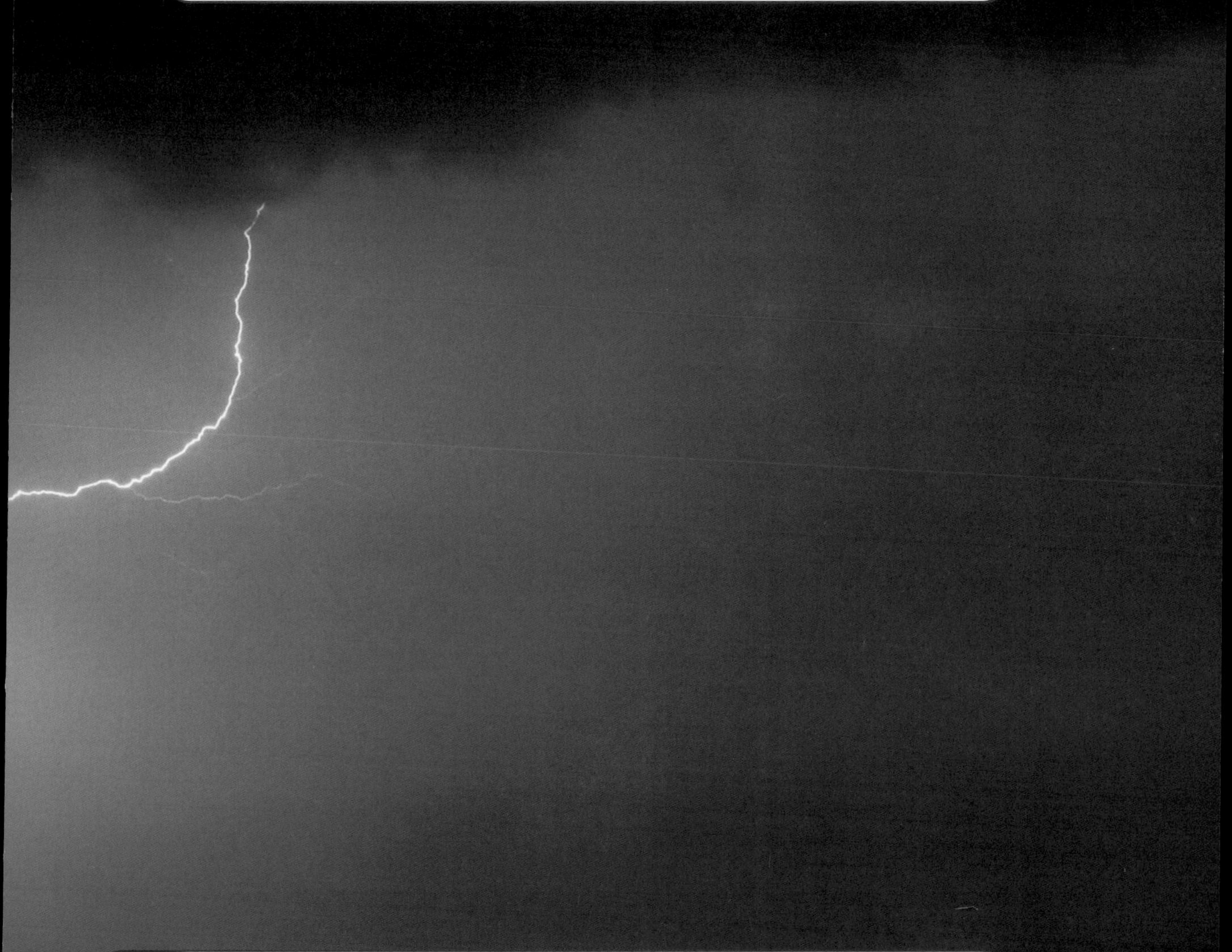

Rich task

Inside a bag are three cards with the letters A, N and D written on them. You are allowed to pick one card from the bag and then replace the card in the bag.

How likely are you to be able to spell the word AND after picking three or more cards?

Probability

This spread will show you how to:
- Understand and use the probability scale
- Calculate and explain the probability of an event

- **Probability** is a measure of how likely an **event** is to happen.
- Probability is measured on a scale from 0 to 1.

$$0 \longrightarrow 1$$

Event:
cannot happen

Event:
certain to happen

1, 2, 3, 4, 5 and 6 are the **outcomes** of throwing a dice.

For all other events the probability is a fraction between 0 and 1.

- Probability of an event happening = $\dfrac{\text{number of favourable outcomes}}{\text{total number of possible outcomes}}$

You can write probability as a fraction, a decimal or a percentage.

Example

There are 30 students in Class 10Z, 18 girls and 12 boys.
All of the students are aged 14 or 15 and all own a mobile phone.
A student is chosen at random from Class 10Z. What is the probability that the student chosen

a is aged 10   **b** owns a mobile phone   **c** is a girl   **d** is a boy?

a This outcome cannot happen. All the students are 14 or 15, so P(aged 10) = 0.
b This outcome is certain to happen. All the students own a mobile phone, so P(owns mobile) = 1.
c There are 18 girls and a total of 30 students, so P(girl) = $\frac{18}{30} = \frac{3}{5}$.
d There are 12 boys and a total of 30 students, so P(boy) = $\frac{12}{30} = \frac{2}{5}$.

P(aged 10) is a short way of writing 'the probability that a student is aged 10'.

The probability that a girl is chosen or that a boy is chosen is certain to happen as the students are all either girls or boys.

Notice that P (girl) + P(boy) = $\frac{18}{30} + \frac{12}{30} = 1$ = P(certainty)

All possible outcomes are accounted for.

- Sum of probabilities of all possible outcomes = 1

Example

A spinner has circles, squares and triangles on its face.
The table gives the probabilities of landing on circle, square and triangle.
Work out the value of x.

Outcome	Circle	Square	Triangle
Probability	0.25	0.625	x

Total probability = 1

$$0.25 + 0.625 + x = 1$$
$$0.875 + x = 1$$
$$\text{So, } x = 0.125$$

1 The probability that a girl chosen at random in Class 10Y has
a cat is 0.47.
What is the probability that a girl chosen at random in Class 10Y
does not have a cat?

2 The probability that a boy chosen at random in Class 10X does not
wear glasses is 0.35.
What is the probability that a boy chosen at random in Class 10X
does wear glasses?

3 A dish contains 5 orange, 3 lemon and 10 strawberry sweets.
One sweet is chosen at random.
What is the probability that the sweet is

  a orange   **b** banana   **c** strawberry
  d lemon   **e** not lemon?

4 A bag contains 7 green, 8 blue and 10 red marbles.
Nine of the marbles are large, 16 are small.
One marble is chosen at random.
What is the probability that the marble is

  a green   **b** not green   **c** large
  d small   **e** red   **f** not blue?

5 Four girls compete to become head girl.
The probabilities of being chosen are shown in the table.

Amy	Beth	Cathy	Debs
0.14	0.32	0.27	x

Work out the value of x.

6 Five boys want to be captain of the cricket team.
The probabilities of their being chosen are shown in the table.

Gerry	Harry	Iain	Jim	Ken
0	0.24	x	0.15	0.21

  a Explain what the probability that Gerry is chosen is 0 means.
  b Work out the value of x.

7 Four teams are left in a football competition.
The probabilities of their winning the competition are shown
in the table.

City	United	Rovers	Rangers
0.18	0.22	x	$2x$

  a Explain the chances of Rangers winning compared to the chances
    of Rovers winning.
  b Work out the value of x.

This spread will show you how to:
- Calculate the probability of mutually exclusive events

Keywords

Mutually exclusive
Event

- Two or more **events** are **mutually exclusive** if they cannot happen at the same time.

For example,

An event is one or more outcomes

When you roll a dice, the events 'an even number' and 'a 3' can both happen, but not at the same time.

When you flip a coin, the events 'head' and 'tail' can both happen, but not at the same time.

When two or more events are mutually exclusive the addition rule (sometimes called the OR rule) is used to find their probability.

- For two mutually exclusive events, A and B, the probability of A or B happening is the sum of P(A) and P(B).
  P(A or B) = P(A) + P(B)

- Probability of an event *not* happening = 1 − Probability an event happens
  For an event A, P(not A) = 1 − P(A)   or   P(A) = 1 − P(not A)

Example

A spinner has 12 equal sides: five green, four blue, two red and one white. The spinner is spun.
a What is the probability that the spinner lands on
   i green or white   **ii** green or blue   **iii** blue or white
  iv blue or red or white   **v** not green?
b Why are the answers to parts **iv** and **v** the same?

p.58

a **i** P(green) = $\frac{5}{12}$ P(white) = $\frac{1}{12}$ P(green or white) = $\frac{5}{12} + \frac{1}{12} = \frac{6}{12}$

   ii P(green) = $\frac{5}{12}$ P(blue) = $\frac{4}{12}$ P(green or blue) = $\frac{5}{12} + \frac{4}{12} = \frac{9}{12}$

  iii P(blue) = $\frac{4}{12}$ P(white) = $\frac{1}{12}$ P(blue or white) = $\frac{4}{12} + \frac{1}{12} = \frac{5}{12}$

  iv P(blue) = $\frac{4}{12}$ P(red) = $\frac{2}{12}$ P(white) = $\frac{1}{12}$

      P(blue or red or white) = $\frac{4}{12} + \frac{2}{12} + \frac{1}{12} = \frac{7}{12}$

   v P(not green) = 1 − P(green) = $1 - \frac{5}{12} = \frac{7}{12}$

b The possible outcomes are green, blue, red and white.
  Part **v**, P(not green), is the same as part **iv**, P(blue or red or white).

1 A spinner has 20 equal sides: 7 have circles, 5 have pentagons,
4 have squares, 3 have triangles, 1 has a rectangle.
The spinner is spun.
What is the probability that the spinner lands on

a a circle

b not a circle

c a rectangle

d not a rectangle

e a circle or a square

f a pentagon or a square

g a triangle or a square or a rectangle

h a circle or a pentagon or a triangle

i not a square

2 A nine sided spinner has its sides numbered 1 to 9.
The spinner is spun.
What is the probability that the spinner lands

a on the number 7

b not on the number 7

c on an odd number

d on a multiple of 3

e on a multiple of 5

f not on a multiple of 5

g on a multiple of 4

h not on a multiple of 4

i on a prime number?

3 The table shows information about the type of pet owned by
students in Class 10A. No student owns more than one pet.

Pet	Cat	Dog	Hamster	Fish	No pet
Number of students	7	8	2	4	9

a How many students are there in the class?

b One student is chosen at random from the class.
Work out the probability that the student chosen will own

  i a cat
  ii a cat or a dog
  iii a dog or a fish

  iv a cat or a hamster **v** no pets
  vi a pet.

4 The table shows information about the number of driving lessons
students in Class 12C had in February one year.

Number of driving lessons	0	1	2	3	More than 3
Number of students	5	2	8	12	3

One student is chosen at random from the class.
Work out the probability that the student chosen had

a 1 driving lesson

b 1 or 2 driving lessons

c 2 or more driving lessons

d 2 or fewer driving lessons.

207

This spread will show you how to:
- Use two-way tables to calculate probabilities

Keywords

Expected
  number
Two-way table

- Probabilities can be found from information given in a **two-way table**.

Example

Each of the students in Class 10Z went abroad last year.
The two-way table shows some information about the countries visited.

	Europe	America	Rest of the world	Total
Girls	13			18
Boys		2		12
Total	22	5		

a Complete the table.

b One student is chosen at random from Class 10Z. Write down the probability that the student
  i visited Europe   **ii** is a boy who visited America.

a

	Europe	America	Rest of the world	Total
Girls	13	**3**	**2**	18
Boys	**9**	2	**1**	12
Total	22	5	**3**	**30**

Use the totals to find the missing numbers.
  $3 + 2 = 5$
  $13 + 9 = 22$

b   **i** Probability that the student visited Europe $= \frac{22}{30}$
  ii Probability that the student is a boy who visited America $= \frac{2}{30}$

You can use the probability of a particular event happening to calculate
the **expected number** of times it will occur.

- Expected number = Total number of trials × Probability of a particular event happening

Example

The foreign countries visited by the students in Class 10Z is typical of the
foreign countries visited by the students in Year 10 at the same school.
There are 240 students in Year 10 at the school.
How many students would you expect to visit
  a Europe   **b** America   **c** America or the Rest of the world?

p.60

a $P(\text{Europe}) = \frac{22}{30}$ so expected number: $240 \times \frac{22}{30} = 176$

b $P(\text{America}) = \frac{5}{30}$ so expected number: $240 \times \frac{5}{30} = 40$

c $P(\text{America or Rest of the world}) = P(\text{America}) + P(\text{Rest of the world})$
  $= \frac{5}{30} + \frac{3}{30} = \frac{8}{30}$

  Expected number: $240 \times \frac{8}{30} = 64$

There are 240
students, so total
number of trials
is 240.

1 The two-way table shows where each of the 32 students in Class 10Y spent their Easter holiday.

	France	Spain	UK	Total
Girls	3			18
Boys		8		
Total	6	12		32

   a Copy and complete the table.

   b One student is chosen at random from Class 10Y. Find the probability that the student

      i visited France   **ii** did not visit France

     iii is a boy   **iv** is a boy who visited Spain

      v visited France or Spain   **vi** is a girl who did not visit France.

2 The two-way table shows some information about the favourite activity of 50 students.

	Orienteering	Paintballing	Quadbiking	Total
Girls	11		4	23
Boys		8		
Total	16			50

   a Copy and complete the table.

   b One student is chosen at random. Find the probability that the student's favourite activity is

      i paintballing   **ii** not paintballing

     iii paintballing or orienteering   **iv** quadbiking or paintballing.

   c These students are a sample chosen from a larger group of 400 students.
     How many of the larger group would you expect

      i to prefer orienteering   **ii** to be girls?

3 The two-way table shows some information about the preferred subject of the 120 students in Year 10.

	Science	Humanities	Other subjects	Total
Girls	29		32	
Boys		3		
Total	56	21		120

   a Copy and complete the table.

   b One student is chosen at random. Find the probability that the student

      i is a girl   **ii** prefers humanities

     iii prefers humanities or science

     iv is a boy who prefers science.

   c This year group is typical of the whole school. There are 600 students in the school.
     How many of the whole school would you expect to prefer

      i science   **ii** humanities?

This spread will show you how to:

- Calculate theoretical probabilities and relative frequencies
- Understand the concepts of experimental and theoretical probability

Keywords

Bias
Experimental
Fair
Theoretical

- You can calculate the **theoretical** probability when an event is **fair** or unbiased.

For example:
A fair coin is thrown.
Probability of a fair coin landing on heads $= \frac{1}{2}$.

One head, two outcomes.

An unbiased dice is rolled.
Probability of an unbiased dice landing on $2 = \frac{1}{6}$.

One '2', six outcomes.

- Theoretical probability is based on equally likely outcomes.
- Theoretical probabilities are calculated on the assumption that the number of favourable outcomes and the number of possible outcomes are as expected.

Example

A fair spinner with 8 sides, three yellow, two red, two blue and one white, is spun. Work out the probability that the spinner lands on

a yellow $\qquad$ **b** red or blue $\qquad$ **c** not white.

....

a P(yellow) $= \frac{3}{8}$ $\qquad$ **b** P(red or blue) $= \frac{4}{8} = \frac{1}{2}$ $\qquad$ **c** P(not white) $= 1 - \frac{1}{8} = \frac{7}{8}$

- You use **experimental** probability when the event is **unfair** or **biased** or when the theoretical outcome is known.

Example

A biased coin is thrown 200 times. It lands on heads 140 times.
a Estimate the probability of this coin landing on heads on the next throw.
b Estimate the number of heads you expect to get when the coin is thrown 500 times.

....

a P(Head) $= \frac{140}{200} = \frac{7}{10}$ $\qquad$ **b** Estimated expected number $= 500 \times \frac{140}{200} = 350$

Example

Tom carries out a survey about the number of people in his village who are left-handed.
He asks 40 people and five of them are left-handed.
a Estimate the probability that a person in the village is left-handed.
b 192 people live in his village. Estimate how many are left-handed.

....

a P(left-handed) $= \frac{5}{40} = \frac{1}{8}$ $\qquad$ **b** Estimated number $= 192 \times \frac{5}{40} = 24$

- The closer experimental probability is to theoretical probability the less likely it is that there is **bias**.

1 The probability that a biased dice will land on a six is 0.35.
The dice is rolled 400 times.
Estimate the number of times the dice will land on a six.

2 The probability that a biased coin will land on tails is 0.24.
The coin is thrown 500 times.
Estimate the number of times the coin will land on tails.

3 A biased coin is thrown 80 times. It lands on heads 50 times.
Estimate the probability that this coin will land on heads on the next throw.

4 A biased four-sided dice is rolled 120 times. The table shows the outcomes.

Score	1	2	3	4
Frequency	22	34	44	20

a Explain why it is twice as likely that the dice will land on 3 as on 1.
b The dice is rolled once more. Estimate the probability that it lands on 2.
c The dice is rolled a further 300 times.
How many of those times would you expect the dice to land on 4?

5 A biased dice is rolled 100 times. The table shows the outcomes.

Score	1	2	3	4	5	6
Frequency	7	20	32	17	13	11

a The dice is rolled once more. Estimate the probability that the dice will land on
i 6   **ii** 3   **iii** 1 or 2
iv 4 or more   **v** a number that is not 5.
b The dice is going to be rolled a further 400 times.
How many times would you expect the dice to land on
i 2   **ii** 4 or 5?

6 There are 240 students in Year 10 at Endeavour School.
a In a survey of 30 students from Year 10, 13 owned an MP3 player.
How many of the whole of Year 10 would you expect to own an MP3 player?
b In a survey of 40 students from Year 10, 19 said they liked peaches.
How many of the whole of Year 10 would you expect to like peaches?
c In a survey of 48 students from Year 10, 5 were left-handed.
How many of the whole of Year 10 would you expect to be left-handed?

Relative frequency

This spread will show you how to:

• Calculate theoretical probabilities and relative frequencies

Keywords

Bias
Relative frequency

• Experimental probability is also known as **relative frequency**.

• Relative frequency is the proportion of successful trials in an experiment.

• The more trials that are carried out the more reliable the estimate of probability.

Example

Rachael has a mixed colours packet of seeds.
She plants 10 seeds each week for 7 weeks.
The table shows the number of purple flowers
that grew in each group of seeds.

Week	1	2	3	4	5	6	7
Number of purple flowers	4	6	5	7	4	6	5

a Work out the relative frequency of a purple
flower.
b Find the best estimate of the probability of getting a purple flower.

a

Week	1	2	3	4	5	6	7
Number of purple flowers	4	6	5	7	4	6	5
Relative frequency	$\frac{4}{10}$	$\frac{10}{20}$	$\frac{15}{30}$	$\frac{22}{40}$	$\frac{26}{50}$	$\frac{32}{60}$	$\frac{37}{70}$

b The best estimate includes all the results.   P(purple flower) $= \frac{37}{70}$

For each
successive week,
find the total
number of purple
flowers and the
total number of
seeds planted.

You can compare the relative frequency of an event with the theoretical
probability. If they are quite different, the experiment may be **biased**.

Example

Dan suspects that a particular dice has a bias towards the number 3.
Dan rolls the dice 30 times and gets these results

4	3	3	3	6	5	1	3	2	5
1	3	4	5	3	3	6	2	5	4
6	3	1	3	6	5	3	2	4	3

a What is the relative frequency of rolling a 3?
b Is the dice biased toward 3? Explain your answer.

a The number 3 is rolled 11 times.   Relative frequency $= \frac{11}{30}$
b Theoretical probability of rolling a '3' is P(3) $= \frac{1}{6}$

In 30 rolls expected number of 3s $= 30 \times \frac{1}{6} = 5$

5 is not close to 11 so the dice does appear to be biased towards 3.

1 Jim carries out an experiment.
He throws a coin 320 times.
The coin lands on tails 114 times.
Is the coin fair? Explain your answer.

2 A spinner has 10 equal sides, 5 black and 5 red.
Dave carries out an experiment.
He spins the spinner 280 times.
The spinner lands on black 133 times.
Is the spinner fair? Explain your answer.

3 Jane carries out an experiment.
She rolls a dice 200 times.
The table shows the results.

Outcome	1	2	3	4	5	6
Frequency	32	34	35	31	35	33

Is the dice fair? Explain your answer.

4 Tom has a four-sided dice.
He rolls the dice 100 times.
The table shows the results.

Outcome	1	2	3	4
Frequency	18	44	19	19

Is the dice fair? Explain your answer.

5 Clara suspects that a coin is biased. She flips the coin and notes how
many heads she gets in each group of 10 flips.
Clara flips the coin 100 times in total.
The table shows her results.

Group of 10 flips	1	2	3	4	5	6	7	8	9	10
Number of heads	4	3	4	2	5	4	4	3	3	2
Relative frequency										

a Copy the table and complete for the relative frequency.
b Write the best estimate of the probability of the coin landing
on heads.
c Is the coin biased? Explain your answer.

Summary

Check out
You should now be able to:

- Understand and use the probability scale
- Calculate the probability of an outcome of an event
- Identify different mutually exclusive outcomes and know that the sum of the probabilities of these outcomes is 1
- Use two-way tables to calculate probabilities
- Calculate theoretical probabilities and relative frequencies
- Compare experimental data and theoretical probabilities

Worked exam question
Here is a 4-sided spinner.
The sides of the spinner are labelled 1, 2, 3 and 4.
The spinner is biased.
The probability that the spinner will land on each of the numbers 2 and 3 is given in the table.
The probability that the spinner will land on 1 is equal to the probability that it will land on 4.

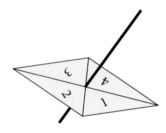

Number	1	2	3	4
Probability	x	0.3	0.2	x

a Work out the value of x. (2)

Sarah is going to spin the spinner 200 times.
b Work out an estimate for the number of times it will land on 2. (2)

(Edexcel Limited 2005)

a
$$\begin{array}{r} 0.3 \\ \underline{0.2} + \\ 0.5 \end{array}$$
$$1 - 0.5 = 0.5$$
$$0.5 \div 2 = 0.25$$

> Show the addition, subtraction and division calculations.

b
$$0.3 \times 200 = 60 \text{ times}$$

> Show this multiplication sum.

Exam questions

1 There are 3 red pens, 4 blue pens and 5 black pens in a box.
Sameena takes a pen, at random, from the box.
a Write down the probability that she takes a black pen. (2)
b Write down the probability that Sameena takes a pen
that is not black. (1)

(Edexcel Limited 2008)

2 The two-way table gives some information about how 100 children travelled
to school one day.

	Walk	Car	Other	Total
Boy	15		14	54
Girl		8	16	
Total	37			100

a Complete the two-way table. (3)

One of the children is picked at random.
b Write down the probability that this child walked to school
that day. (1)

(Edexcel Limited 2009)

3 A DIY store bought 1750 boxes of nails.
Barry took 25 of these boxes and counted the number of nails in each.
The table shows his results.

Number of nails	Number of boxes
14	2
15	9
16	8
17	4
18	2

The numbers of nails in the 25 boxes are typical of the number of nails
in the 1750 boxes.

Work out an estimate for how many of the 1750 boxes
contain 16 nails. (3)

(Edexcel Limited 2006)

Decimal calculations

There are lots of professions in which it is vital to perform mental calculations as a check on the answer they have calculated either manually or by using a calculator. These include doctors working out the dose of medicine to give a patient, pilots checking the fuel required for a flight or civil engineers calculating the amount of material required for the construction of a building.

What's the point?
Standard written methods are used to perform more complex mathematical calculations. However it is vital that at all times mental calculations (approximations) are used to check that any answers are sensible

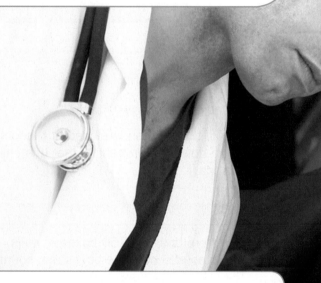

Check in

You should be able to

- **use mental methods for addition and subtraction**
1   Write notes to show how you could carry out these additions and subtractions mentally.
    a 23 + 97   **b** 234 − 96   **c** 46 + 44   **d** 973 − 708

- **use written methods for addition and subtraction of integers**
2   Show how you could use a column method to work these out.
    a 3775 + 663   **b** 2886 − 909   **c** 3775 − 2918   **d** 5549 + 8675

- **use mental methods for multiplication and division**
3   Write notes to show how you could carry out these multiplications and divisions mentally.
    a 21 × 7   **b** 101 × 15   **c** 8320 ÷ 8   **d** 7350 ÷ 21

- **use written methods for multiplication and division of integers**
4   Show how you can calculate these multiplications and divisions using a written method.
    a 325 × 36   **b** 4356 × 18   **c** 3485 ÷ 17   **d** 889 × 76

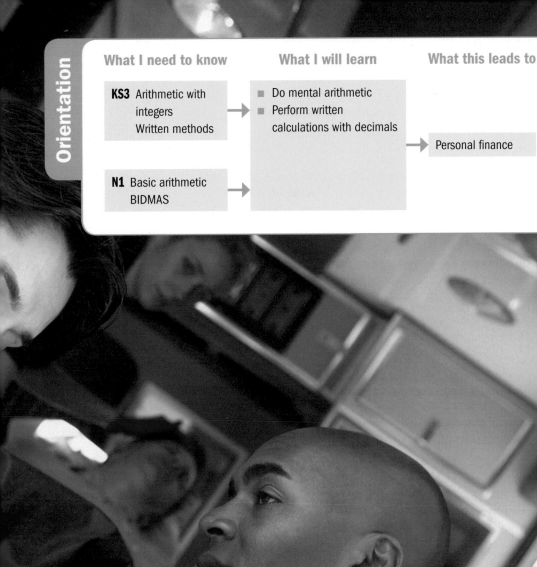

What I need to know	What I will learn	What this leads to

KS3 Arithmetic with integers
Written methods

- Do mental arithmetic
- Perform written calculations with decimals

Personal finance

N1 Basic arithmetic
BIDMAS

How many hamsters would you need to make an elephant?
How much food would an elephant-sized hamster need to eat?
How big would the cage for a elephant-sized hamster need to be?

This spread will show you how to:

- Understand where to place the decimal point in calculations
- Use mental methods to calculate with decimal numbers

Keywords
Compensation
  method
Estimate

You already know several ways of adding and subtracting whole numbers, such as 36 + 59 or 326 − 138, in your head.
You can use the same techniques with decimal numbers.

- Start by working out an **estimate** in your head so that you can check that your final answer is reasonable.

  $15.1 - 3.8 \approx 15 - 4$
  $= 11$

Then calculate the exact answer. One way is using the **compensation method**.

  $15 - 4 = 11$
  Add 0.1 (for 15.1) = 11.1
  Add 0.2 (for 3.8) = 11.3
  Answer is 11.3

Or treat the numbers as whole numbers, and adjust the place value afterwards. For 15.1 − 3.8, work out 151 − 38 = 113, and adjust this to 11.3.

  $151 - 38 = 113$
  so $15.1 - 3.8 = 11.3$

Example

Work these out in your head.
a 4.7 + 9.8   **b** 16.9 − 7.3   **c** 108.4 − 37.1
d 6.07 + 5.3   **e** 13.9 − 6.75

....

a Initial estimate: 5 + 10 = 15.
One method is: 4.7 + 10 = 14.7, and then 14.7 − 0.2 = 14.5.
4.7 + 9.8 = 14.5

b Initial estimate: 17 − 7 = 10.
One method is: 169 − 73 = 169 − 69 − 4 = 96.
16.9 − 7.3 = 9.6

c Initial estimate: 100 − 40 = 60.
One method is to adjust the initial approximation:
60 + 8.4 + 2.9 = 71.3.
108.4 − 37.1 = 71.3

d Initial estimate: 6 + 5 = 11.
Now add the figures after the decimal point (0.07 and 0.3).
6.07 + 5.3 = 11.37

e Initial estimate: 14 − 7 = 7.
The difference between 6.75 and 14 is 7.25, so the difference between 6.75 and 13.9 is 0.1 less.
13.9 − 6.75 = 7.15

There is more than one way to solve each of the these problems.

1   Write the answers to these additions.
   a 0.1 + 0.7   **b** 0.2 + 0.3   **c** 0.3 + 0.1
   d 0.7 + 0.4   **e** 0.3 + 0.1   **f** 0.5 + 0.5

2   Use your answers to question **1** to write the answers to these
   additions.
   a 5.1 + 0.7   **b** 6.2 + 0.3   **c** 7.3 + 0.1
   d 3.7 + 0.4   **e** 11.3 + 0.1   **f** 9.5 + 0.5

3   Work out these in your head and write the answers.
   a 4.7 + 5.3   **b** 4.7 + 5.4   **c** 3.6 + 6.7
   d 6.8 + 4.3   **e** 7.5 + 8.9   **f** 2.7 + 4.8

4   Work out these in your head and write the answers.
   a 3.55 + 4.22   **b** 2.13 + 3.12   **c** 3.18 + 0.42
   d 3.72 + 0.18   **e** 1.42 + 0.71   **f** 8.39 + 4.65

5   Work out these in your head and write the answers.
   a 3.35 + 0.8   **b** 0.15 + 6.7   **c** 0.7 + 3.88
   d 6.92 + 3.5   **e** 5.34 + 6.8   **f** 1.44 + 0.6

6   Write the answers to these subtractions.
   a 1 − 0.3   **b** 1 − 0.4   **c** 1 − 0.7
   d 4 − 0.6   **e** 11 − 0.8   **f** 15 − 0.9

7   Use your answers to question **6** to write the answers to these
   subtractions.
   a 1.1 − 0.3   **b** 1.1 − 0.4   **c** 1.5 − 0.7
   d 4.7 − 0.6   **e** 11.5 − 0.8   **f** 15.3 − 0.9

8   Work out these in your head and write the answers.
   a 2.16 − 1.42   **b** 1.51 − 0.46   **c** 6.39 − 4.88
   d 15.46 − 8.32   **e** 5.17 − 4.09   **f** 4.29 − 3.65

9   Write the answers to these subtractions.
   a 2 − 0.38   **b** 4 − 0.49   **c** 1 − 0.55
   d 5 − 0.63   **e** 15 − 0.48   **f** 12 − 0.34

10 Use your answers to question **9** to write the answers to these
   subtractions.
   a 2.1 − 0.38   **b** 4.1 − 0.49   **c** 1.2 − 0.55
   d 5.3 − 0.63   **e** 15.5 − 0.48   **f** 12.7 − 0.34

N5.2 Written methods for adding and subtracting decimals

This spread will show you how to:

- Understand where to place the decimal point in calculations
- Use mental and written methods to calculate with decimal numbers

p.14

Keywords
Column
Decomposition
Digit

When adding or subtracting decimal numbers with more than one decimal place, use a written method to ensure accuracy.

Example

Calculate these using a written method.

a 102.773 + 28.47 **b** 26.44 − 1.105

a Initial estimate: 100 + 30 = 130

```
  1 0 2 . 7 7 3
+   2 8 . 4 7 0
  1 3 1 . 2 4 3
```

b Initial estimate: 30 − 1 = 29

```
  2 6 . 4 ³4̸ ¹⁰0
−   1 . 1  0  5
  2 5 . 3  3  5
```

You should still use an estimate to help check your answer.

Make sure that you line up the numbers correctly; put the decimal points above each other.

The 'carrying' and **decomposition** of numbers works in the same way as with whole numbers.

When the numbers in the calculation have different numbers of decimal digits, it can be useful to add extra zeros at the end of the number with fewer digits (shown above in red).

- When you use a written method for adding or subtracting decimals, you should estimate first.

Example

Here are some incorrect attempts at adding and subtracting with decimals. Try to decide what went wrong in each case.

a 10.03 − 2.55

```
  10.⁹0̸¹3
 − 2. 55
   8. 48  ✗
```

b 38.53 + 2.474

```
 38.53
 2.474
 6.327  ✗
 1 1
```

c 100.773 − 28.782

```
  100.773
 −28.782
  128.011  ✗
```

a Initial estimate:
10 − 3 = 7

The top number has not been split up (decomposed) properly. Note that this calculation is actually quite easy to do mentally.

The correct answer is 7.48.

b Initial estimate:
39 + 2 = 41

In this example the decimal points, and therefore all of the columns, were not properly aligned.

The correct answer is 41.004.

c Initial estimate:
100 − 30 = 70

Here the smallest digit in each column has been subtracted from the bigger one. This seems silly, but it's easy to do when in a hurry!

The correct answer is 71.991

1 Calculate these using a mental method.
  a 4.3 + 8.1   **b** 6.4 + 5.6   **c** 9.2 + 3.9
  d 12.7 + 9.8   **e** 14.3 + 8.8   **f** 16.2 + 9.9

2 Work out each of the calculations from question **1**, using a standard written method.
  If you do not get the same answers by both methods, check to find your mistake.

3 Some of the calculations below are easy to do mentally.
  Others are best performed using a written method.
  Find the answer to each, showing your method clearly.
  a 5.9 + 7.1   **b** 0.673 + 1.198   **c** 94.834 + 106.487
  d 16.7 + 28.3   **e** 36.87 + 2.1   **f** 17.71 + 3.98

4 Work out these calculations using a standard written method.
  Remember to write down an estimate first.
  a 31.45 + 108.88   **b** 182.7 + 59.6   **c** 81.927 + 16.88
  d 104.7 + 98.89   **e** 57.784 + 103.218   **f** 61.386 + 40.614

5 Use a calculator to check your answers to question **4**.

6 Work out these using a mental method.
  a 8.4 − 6.2   **b** 9.7 − 0.6   **c** 17.9 − 2.9
  d 7.2 − 2.3   **e** 15.3 − 6.9   **f** 17.8 − 14.9

7 Work out the calculations from question **6**, using a standard written method.
  Check that you get the same answers by both methods. If not, find and correct your mistake.

8 Work out these calculations using an appropriate method (written or mental). Show your method clearly.
  a 5.8 − 3.2   **b** 16.73 − 8.87   **c** 9.6 − 3.7
  d 109.54 − 17   **e** 2.37 − 1.4   **f** 26.25 − 1.98

9 Work out these calculations using a standard written method.
  Remember to write an estimate before you do the calculation.
  a 21.864 − 7.968   **b** 104.87 − 85.42   **c** 417.48 − 57.69
  d 24.503 − 16.82   **e** 19.21 − 18.884   **f** 102.01 − 90.59

10 Use a calculator to check your answers to question **9**.

11 Work out these calculations using an appropriate method, and show your working clearly.
  a 8.6 − 4.5   **b** 26.4 + 13.8   **c** 18 − 6.712
  d 15.808 − 9.84   **e** 4.008 − 3.116   **f** 4.109 − 3.64

This spread will show you how to:

- Multiply and divide decimal numbers
- Understand where to place the decimal point in calculations

Keywords
Estimate
Place value

p.12

You already know some methods for mental multiplication and division of whole numbers. For example,

$$7 \times 14 = 7 \times (10 + 4) = 7 \times 10 + 7 \times 4 = 70 + 28 = 98.$$

You can extend these methods to work with decimal numbers.

- One basic number fact can be extended to many different decimal calculations. For example, $4 \times 3 = 12$ leads to $4 \times 0.3 = 1.2$, $0.4 \times 3 = 1.2$, $0.3 \times 0.4 = 0.12$, ...

Start with a basic calculation using whole numbers, and then adjust the **place value**.

Example

Calculate mentally   **a** 7×19 **b** 0.7×19 **c** 0.007×190

a $7 \times 20 = 7 \times 2 \times 10 = 14 \times 10 = 140$.
   So, $7 \times 19 = 140 - 7 = 133$.
b The original 7 (in 7×19) has changed to 0.7, so $0.7 \times 19 = 13.3$.
   - A rough estimate is another way to get the correct answer.
     $0.7 \times 19 \approx 1 \times 20$. The answer must be 13.3, since the alternatives (1.33 or 133) are much further from 20.
c Compare this to part **a**. 0.007, not 7, so answer is 1000 times smaller. 190, not 19, so answer is 10 times bigger. So, overall, answer is 100 times smaller. The answer is 1.33.
   - An estimate confirms this:
     $0.007 \times 190 \approx 0.007 \times 200 = 0.007 \times 100 \times 2 = 0.7 \times 2 = 1.4$

Often you know what the digits in the answers are, but you need to decide on the place value. A rough estimate is usually good enough to decide.

Example

Given that $238 \times 17 = 4046$, find the value of
a $2.38 \times 17\ 000$ **b** $404.6 \div 170$

a $2.38 \times 17\ 000 \approx 2 \times 20\ 000$, so the answer must be 40 460.
b The original calculation gives $4046 \div 17 = 238$.
   $404.6 \div 170 \approx 400 \div 200 = 2$. The answer must be 2.38.

Example

Davinder buys 15 bottles of cola for a party. Each bottle costs 79p. Work out mentally the total cost.

Estimate: $15 \times 80 = 10 \times 80 + 5 \times 80$
              $= 800 + 400 = 1200$
        $15 \times 79 = 15 \times 80 - 15 \times 1$
              $= 1200 - 15 = 1185$
The answer is £11.85.

1 Write the answer to the calculation 4×12.
   Use this answer to work out
   a 0.4×12 **b** 4×1.2 **c** 0.4×1.2
   d 40×1.2 **e** 400×0.12

2 Use a mental method to work out 7×13. Then use your answer to
   work out
   a 0.7×13 **b** 7×1.3 **c** 0.7×1.3
   d 70×1.3 **e** 700×0.13

3 Write the answers to
   a 14×3 **b** $64 \div 4$ **c** $35 \div 5$
   d $208 \div 2$ **e** $48 \div 24$ **f** 4×7

4 Use your answers from question **3** to work out
   a 1.4×3 **b** $6.4 \div 4$ **c** $3.5 \div 5$
   d $20.8 \div 2$ **e** $4.8 \div 2.4$ **f** 0.4×0.7

5 Any multiplication (such as $2 \times 3 = 6$) is part of a larger family of
   related facts: the equivalent multiplication $3 \times 2 = 6$, and the divisions
   $6 \div 2 = 3$ and $6 \div 3 = 2$. Write three other related facts for each of
   these multiplications.
   a $7 \times 9 = 63$ **b** $6 \times 8 = 48$ **c** $13 \times 7 = 91$
   d $15 \times 18 = 270$ **e** $5 \times 3.5 = 17.5$ **f** $2.4 \times 3.9 = 9.36$

6 You can use your knowledge of place value to extend families of
   multiplication facts further. For example, starting with $8 \times 7 = 56$, you
   can write $7 \times 80 = 560$, $56 \div 7 = 8$, $5.6 \div 7 = 0.8$, $5.6 \div 8 = 0.7$,
   $5.6 \div 80 = 0.07$, and so on.
   Use these ideas to write some number facts related to these
   calculations, and check your answers with a calculator.
   a $4 \times 5 = 20$ **b** $7 \times 9 = 63$ **c** $32 \div 8 = 4$
   d $24 \div 8 = 3$ **e** $81 \div 9 = 9$

7 Given that $5 \times 9 = 45$, calculate
   a 0.5×9 **b** 50×90 **c** 0.5×90 **d** 0.5×0.9
   e $45 \div 9$ **f** $450 \div 5$ **g** $450 \div 0.9$

8 Given that $38 \times 91 = 3458$, calculate
   a 3.8×91 **b** 38×9.1 **c** $3458 \div 91$ **d** 0.38×910
   e $345.8 \div 38$ **f** $0.3458 \div 38$

9 Given that $2.91 \times 350 = 1018.5$, write the value of
   a 29.1×350 **b** 29.1×3.5 **c** 0.0291×3.5
   d $101.85 \div 291$ **e** $10.185 \div 2.91$ **f** $1.0185 \div 350$

10 Given that $4.7 \times 6.3 = 29.61$, write the answer to
   a 0.047×630 **b** 47×0.0063 **c** $2961 \div 0.47$

Written methods for multiplying and dividing decimals

This spread will show you how to:

p.14

- Use written methods to calculate with decimal numbers
- Understand where to place the decimal point in calculations

Keywords
Place value

You already know written methods for multiplying and dividing whole numbers. You can use the same methods with decimal numbers.

- Always start with an estimate.
- Carry out the calculation.
- Use the estimate to position the decimal point in the answer.

Example

Work out $34.27 \div 2.3$

Rhian uses repeated subtraction.

$34.27 \div 2.3$ Use only whole
- Estimate: $30 \div 2 = 15$ numbers in the
- Work out $3427 \div 23$ calculation

$$
\begin{array}{ll}
3427 & \\
2300 & (100 \times 23) \\
\hline
1127 & 10 \times 23 = 230 \\
920 & 20 \times 23 = 460 \\
& (40 \times 23 = 920) \\
\hline
207 & (9 \times 23) = \begin{array}{c} 230 - 23 \\ 207 \end{array} \\
207 & \\
\hline
- &
\end{array}
$$

$3427 \div 23 = 100 + 40 + 9 = 149$
So, $34.27 \div 2.3 = 14.9$

Tom uses long division.

$34.27 \div 2.3$
- Estimate: $30 \div 2 = 15$
- Work out $3427 \div 23$

$$
\begin{array}{r}
149 \\
23\overline{)3427} \\
23 \\
\hline
112 \\
92 \\
\hline
207 \\
207 \\
\hline
-
\end{array}
$$

$3427 \div 23 = 149$
So, $34.27 \div 2.3 = 14.9$

Rhian and Tom both use a written method to calculate $3427 \div 23$, and an estimate to position the decimal point in the answer, so that the digits have the correct place values.

Example

Solve $2.93 = 18.5$

Rhian uses grid multiplication.

2.93×18.5
- Estimate: $3 \times 20 = 60$
- Work out 293×185

	200	90	3	
100	20000	9000	300	29300
80	16000	7200	240	23440
5	1000	450	15	1465
				54205

$293 \times 185 = 54205$
$\Rightarrow 2.93 \times 18.5 = 54.205$

Tom uses long multiplication.

2.93×18.5
- Estimate: $3 \times 20 = 60$
- Work out 293×185

$$
\begin{array}{r}
293 \\
\times 185 \\
\hline
1465 \\
23440 \\
29300 \\
\hline
54205
\end{array}
$$

$293 \times 185 = 54205$
$\Rightarrow 2.93 \times 18.5 = 54.205$

1 Use a mental method to work out
   a 14×7 **b** 19×8 **c** 21×13
   d 17×19 **e** 11×28

2 Now use a written method to work out the answers for question **1**.
   Check that you get the same answers with both methods.

3 Use your answers to questions **1** and **2** to write the answers to
   a 1.4×7 **b** 19×0.8 **c** 2.1×1.3
   d 1.7×0.019 **e** 1.1×0.28

4 Use a mental method to work out
   a $320 \div 4$ **b** $180 \div 15$ **c** $276 \div 23$
   d $357 \div 17$ **e** $440 \div 20$

5 Now use a written method to work out the answers for question **4**.
   Check that you get the same answers with both methods.

6 Use your answers to questions **4** and **5** to write the answers to
   a $3.2 \div 4$ **b** $18 \div 15$ **c** $2.76 \div 2.3$
   d $0.357 \div 1.7$ **e** $0.44 \div 2$

7 Use a written method to work out
   a 4.7×5.3 **b** 1.53×2.8 **c** 21.6×4.9
   d 33.65×3.89 **e** 21.58×1.99

8 Use a calculator to check your answers to question **7**.

9 Use a written method to work out
   a $34.83 \div 9$ **b** $5.425 \div 7$ **c** $7.328 \div 8$
   d $451.8 \div 60$ **e** $54.39 \div 3$

10 Use a calculator to check your answers to question **9**.

11 Use a written method to work out
   a $58.65 \div 17$ **b** $66.4 \div 16$ **c** $185.76 \div 24$
   d $7.752 \div 1.9$ **e** $3.055 \div 1.3$

12 Use a calculator to check your answers to question **11**.

13 Use a written method to work out these, giving your answers
   correct to two decimal places.
   a $14.73 \div 2.8$ **b** $51.99 \div 1.8$ **c** $193.8 \div 0.14$
   d $1013 \div 5.77$ **e** $23.78 \div 0.83$

14 Use a calculator to check your answers to question **13**.

Check out

You should now be able to:

- Use mental and written methods to calculate with decimals, understanding where to place the decimal point
- Use one calculation to find the answer to another

Worked exam question

Using the information that

$4.8 \times 34 = 163.2$

write down the value of

a 48×34 (1)

b 4.8×3.4 (1)

c $163.2 \div 48$ (1)

(Edexcel Limited 2008)

a

1632

> 48 × 34 is approximately 50 × 30
> 50 × 30 = 1500
> 1632 is the same order of size as 1500

b

16.32

> 4.8 × 3.4 is approximately 5 × 3
> 5 × 3 = 15
> 16.32 is the same order of size as 15

c

3.4

> 163.2 ÷ 48 is approximately 150 ÷ 50
> 150 ÷ 50 = 3
> 3.4 is the same order of size as 3

> There are no marks awarded for the working out, but the working out will help you give answers of the correct order of size.

Exam questions

A02

1 Laura has a job as a sales assistant.
She is paid £6.20 for each hour she works.
Last week Laura worked for 35 hours.

Work out Laura's total pay for last week. (3)

2 The cost of a calculator is £6.79
Work out the cost of 28 of these calculators. (3)

(Edexcel Limited 2005)

3 Using the information that
$19 \times 24 = 456$
write down the value of
a 19×240 (1)
b 19×2.4 (1)
c $456 \div 190$ (1)

(Edexcel Limited 2007)

A03

4 Sue buys 17 identical boxes of chocolates for £64.43
Work out the cost of 27 of these boxes of chocolates (3)

5 Use the information that
$322 \times 48 = 15\,456$
to find the value of
a 3.22×4.8 (1)
b 0.322×0.48 (1)
c $15\,456 \div 4.8$ (1)

(Edexcel Limited 2009)

Equations

Mathematicians use algebra to turn real world problems into mathematical equations. Formula 1 engineers use complex mathematical equations to predict the effect on performance of their cars when they make technical modifications.

What's the point?
Learning how to write and solve equations allows you to understand the real world and make predictions.

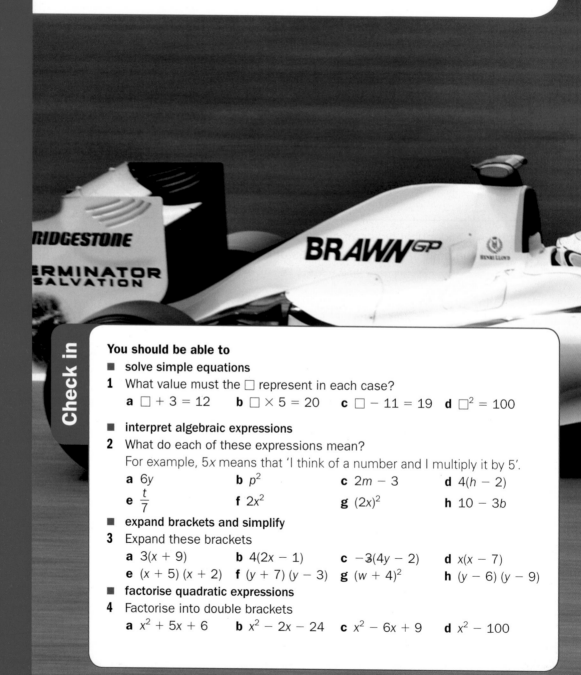

Check in

You should be able to
- **solve simple equations**

1 What value must the □ represent in each case?

   a $□ + 3 = 12$ **b** $□ \times 5 = 20$ **c** $□ - 11 = 19$ **d** $□^2 = 100$

- **interpret algebraic expressions**

2 What do each of these expressions mean?

   For example, $5x$ means that 'I think of a number and I multiply it by 5'.

   a $6y$ **b** p^2 **c** $2m - 3$ **d** $4(h - 2)$

   e $\dfrac{t}{7}$ **f** $2x^2$ **g** $(2x)^2$ **h** $10 - 3b$

- **expand brackets and simplify**

3 Expand these brackets

   a $3(x + 9)$ **b** $4(2x - 1)$ **c** $-3(4y - 2)$ **d** $x(x - 7)$

   e $(x + 5)(x + 2)$ **f** $(y + 7)(y - 3)$ **g** $(w + 4)^2$ **h** $(y - 6)(y - 9)$

- **factorise quadratic expressions**

4 Factorise into double brackets

   a $x^2 + 5x + 6$ **b** $x^2 - 2x - 24$ **c** $x^2 - 6x + 9$ **d** $x^2 - 100$

Exam questions

A02

1 Laura has a job as a sales assistant.
She is paid £6.20 for each hour she works.
Last week Laura worked for 35 hours.

Work out Laura's total pay for last week. (3)

2 The cost of a calculator is £6.79
Work out the cost of 28 of these calculators. (3)

(Edexcel Limited 2005)

3 Using the information that
$19 \times 24 = 456$
write down the value of
a 19×240 (1)
b 19×2.4 (1)
c $456 \div 190$ (1)

(Edexcel Limited 2007)

A03

4 Sue buys 17 identical boxes of chocolates for £64.43
Work out the cost of 27 of these boxes of chocolates (3)

5 Use the information that
$322 \times 48 = 15\,456$
to find the value of
a 3.22×4.8 (1)
b 0.322×0.48 (1)
c $15\,456 \div 4.8$ (1)

(Edexcel Limited 2009)

Equations

Mathematicians use algebra to turn real world problems into mathematical equations. Formula 1 engineers use complex mathematical equations to predict the effect on performance of their cars when they make technical modifications.

What's the point?

Learning how to write and solve equations allows you to understand the real world and make predictions.

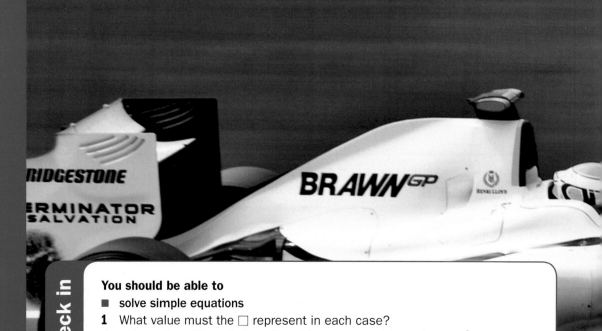

You should be able to

- **solve simple equations**
1   What value must the □ represent in each case?
    a □ + 3 = 12   **b** □ × 5 = 20   **c** □ − 11 = 19   **d** □² = 100

- **interpret algebraic expressions**
2   What do each of these expressions mean?
    For example, 5x means that 'I think of a number and I multiply it by 5'.
    a 6y **b** p^2 **c** 2m − 3   **d** 4(h − 2)
    e $\frac{t}{7}$ **f** 2x^2 **g** $(2x)^2$ **h** 10 − 3b

- **expand brackets and simplify**
3   Expand these brackets
    a 3(x + 9)   **b** 4(2x − 1)   **c** −3(4y − 2)   **d** $x(x$ − 7)
    e (x + 5) (x + 2)   **f** (y + 7) (y − 3)   **g** $(w + 4)^2$ **h** (y − 6) (y − 9)

- **factorise quadratic expressions**
4   Factorise into double brackets
    a x^2 + 5x + 6   **b** x^2 − 2x − 24   **c** x^2 − 6x + 9   **d** x^2 − 100

What I need to know

KS3 Experience of solving equations

A2 Expand brackets

What I will learn

- Use inverse operations with algebra
- Solve linear equations including those with fractions and brackets
- Solve quadratic equations by factorisation

What this leads to

A5 Solve simultaneous equations

A6 Plot and interpret further equations

A-level
Maths, Chemistry, Physics

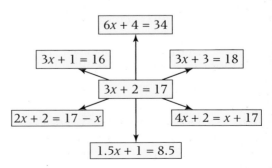

In this diagram the equation $3x + 2 = 17$ has been changed in different ways but all of these ways still give the same solution of $x = 5$. Describe each change to the equation.
Continue each change for at least one more step.
Invent some changes of your own.

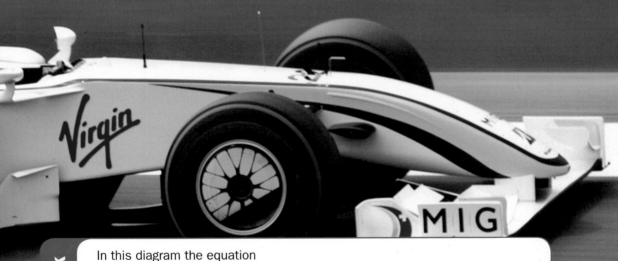

$6x + 4 = 34$

$3x + 1 = 16$ $3x + 3 = 18$

$3x + 2 = 17$

$2x + 2 = 17 - x$ $4x + 2 = x + 17$

$1.5x + 1 = 8.5$

Working with inverse operations

This spread will show you how to:

- Understand and use inverse operations
- Use function machines

Keywords

Inverse
Operation
Variable

- Addition, subtraction, multiplication and division are all **operations**.

- An operation has an **inverse**, for example subtraction is the inverse of addition.

An operation acts on a number or a **variable**.

The inverse operation changes the variable back to its original value.

$$a \rightarrow \boxed{\times 3} \rightarrow 3a \rightarrow \boxed{\div 3} \rightarrow a$$

x and *y* are examples of variables.

If two or more operations act on a variable then you do their inverse operations in reverse order to get back to the original value.

$$x \rightarrow \boxed{\times 2} \rightarrow 2x \rightarrow \boxed{-5} \rightarrow 2x - 5$$
$$x \leftarrow \boxed{\div 2} \leftarrow 2x \leftarrow \boxed{+5} \leftarrow 2x - 5$$

The operations are 'multiply by 2' and 'subtract 5'.

The inverse operations are 'add 5' and 'divide by 2'.

Example

Find the value of each letter.

a	5			*x*
b	−12	$\boxed{\times 3}$	$\boxed{-7}$	*y*
c	$\frac{1}{2}$			*z*
d	*p*			53

....

a $5 \xrightarrow{\times 3} 15 \xrightarrow{-7} 8$ So $x = 8$

b $-12 \xrightarrow{\times 3} -36 \xrightarrow{-7} -43$ So $y = -43$

c $\frac{1}{2} \xrightarrow{\times 3} 1\frac{1}{2} \xrightarrow{-7} -5\frac{1}{2}$ So $z = -5\frac{1}{2}$

d $53 \xrightarrow{+7} 60 \xrightarrow{\div 3} 20$ So $p = 20$

To find *p*, do the inverse operations in the reverse order.

Example

Find the starting number in each case.

a I think of a number, subtract 8 and square root it. The answer is 5.
b I think of a number, multiply by 2, subtract 4 and cube it. The answer is −216.

....

a $5 \xrightarrow{\text{square}} 25 \xrightarrow{+8} 25 + 8 = 33$ The starting number is 33.

b $-216 \xrightarrow[\text{root}]{\text{cube}} -6 \xrightarrow{+4} -2 \xrightarrow{\div 2} -1$ The starting number is −1.

Undo the operations in the reverse order.

1 Copy and complete these diagrams.

a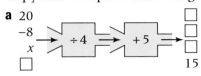

b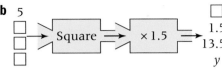

c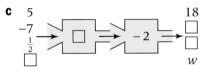

d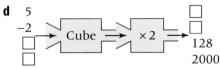

2 Explain why 'adding 10%' cannot be undone using the inverse operation 'subtracting 10%'.

Problem

A03

3 In each case, use inverse operations to find the starting number.
  a I think of a number, double it and subtract 4. This gives me 7.
  b I think of a number, square it, multiply by 3 and get 75.
  c I think of a number, add 11, cube root it and get 2.
  d I think of a number, halve it, treble it then subtract 6. I get 54.
  e I think of a number, multiply it by $\frac{1}{4}$, square root it and get 5.

4 Is there another starting number for question **3b**?
   Explain your answer.

5 a Write, in order, the operations that have acted on x.
   i $x + 3 = 17$ **ii** $2x - 1 = 17$ **iii** $\dfrac{x + 5}{2} = 24$ **iv** $3x^2 = 48$
  b Using the inverse operations, in reverse order, find the value of x
   in each case.

6 Look at this function machine.

Start	× 2	− 9	÷ 5	+ 2	÷ $\frac{1}{2}$	Finish

  a What value do you end up with if (-3) enters the machine?
  b What value did you start with if these values leave the machine?
   i 6   **ii** 8.4   **iii** $\frac{1}{4}$

7 a Invent your own function machine that uses five operations to
   convert a value of 5 into 53.
  b Use your machine to describe what value you started with if the
   output is 100.

Solving one-sided equations

This spread will show you how to:

- Set up and solve simple equations

Keywords

Equation
Operation
Solve

- An **equation** is a statement with an equals sign. For example $2x - 4 = 18$ is an equation.

- To **solve** an equation, do the same **operation** to both sides. For example

$$2x - 4 = 18 \quad \text{Add 4 to both sides.}$$
$$2x = 22 \quad \text{Divide both sides by 2.}$$
$$x = 11$$

This equation is only true when $x = 11$.

Undo the operations in reverse order.

Example

Solve these equations.

a $\dfrac{3x - 5}{2} = 8$ **b** $4(y + 3) = 40$ **c** $17x = 35$

....

a $\dfrac{3x - 5}{2} = 8$ Multiply both sides by 2.

$3x - 5 = 16$ Add 5 to both sides.

$3x = 21$ Divide both sides by 3.

$x = 7$

b $4(y + 3) = 40$ Divide both sides by 4.   **or** $4(y + 3) = 40$ Expand the bracket.

$y + 3 = 10$ Subtract 3 from both sides. $4y + 12 = 40$ Subtract 12 from both sides.

$y = 7$ $4y = 28$ Divide both sides by 7.

$y = 7$

c $17x = 35$ Divide both sides by 17.

$x = \dfrac{35}{17}$ Change to a mixed number.

$= 2\dfrac{1}{17}$ Avoid decimals. $\dfrac{1}{17}$ is a recurring decimal and has to be rounded, so a decimal answer is not exact.

Example

The perimeter of this rectangle is 34 cm. Find x.

$3x + 1$

x

....

$\text{perimeter} = x + x + 3x + 1 + 3x + 1$

$= 8x + 2$

$\text{so } 8x + 2 = 34$

$8x = 32$

$x = 4$

1 Solve these one-sided equations.

a $3x - 7 = 8$ **b** $4(x - 1) = 20$ **c** $\dfrac{x}{2} + 8 = 13$

d $\dfrac{2x - 8}{4} = -3$ **e** $2(x^2 + 9) = 68$ **f** $2\left(\dfrac{x + 1}{2} - 3\right) = 8$

g $\dfrac{3x - 5}{2} = 10$ **h** $x^3 - 4 = 12$ **i** $\sqrt{x} - 1 = 9$

2 Copy and complete this crossword by solving the equations in the clues.

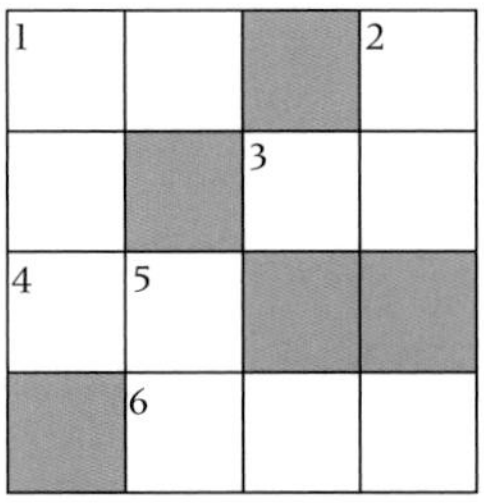

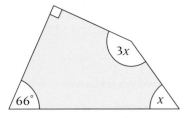

Across	**Down**
1 $2(x + 2) = 30$	**1** $3x + 15 = 330$
3 $\frac{x}{2} - 5 = 15$	**2** $x - 10 = 30$
4 $\frac{3x - 11}{2} = 71$	**5** $x^2 + 1 = 170$
6 $3(x - 100) = 675$	

3 The perimeter of this shape is 30 mm. Find the length of each side.

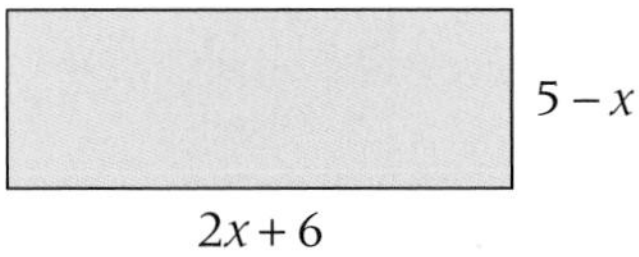

$5 - x$

$2x + 6$

4 This shape is a quadrilateral. Work out the value of x.

$3x$

$66°$

x

The interior angle sum of a quadrilateral is 360°.

5 a Given that $y = 4x - 8$, work out the value of x when $y = 12$.
   b Repeat for $y = -4$.

A03 Problem

6 The diagram shows an isosceles triangle.
   Given that the equal angles are 10 less than
   double the third angle:
   a Write an expression for each equal angle.
   b Write an equation connecting the angles.
   c Solve the equation, using your answer to find the size of
      each angle.

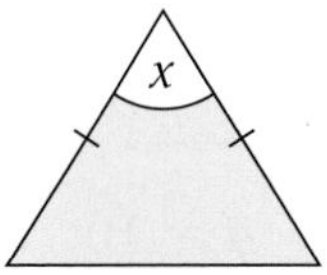

x

90

Solving double-sided equations

This spread will show you how to:
- Set up and solve simple equations

When you solve an equation with **unknowns** on both sides

Subtract the smaller algebraic term from both sides, for example

$5x - 3 = 3x + 1$	Subtract $3x$ from both sides.
$2x - 3 = 1$	This is now a **one-sided** equation.

Solve the one-sided equation by using inverse operations.

$2x - 3 = 1$	Add 3 to both sides.
$2x = 4$	Divide both sides by 2.
$x = 2$	

Example

Solve

a $3x + 7 = 5x - 1$ **b** $5 - 6y = 2y - 3$

....

a $3x + 7 = 5x - 1$	Subtract $3x$ from both sides ($3x$ is smaller than $5x$).
$7 = 2x - 1$	Add 1 to both sides.
$8 = 2x$	Divide both sides by 2.
$4 = x$	
$x = 4$	

b $5 - 6y = 2y - 3$	Subtract $-6y$ from both sides ($-6y$ is smaller than $2y$):
	Subtracting $-6y$ is the same as adding $6y$.
$5 = 8y - 3$	Add 3 to both sides.
$8 = 8y$	Divide both sides by 8.
$y = 1$	

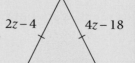

Example

The diagram shows an isosceles triangle.
Find the length of the equal sides.

....

$2z - 4 = 4z - 18$	Write an equation and solve it.
$-4 = 2z - 18$	After subtracting $2z$ from both sides.
$14 = 2z$	After adding 18 to both sides.
$7 = z$	

Labels on triangle: $2z - 4$ $4z - 18$

Hence, $2z - 4 = 14 - 4 = 10$ and $4z - 18 = 28 - 18 = 10$
The equal sides of the triangle are 10 units long.

1 Solve these equations
   a $2x + 4 = x + 3$ **b** $10 - 3x = 7x - 10$
   c $8 - 3x = 5 - 2x$ **d** $4(7 + 2z) = 15 - 8z$

2 In each row of equations, one has a different solution from the other two. Find the odd one out.
   a $2a + 7 = 4a + 1$ $6a - 2 = 2a + 6$ $10 + 2a = 7a - 5$
   b $10 - 3b = 6b + 1$ $15 - 2b = 14 - b$ $3b - 14 = b$
   c $2c + 2 = 4c - 1$ $8c - 7 = 6c - 4$ $1 - 4c = 2 - 8c$
   d $2d + d + 8 = 3d + 2d - 4$ $5d + 7d - 3 = 10 - d$ $15 - d - d = 9 - d$

3 In each case, use the information to write an equation and solve it to find the starting number.
   a I think of a number, multiply it by 8 and subtract 2. I get the same answer as when I multiply this number by 2 and add 10.
   b I think of a number, multiply it by 5 and add 3. I get the same answer as when I multiply it by 2 and subtract it *from* 24.
   c Taking double a number from 11 is equal to taking treble that number from 14.

4 In each case, use the information to form an equation and solve it.
   a The angles in a triangle are $x°$, $x + 20°$ and $x + 40°$. Find the angles of the triangle.
   b The perimeters of these two shapes are equal. What are the dimensions of each shape?

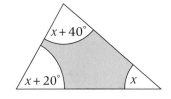

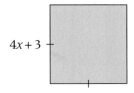 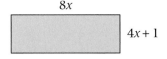

   c Abdul is 180 cm tall and Mark is $10x$ cm tall.
   Alice is 164 cm tall and Miranda is $9x$ cm tall.
   The difference in height between the two boys is equal to the difference in height between the two girls.
   How tall are Mark and Miranda?

Solving equations with fractions

This spread will show you how to:

• Solve equations involving fractions

Keywords
Cross multiply
Reciprocal

• When you solve an equation involving fractions, clear the fractions first.

You can use the fact that $x \times \frac{1}{x} = 1$ to help you.

$\frac{1}{x}$ is called the **reciprocal** of x.

• Any non-zero number multiplied by its reciprocal is 1.

Example

Solve the equation $\frac{15}{x} = \frac{3}{7}$

··

$\frac{15}{x} = \frac{3}{7}$ Multiply both sides by the common denominator $7x$ to clear the fractions.

$\frac{15}{x} \times 7x = \frac{3}{7} \times 7x$ Cancel common factors.

$15 \times 7 = 3 \times x$,   so $105 = 3x$,   so $x = 35$

$7x$ is the reciprocal of $\frac{1}{7x}$.

When the equation has just a single fraction on each side you can **cross multiply** to clear the fractions. For example,

$\frac{15}{x} \diagdown \frac{3}{7}$ $15 \times 7 = x \times 3 \rightarrow 105 = 3x \rightarrow x = 35$ as above

Cross multiply:
LH numerator ×
RH denominator
and vice versa.

Example

Solve these equations

a $\frac{3}{x} = \frac{4}{9}$ **b** $\frac{12}{p} + 9 = 28$

···

a $\frac{3}{x} = \frac{4}{9}$ Simple fraction each side, so cross multiply.

$3 \times 9 = 4x$ Divide both sides by 4.

$27 = 4x$

$x = \frac{27}{4} = 6\frac{3}{4}$

b $\frac{12}{p} + 9 = 28$ Subtract 9 from both sides.

$\frac{12}{p} = 19$ Cross multiply $(19 = \frac{19}{1})$.

$12 \times 1 = 19p$ Divide both sides by 19.

$p = \frac{12}{19}$

It is tempting to use cross multiplication straightaway but you cannot cross multiply until you have a simple fraction on each side.

Example

I divide 15 by a certain number and get $\frac{3}{4}$.
Write an equation and solve it to find the starting number.

···

If x is the missing number then: $\frac{15}{x} = \frac{3}{4}$ Cross multiply.

$15 \times 4 = 3x$

$60 = 3x$, so $x = 20$

1 Solve these equations by cross multiplying.

a $\dfrac{7}{x} = 21$ **b** $15 = \dfrac{5}{x}$ **c** $\dfrac{4}{y} = 3$ **d** $\dfrac{7}{p} = 8$

Think of
21 as $\dfrac{21}{1}$

e $\dfrac{10}{x} = -2$ **f** $11 = \dfrac{5}{y}$ **g** $-3 = \dfrac{7}{y}$ **h** $-\dfrac{3}{x} = -9$

2 Solve these equations

a $\dfrac{5}{x} + 9 = 10$ **b** $\dfrac{10}{p} + 7 = 8$ **c** $\dfrac{x}{4} + 3 = 10$ **d** $-2 = 1 + \dfrac{3}{x}$

3 Solve these equations

a $\dfrac{16}{x} + 4 = 2$ **b** $\dfrac{12}{2y} - 3 = 5$ **c** $\dfrac{6}{3p} - 1 = 10$ **d** $\dfrac{15}{2x} + 4 = -2$

4 Solve these equations

a $\dfrac{x+1}{3} = \dfrac{x-1}{4}$ **b** $\dfrac{2y-1}{3} = \dfrac{y}{2}$ **c** $\dfrac{5}{w+5} = \dfrac{15}{w+7}$ **d** $\dfrac{3}{x-1} = \dfrac{9}{2x-1}$

5 Write an equation and solve it to find the starting number in each case.

a I think of a number, divide it into 16 and I get 10.

b I think of a number, add 4, divide it into 12 and get 7.

c I think of a number, take 3 and divide it into 11. This gives me the same answer as when I take the same number and divide it into 8.

6 Use the formula $P = \dfrac{180}{n+1}$ to find

a the value of P when $n = 4$

b the value of n when $P = 12$

c the value of n when $P = 2\dfrac{2}{9}$.

.78

A03 Problem

7 The means of each set of expressions are equal. Use this information to find x and, hence, the value of each expression.

Set 1

$2x - 1$	$3x + 2$
$5x + 4$	$7x$
$6x - 4$	$10 - 2x$

Set 2

$3x - 7$	$5x + 8$
$13 - x$	
$2(3x + 1)$	$12 + 4x$

To find the
mean add all
the expressions
and divide by
the number of
expressions.

Solving harder equations

This spread will show you how to:

- Solve linear equations after simplifying them

Keywords
Expand
Negative term

When solving an equation with brackets, **expand** the brackets first.

$4(x + 2) = 7(x - 1)$ Expand the brackets.

$\quad 4x + 8 = 7x - 7$ Subtract $4x$ from both sides.

$\qquad\quad 8 = 3x - 7$ Add 7 to both sides.

$\qquad 15 = 3x$ Divide both sides by 3.

$\qquad\quad x = 5$

An equation with a negative x-term is easier to solve if you get rid of the **negative term** first.

$5 - 8x = 45$ Add $8x$ to both sides to remove the $-8x$.

$\quad 5 = 45 + 8x$ Subtract 45 from both sides.

$-40 = 8x$ Divide both sides by 8.

$\quad x = -5$

Example

Solve $(x + 3)(x + 5) = (x + 2)^2$

$\qquad (x + 3)(x + 5) = (x + 2)^2$

$x^2 + 3x + 5x + 15 = (x + 2)(x + 2)$ Remove brackets first (FOIL): *multiply* the terms.

$\quad x^2 + 8x + 15 = x^2 + 2x + 2x + 4$ Collect like terms.

$\quad x^2 + 8x + 15 = x^2 + 4x + 4$ Subtract x^2 from each side.

$\qquad\quad 8x + 15 = 4x + 4$ Subtract $4x$ from both sides.

$\qquad\quad 4x + 15 = 4$ Subtract 15 from both sides.

$\qquad\qquad\quad 4x = -11$

$\qquad\qquad\quad\ x = \frac{-11}{4} \text{ or } -2\frac{3}{4}$

Don't use a calculator – fractions are exact. Decimals that need rounding, make the **solution** less accurate.

Example

The square and the rectangle have the same area.
Find the value of x.

Area of square = Area of rectangle

$\qquad x^2 = (x + 5)(x - 2)$ Remember brackets first (FOIL).

$\qquad x^2 = x^2 + 5x - 2x - 10$ Collect like terms.

$\qquad x^2 = x^2 + 3x - 10$ Subtract x^2 from each side.

$\qquad\ 0 = 3x - 10$ Add 10 to both sides.

$\quad 10 = 3x$

$\quad \frac{10}{3} = x$

$\qquad x = 3\frac{1}{3}$

1 Solve these equations

  a $4x - 9 = 15$

  c $3x + 9 = 2x + 15$

  e $10 - 3t = 8t - 12$

  b $\frac{y}{7} + 2 = 4$

  d $5p - 9 = 3p + 7$

  f $3 - 7b = 9 - 10b$

2 Solve these equations by first expanding brackets and/or collecting like terms.

  a $3x + 7x - 3 + 9 = 17$

  c $2(3 - 8z) = 4(5 - 6z)$

  e $3a - (9 - 7a) = 34$

  g $(x + 3)(x + 4) = (x + 7)(x - 2)$

  b $3(2y - 1) = 4(7y + 6)$

  d $2a + 3(2a - 7) = 20$

  f $6 - (3x - 9) = -2(5 - x)$

  h $(y - 7)^2 = (y + 5)^2$

3 Solve these equations containing negative terms.

  a $10 - 4x = 2$　　**b** $20 - 3y = 11$　　**c** $15 - 8z = 12$

  d $14 - 7a = 19$　　**e** $18 = 17 - b$　　**f** $36 = 2(3 - 6c)$

A03 Problem

4 In each case, write an equation to represent the information given and solve the equation to find the starting number.

  a I think of a number, treble it and subtract this *from* 40. My answer is 22.

  b I think of a number, add 6 and double the result. This gives me the same answer as when I subtract 5 from the number and multiply the result by 6.

  c I think of a number, add 7 and square it. This gives me the same answer as when I take 5 from the number and square it.

5 In each case, form an equation to represent the information given and solve it to find x.

  a

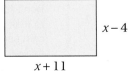

$x - 4$

$x + 11$

$x + 3$

The area of the square and rectangle are equal.

  b

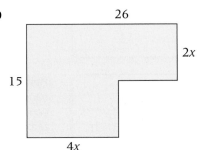

26

$2x$

15

$4x$

The lengths of the unmarked sides are equal.

Introducing quadratic equations

This spread will show you how to:
- Solve quadratic equations by factorisation

Keywords
Factorise
Product
Quadratic
Solution
Sum

- A **quadratic** equation contains an x^2 term as the highest power, for example
$$x^2 + 5x + 6 = 0$$

Many quadratic equations can be solved by **factorising**.

For example, solve the equation $x^2 + 5x + 6 = 0$.

x^2 term, so equation is a quadratic.

p.132

$x^2 + 5x + 6 = 0$ The two factors will each start with x.

$(x + \square)(x + \square) = 0$ Now find two numbers with a **sum** of 5 (for $5x$) and a **product** of 6.

$(x + 3)(x + 2) = 0$ Check; $(x + 3)(x + 2) = x^2 + 3x + 2x + 6 = x^2 + 5x + 6$.

The two factors have a product of zero. This means that at least one of them is equal to zero.

Either $x + 3 = 0$ or $x + 2 = 0$

If $x + 3 = 0$, $x = -3$ and if $x + 2 = 0$, $x = -2$ Check $(-3)^2 + 5(-3) + 6 = 9 - 15 + 6 = 0$

The **solutions** are $x = -3$ and $x = -2$.   and $(-2)^2 + 5(-2) + 6 = 4 - 10 + 6 = 0$

- The key to solving quadratic equations is that the product of the two factors is zero, so you can say that one or other (or both) of the factors is zero. This leads to the two solutions.

Example

Solve these equations

a $x^2 - 7x + 12 = 0$ **b** $x^2 + 3x = 0$

. . . .

a $x^2 - 7x + 12 = 0$ To factorise, you need two negatives to

$(x - 3)(x - 4) = 0$ multiply to make $+12$ and add to make -7.

   Either $x - 3 = 0$ or $x - 4 = 0$

           $x = 3$ $x = 4$

b $x^2 + 3x = 0$

   $x(x + 3) = 0$

   Either $x = 0$ or $x + 3 = 0$

                        $x = -3$

Don't forget to look out for common factors as well as double bracket style factorisation.

Example

Solve $x^2 = 2x + 15$

. . . .

$x^2 - 2x - 15 = 0$ Find two numbers with sum -2 and product -15.

$(x + 3)(x - 5) = 0$

   Either $x + 3 = 0$ or $x - 5 = 0$

           $x = -3$ $x = 5$

First make the left hand side of the equation equal to zero.

1 Here are six equations

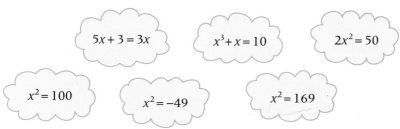

$5x + 3 = 3x$ $x^3 + x = 10$ $2x^2 = 50$

$x^2 = 100$ $x^2 = -49$ $x^2 = 169$

   a Write the four quadratic equations.
   b From the quadratic equations, find three that have two solutions.
     Write these three equations and their solutions.

2 Factorise these quadratic expressions
   a $x^2 + 8x + 12$ **b** $x^2 + 11x + 24$ **c** $x^2 + 13x + 36$
   d $x^2 + 16x + 55$ **e** $x^2 - 7x + 12$ **f** $x^2 - 10x + 24$
   g $x^2 + 3x - 28$ **h** $x^2 - 2x - 15$

3 Solve these quadratic equations by factorising them into
   double brackets.
   a $x^2 + 7x + 12 = 0$ **b** $x^2 + 8x + 12 = 0$ **c** $x^2 + 10x + 25 + 0$
   d $x^2 + 5x - 14 = 0$ **e** $x^2 - 4x - 5 = 0$ **f** $x^2 - 5x + 6 = 0$
   g $x^2 - 10x + 21 = 0$ **h** $x^2 = 3x + 40$

4 Solve these quadratic equations by factorising them into a
   single bracket.
   a $x^2 - 8x = 0$ **b** $x^2 + 4x = 0$ **c** $x^2 - 6x = 0$
   d $y^2 + 5y = 0$ **e** $x^2 = 9x$ **f** $x^2 - 12x = 0$
   g $2x^2 + 8x = 0$ **h** $6x - x^2 = 0$

5 Explain why $x^2 + 7x + 11 = 0$ cannot be solved by factorisation.

6 **a** Vicky is trying to solve $(x + 4)(x - 2) = 7$.
     Here is her attempt.

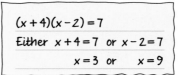

     $(x + 4)(x - 2) = 7$
     Either $x + 4 = 7$ or $x - 2 = 7$
               $x = 3$ or $x = 9$

     What is wrong with her method?
   b Can you solve this equation correctly to
     show that the two solutions should really
     be $x = 3$ and $x = -5$?

7 Simplify and then solve this equation
    $x^2 + 4x - 5 = 2x(x - 1)$

Summary

Check out

You should now be able to:

- Set up and solve simple equations
- Solve linear equations where the unknown appears on either side or on both sides of the equation
- Solve linear equations after simplifying them
- Solve quadratic equations by factorisation

Worked exam question

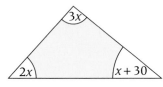

Diagram NOT
accurately drawn

The diagram shows a triangle.

The sizes of the angles, in degrees, are $3x$, $2x$ and $x + 30$

Work out the value of x.   (3)

(Edexcel Limited 2008)

The angles in a triangle add to 180°.

$3x + 2x + (x + 30) = 180$
$3x + 2x + x + 30 = 180$
$\qquad 6x + 30 = 180$
$\qquad 6x = 180 - 30$
$\qquad 6x = 150$
$\qquad x = 25°$
$3x = 75°$
$2x = 50°$
$x + 30 = 55°$

Form the equation

Show each line of working out.

Check:
$75° + 50° + 55° = 180°$

Exam questions

1 **a** Solve $\quad 6x - 7 = 38$ (2)
   b Solve $\quad 4(5y - 2) = 40$ (3)
(Edexcel Limited 2007)

2 **a** Solve $\quad 4y + 3 = 2y + 9$ (2)
   b Solve $\quad 5(t - 3) = 8$ (2)
(Edexcel Limited 2007)

3 Solve $\quad 6x - 5 = 2x + 9$ (3)
(Edexcel Limited 2006)

4

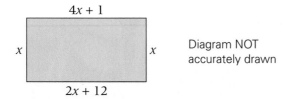

Diagram NOT accurately drawn

The diagram shows a rectangle.
All the measurements are in centimetres.

a Explain why $\quad 4x + 1 = 2x + 12$ (1)
b Solve $\quad 4x + 1 = 2x + 12$ (2)
c Use your answer to part **b** to work out the perimeter of the rectangle. (2)
(Edexcel Limited 2009)

5 **a** Factorise $\quad x^2 + 6x + 8$ (2)
   b Solve $\quad x^2 + 6x + 8 = 0$ (1)
(Edexcel Limited 2006)

Functional Maths 5: Art

Graffiti artists often sketch their designs before projecting them onto the surface. They sometimes use grids or parts of their body as measuring tools to help them copy the proportions accurately.

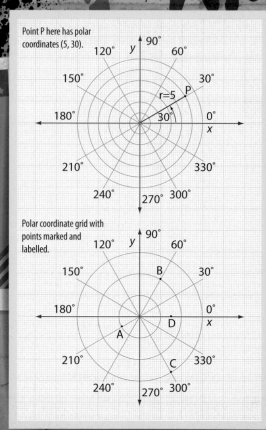

1. A graffiti artist projects an image from a sketchpad of length 20cm and width 14.8cm onto a wall of length 6m.

   a. What scale factor is being used?
   b. What is the width of the graffiti wall?

   The artist's hand-span is 150mm.

   c. What are the dimensions of the wall in terms of hands?
   d. If a 1cm square grid was used in the sketch, what size would the grid squares be on the graffiti wall?
   e. What effect does the enlargement have on the area of the graffiti image?

Polar coordinates ★★★★

Graffiti designs often involve arcs and spirals.
You can plot these on a polar grid using polar coordinates.

Polar coordinates describe the position of a point, **P**, in terms of **r**, the distance P is from the origin and ϑ, the angle that the line **OP** makes with the horizontal axis. **P** has polar coordinates (**r**, ϑ).

2. a. Match points **A**, **B** and **C** with their polar coordinates:

   i. (2, 60)   ii. (1, 210)   iii. (3, 300)

   b. Write down the polar coordinates of point **D**.

Plot and label some other points on a polar coordinate grid.

Point P here has polar coordinates (5, 30).

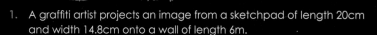

Polar coordinate grid with points marked and labelled.

Crop circles are geometric patterns that are displayed in crop fields. They are often based on circles and spirals.

A circle has a fixed radius.

The radius of a spiral changes as the angle changes.

3. This diagram shows a circle (red) and a spiral (blue) through 360°.

a. What is the radius of the circle?
The circle has an equation of the form
$r = a$ where a is a constant.

b. What is the value of a?
The radius of the spiral increases by 0.025 units every degree.

c. How many units does the radius increase by in total?
The spiral has an equation of the form
$r = k\vartheta$ where k is a constant.

d. What is the value of k?

4. Use a polar coordinate grid to sketch

a. a circle with the equation $r = 5$
b. a semi-circle (starting at 90°) with equation $r = 3$
c. a spiral, starting at the origin, with equation $r = 0.01\vartheta$

Polar coordinate grid with red circle and blue spiral.
Circle $r = a$
Spiral $r = 0.025$

90°
180°
270°
0   a

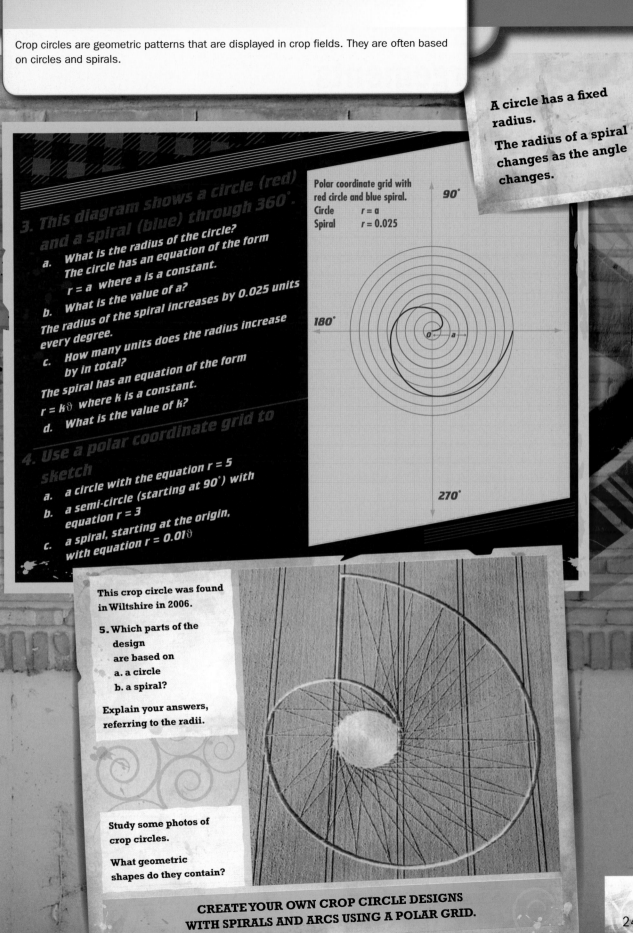

This crop circle was found in Wiltshire in 2006.

5. Which parts of the design are based on
a. a circle
b. a spiral?

Explain your answers, referring to the radii.

Study some photos of crop circles.

What geometric shapes do they contain?

CREATE YOUR OWN CROP CIRCLE DESIGNS WITH SPIRALS AND ARCS USING A POLAR GRID.

From giant billboard posters to making model planes, enlarging an object is a very visible part of the modern world. Whether you are using a map or zooming in on a computer page you are making use of enlargements.

What's the point?
Using the similarity properties of an enlargement allows you to calculate properties of the object from those of its image and visa versa.

You should be able to

■ **plot points**

1 Draw a grid with axes from −3 to +5.
  Draw these points on the grid.

  a (1, 3)   **b** (3, −2)   **c** (4, 5)   **d** (5, 2)
  e (−3, 3)   **f** (0, 4)   **g** (−1, 0)   **h** (−3, −2)

■ **simplify ratios**

2 Write these ratios in their simplest form.

  a 4 : 12   **b** 5 : 30   **c** 6 : 9   **d** 15 : 6

■ **solve equations involving fractions**

3 Solve these equations.

  a $\frac{x}{2} = 7$ **b** $\frac{x}{5} = 3$ **c** $\frac{x}{6} = \frac{5}{3}$ **d** $\frac{x}{4} = \frac{9}{2}$

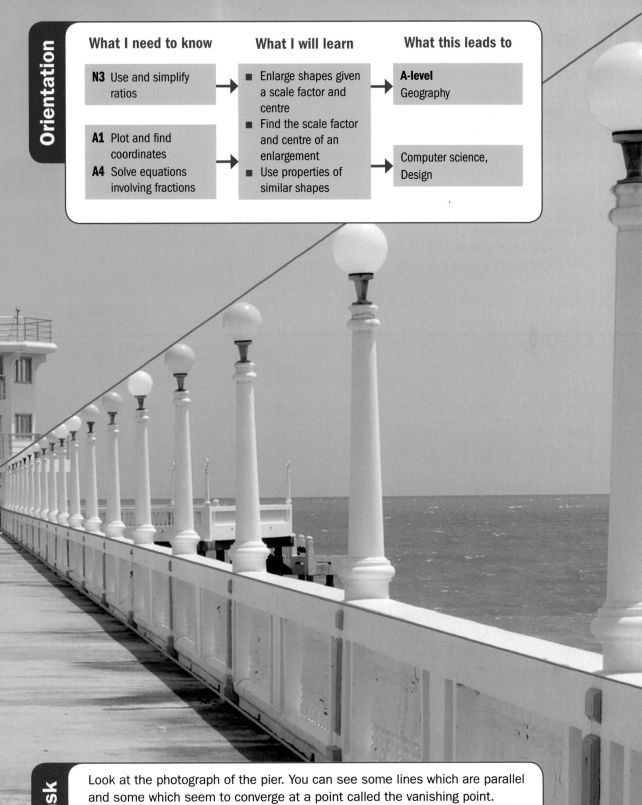

What I need to know

N3 Use and simplify ratios

A1 Plot and find coordinates

A4 Solve equations involving fractions

What I will learn

- Enlarge shapes given a scale factor and centre
- Find the scale factor and centre of an enlargement
- Use properties of similar shapes

What this leads to

A-level
Geography

Computer science, Design

Rich task

Look at the photograph of the pier. You can see some lines which are parallel and some which seem to converge at a point called the vanishing point. Investigate the heights of the lampposts and their distances from the vanishingpoint.

Enlargement

This spread will show you how to:

- Enlarge objects, given a centre of enlargement and scale factor, including fractional scale factors
- Recognise that enlargements preserve angle but not length
- Understand how enlargement affects perimeter

Keywords
Centre of enlargement
Enlargement
Scale factor
Similar

To enlarge a shape, you multiply all the lengths by the same scale factor.

- In an **enlargement**
  - corresponding angles are the same
  - corresponding lengths are in the same ratio.

You draw an enlargement from a **centre of enlargement**.

In the diagram, △PQR is enlarged by **scale factor** 2 from the centre, C, (0, 1).

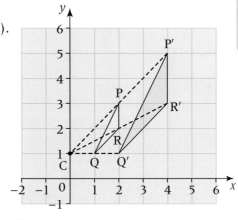

The image, P'Q'R', is **similar** to the object, PQR.

All the distances are × 2 so CQ' = 2CQ, CP' = 2CP, CR' = 2CR.

Lengths on the image are 2 × corresponding lengths on the object so P'R' = 2PR and so on.

- Perimeter of image = scale factor × perimeter of object.

Example

a Enlarge the white triangle by scale factor 3, centre (2, 1).
b How much larger is the perimeter of the image than the perimeter of the object?

a

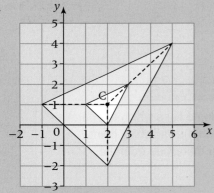

Lengths on the image are 3 × the corresponding lengths on the object.

The distance from C to a point on the image is 3 × the distance from C to the corresponding point on the object.

b Perimeter of image = 3 × perimeter of object.

1 a Copy this diagram, but extend both axes to 16.
   b Enlarge triangle T by scale factor 2, centre (0, 0). Label the image U.
   c Enlarge triangle T by scale factor 3, centre (0, 0). Label the image V.

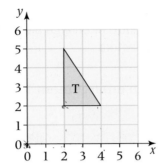

2 a Copy this diagram, but extend both axes from −7 to 7.
   b Enlarge kite K by scale factor 2, centre (1, 3). Label the image L.
   c Enlarge kite K by scale factor 4, centre (1, 3). Label the image M.

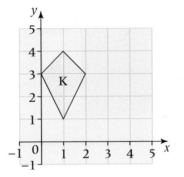

3 a Draw a grid with an *x*-axis from −6 to 6 and a *y*-axis from −3 to 15. Plot the points (1, 3) (1, 5) (3, 7) (4, 6). Join them to make quadrilateral Q.
   b Enlarge quadrilateral Q by scale factor 2, centre (4, 8). Label the image R.
   c Enlarge quadrilateral Q by scale factor 2, centre (0, 6). Label the image S.
   d How much larger is the perimeter of R than the perimeter of Q?

4 a Draw a grid with *x*- and *y*-axes from 0 to 13. Plot the points (1, 3) (4, 4) (2, 1). Join them to make triangle A.
   b Enlarge triangle A by scale factor 3, centre (0, 0). Label the image B.
   c Enlarge triangle A by scale factor 2, centre (2, 1). Label the image C.
   d How much larger is the perimeter of B than the perimeter of A?

DID YOU KNOW?

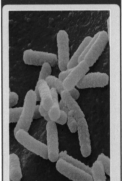

The E. coli. bacteria in this picture has been enlarged by a scale factor of 5000.

Unit 3

249

This spread will show you how to:

- Enlarge objects, given a centre of enlargement and scale factor, including fractional scale factors

Keywords
Enlargement
Scale factor

A map or a scale drawing is an **enlargement** by a fractional **scale factor**.

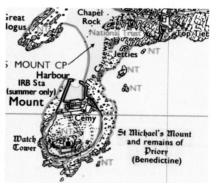

This map is an enlargement of St Michael's Mount by scale factor $\frac{1}{25\,000}$.

- Enlargement by a scale factor less than 1 produces a smaller image.

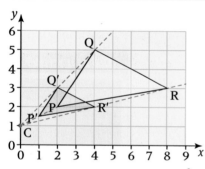

In the diagram $\triangle PQR$ is enlarged by scale factor $\frac{1}{2}$, centre $(0, 1)$.

All the distances are multiplied by $\frac{1}{2}$ so $CQ' = \frac{1}{2}CQ$, $CP' = \frac{1}{2}CP$, $CR' = \frac{1}{2}CR$.

Lengths on the image are half the corresponding lengths on the object so $P'R' = \frac{1}{2}PR$ and so on.

Enlargement by scale factor $\frac{1}{2}$ is the inverse of enlargement by scale factor 2.

Example

Enlarge triangle A by scale factor $\frac{1}{2}$, centre $(-4, 2)$.

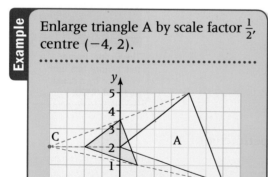

Example

Enlarge triangle B by scale factor $\frac{1}{3}$ centre $(-2, -2)$.

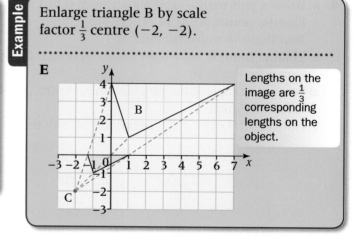

Lengths on the image are $\frac{1}{3}$ corresponding lengths on the object.

1 a Copy this diagram.
  b Enlarge triangle T by scale factor $\frac{1}{2}$, centre (0, 0). Label the image U.
  c Enlarge triangle T by scale factor $\frac{1}{2}$, centre (2, 2). Label the image V.

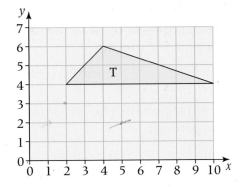

2 a Copy this diagram.
  b Enlarge kite K by scale factor $\frac{1}{3}$, centre (0, 0). Label the image L.
  c Enlarge kite K by scale factor $\frac{1}{3}$, centre (3, 3). Label the image M.

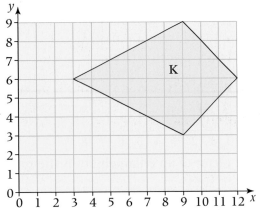

3 a Draw a grid with an x-axis from -3 to 5 and a y-axis from -2 to 10.
  Plot the points (1, 3) (1, 6) (3, 9) (4, 6).
  Join them to make quadrilateral Q.
  b Enlarge quadrilateral Q by scale factor $\frac{1}{3}$, centre (4, 3).
  Label the image R.
  c Enlarge quadrilateral Q by scale factor $\frac{1}{2}$, centre $(-2, -2)$.
  Label the image S.

4 a Draw a grid with x-axis from -6 to 6 and y-axis from 0 to 11.
  Plot the points $(-6, 6)$ (0, 6) (3, 3).
  Join them to make triangle A.
  b Enlarge triangle A by scale factor $\frac{1}{3}$, centre (0, 0).
  Label the image B.
  c Enlarge triangle A by scale factor $\frac{1}{2}$, centre (0, 10).
  Label the image C.

Unit 3

Describing an enlargement

This spread will show you how to:

- Describe an enlargement by giving the scale factor and centre of enlargement
- Understand, identify and use scale factors

To describe an **enlargement** you give the **scale factor** and the **centre of enlargement**.

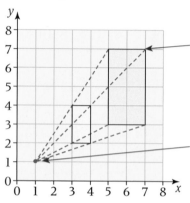

Construction lines join corresponding vertices on the object and image.

The construction lines meet at the centre of enlargement.

This is an enlargement of scale factor $\frac{1}{2}$, centre (1, 1).

The scale factor of an enlargement is the ratio of corresponding sides.

- Scale factor = $\dfrac{\text{length of image}}{\text{length of original}}$

You can write this as a ratio length of image : length of object.

Example

Describe fully the single transformation that maps triangle PQR onto P′Q′R′.

The construction lines meet at (−3, −2)

Scale factor = $\frac{P'R'}{PR}$ = $\frac{6}{2}$ = 3.

The transformation that maps PQR onto P′Q′R′ is an enlargement, centre (−3, −2), scale factor 3.

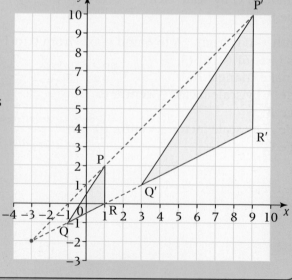

1 Describe fully the single transformation that maps
  a triangle P onto triangle Q
  b triangle Q onto triangle P.

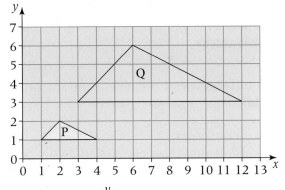

2 a Describe fully the single transformation that maps
  i rectangle R onto rectangle S
  ii rectangle S onto rectangle R.
  b The perimeter of rectangle R is 8 units. What is the perimeter of rectangle S?

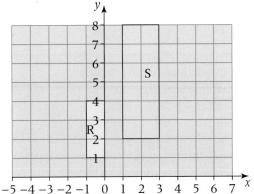

3 Describe fully the single transformation that maps
  a triangle W onto triangle X
  b triangle X onto triangle W.

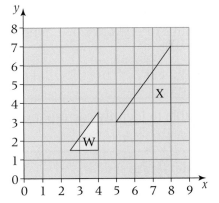

4 a Describe fully the single transformation that maps
  i triangle B onto triangle A
  ii triangle B onto triangle C
  iii triangle A onto triangle C.
  b How many times bigger is the perimeter of triangle C than triangle A?

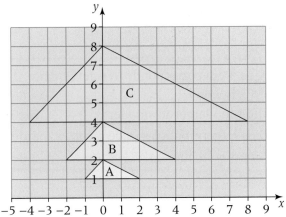

Similar shapes

This spread will show you how to:

● Understand similarity of 2-D shapes, using this to find missing lengths and angles

Keywords
Enlargement
Ratio
Similar

In an **enlargement** the object and the image are mathematically **similar**.

● In similar shapes
 – corresponding pairs of angles are equal
 – corresponding pairs of sides are in the same ratio.

Any two circles are similar.
Any two squares are similar.

You can use the **ratio** between similar shapes to find missing side lengths.

Example

These quadrilaterals are similar. Find the side lengths s and t.

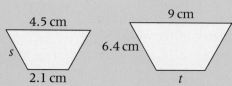

....

Side 4.5 cm corresponds to side 9 cm.
As a fraction,

the ratio larger to smaller is $\frac{9}{4.5} = 2$ the ratio smaller to larger is $\frac{4.5}{9} = \frac{1}{2}$

Side s corresponds to side 6.4 cm.   Side t corresponds to side 2.1 cm.

$\frac{t}{6.4} = \frac{1}{2}$ $\frac{t}{2.1} = 2$

$s = \frac{1}{2} \times 6.4 = 3.2$ cm $t = 2.1 \times 2 = 4.2$ cm

Choose the ratio that gives a fraction with the side you want on top.

You need to be able to identify corresponding sides when one shape is upside down.

Example

These quadrilaterals are similar. Find the side lengths x and y.

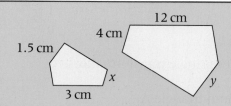

....

Side 3 cm corresponds to side 12 cm.
As a fraction,

the ratio larger to smaller is $\frac{3}{12} = \frac{1}{4}$ the ratio smaller to larger is $\frac{12}{3} = 4$

Side x corresponds to side 4 cm.   Side y corresponds to side 1.5 cm.

Side x is smaller. $4 \times \frac{1}{4} = 1$ Side y is larger. $1.5 \times 4 = 6$

$x = 1$ cm $y = 6$ cm

1 These two trapeziums are similar.
Find the lengths *a* and *b*

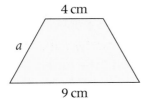
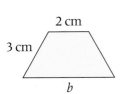

4 cm

2 cm

3 cm

a

b

9 cm

2 These two quadrilaterals are similar.
Find the lengths *c* and *d*

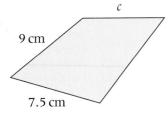

c

9 cm

2 cm

3 cm

7.5 cm

d

3 These two pentagons are similar.
Find the lengths *e* and *f*.

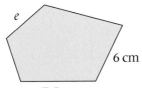

2 cm

e

f

3 cm

6 cm

7.5 cm

4 These two parallelograms are similar.
Find the perimeter of the smaller parallelogram.

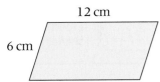

12 cm

5 cm

6 cm

5 These two isosceles triangles are similar.
  a Write the ratio larger to smaller
    for these triangles.
  b Find the length *x*.
  c Find the perimeter of each triangle.
  d Write the ratio of their perimeters.
    What do you notice?

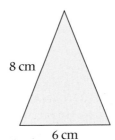

8 cm

x

6 cm

4 cm

Similar triangles

This spread will show you how to:

- Understand similarity of 2-D shapes, using this to find missing lengths and angles
- Use common sense to check answers to geometric problems

Keywords
Ratio
Similar

Two shapes are **similar** if one is the enlargement of the other.

- In similar triangles
  - corresponding pairs of angles are equal
  - corresponding pairs of sides are in the same **ratio**.

If a line is drawn parallel to one side of a triangle, the smaller triangle and the larger triangle are similar.

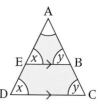

Corresponding angles are equal (using properties of angles in parallel lines), so the triangles ABE and ACD are similar.

p.88

Example

Find the length CD in this triangle.

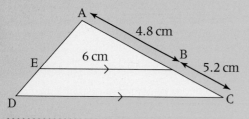

Triangle ACD is similar to triangle ABE.

$$\frac{CD}{BE} = \frac{AC}{AB}$$
$$\frac{CD}{6} = \frac{10}{4.8}$$
$$CD = \frac{6 \times 10}{4.8} = 12.5 \text{ cm}$$

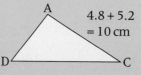

4.8 + 5.2 = 10 cm

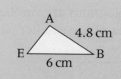

4.8 cm

6 cm

CD corresponds to BE, and AC corresponds to AB.

Check that your answer makes sense. Did you expect a longer or shorter length?

Example

Find the length QS in this triangle.

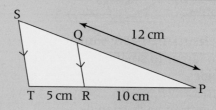

Triangle PQR is similar to triangle PST.

$$\frac{PS}{PQ} = \frac{PT}{PR}$$
$$\frac{PS}{12} = \frac{15}{10}$$
$$PS = \frac{12 \times 15}{10}$$
$$PS = 18 \text{ CM}$$
$$QS = PS - PQ$$
$$= 18 - 12 = 6 \text{ cm}$$

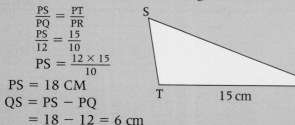

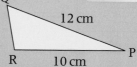

PQ corresponds to PS and PR corresponds to PT.

1 In the diagram,
BE is parallel to CD.
Find the lengths
a AB
b AD

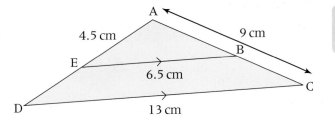

Start by
sketching the
two triangles.

2 In the diagram, QT is parallel to RS.
Find the lengths
a PT
b QR

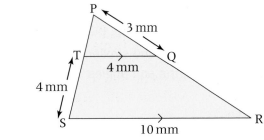

3 In the diagram, VZ is parallel to WY.
Find the lengths
a WY
b VW
c VX

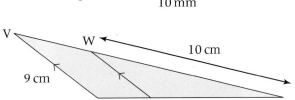

4 In the diagram, KL is parallel to NM.
a Find the lengths MN and JN.
b Find the perimeter of the
trapezium KLMN.

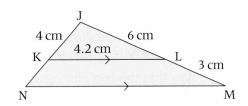

5 EB is parallel to DC.
Work out the perimeter of
a triangle ABE
b trapezium EBCD.

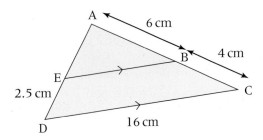

Check out

You should now be able to:

- Describe and transform 2-D shapes using enlargements
- Enlarge 2-D shapes using a centre of enlargement and a positive, fractional or negative scale factor
- Identify shapes that are similar
- Understand similarity of 2-D shapes, using this to find missing lengths and angles

Worked exam question

 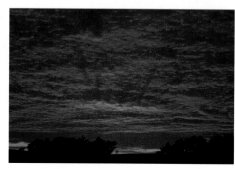

Photographs NOT accurately drawn

The smaller photograph measures 4 inches by 6 inches.
The larger photograph measures 8 inches by 10 inches.

Show that the two photographs are not mathematically similar.   (3)

$$\frac{10}{6} = 1.667 \text{ and } \frac{8}{4} = 2$$

OR

$$\frac{6}{10} = 0.6 \text{ and } \frac{4}{8} = 0.5$$

OR

$$\frac{4}{6} = 0.667 \text{ and } \frac{8}{10} = 0.8$$

OR

$$\frac{6}{4} = 1.5 \text{ and } \frac{10}{8} = 1.25$$

The scale factors are different.

> Two calculations need to be shown whichever method is chosen.

> You need to convert the fractions to decimals to calculate the scale factors.

Exam questions

1

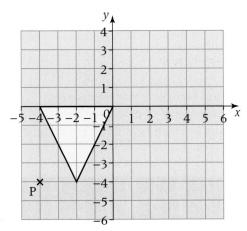

On an accurate copy, enlarge the shaded triangle by scale factor $1\frac{1}{2}$, centre P. (3)

(Edexcel Limited 2004)

2 The diagram shows two quadrilaterals that are mathematically similar.

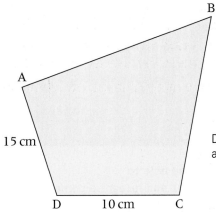

Diagram NOT accurately drawn.

In quadrilateral *PQRS*, *PQ* = 8 cm, *SR* = 4 cm.
In quadrilateral *ABCD*, *AD* = 15 cm, *DC* = 10 cm.
Angle *PSR* = angle *ADC*.
Angle *SPQ* = angle *DAB*.

a Calculate the length of *AB*. (2)

b Calculate the length of *PS*. (2)

(Edexcel Limited 2007)

Factors and powers

When you use the internet to pay for goods you need to know that your financial details are safe. To make these details secure they are turned into a secret code, encrypted, using the product of two very large prime numbers. Whilst the product is publicly known, the difficulty in factorising it makes the message safe.

What's the point?

Prime numbers are those numbers that only have two factors: 1 and themselves. Since all numbers can be broken down into a unique product of prime numbers, mathematicians believe that studying prime numbers will lead to a better understanding of all numbers.

Check in

You should be able to

■ **find factors and recognise prime numbers**

1   **a** Write all of the factors of these numbers.

   i 6   **ii** 12   **iii** 28   **iv** 36

   b Write all of the prime numbers between 1 and 50.

■ **write and interpret numbers given in power form**

2   Write these multiplications in power form.

   For example, $4 \times 4 \times 4 = 4^3$

   a 3×3 **b** $4 \times 4 \times 4 \times 4 \times 4$ **c** $6 \times 6 \times 6$ **d** $5 \times 5 \times 5 \times 5$

3   Write these powers of numbers as multiplications in expanded form.

   For example, $5^3 = 5 \times 5 \times 5$

   a 3^3 **b** 6^2 **c** 4^5 **d** 8^3 **e** 7^4 **f** 5^5

4   Evaluate

   a 2^4 **b** 3^3 **c** 4^2 **d** 5^3 **e** 2^7 **f** 7^1

Orientation

What I need to know	What I will learn	What this leads to
KS3 Recognise prime numbers	■ Find prime factorisations, the HCF and LCM ■ Order fractions ■ Use indices and the index laws ■ Write numbers in standard form	**A-level** Maths, Sciences
N1 Identify significant figures **N2** Use fractions **A1** Use indices		Cryptography

Rich task

Throughout history people have tried to find a mathematical formula that will generate prime numbers. Here are some of the formulae that have been used.

$6n + 1$

$6n - 1$

$n^2 + n + 41$

$2^n - 1$

Investigate these formulae and see if they really work.

Factors and primes

This spread will show you how to:

• Use the concepts of factors, prime numbers and prime factor decomposition

Keywords
Factor
Prime
Prime factor
  decomposition
Product

You can write a number as a **product** of **factors** in different ways.

• A **prime** number is a number with only two factors – itself and 1.
$17 = 17 \times 1$

This means 1 is not a prime number

• Any number greater than 1 can be written as a unique product of its prime factors. This is called the **prime factor decomposition** of the number.
$56 = 2 \times 2 \times 2 \times 7 = 2^3 \times 7$

The factor tree method is a good way to find factors systematically.

Example

Express 84 and 112 as a product of their prime factors.

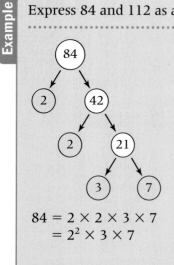

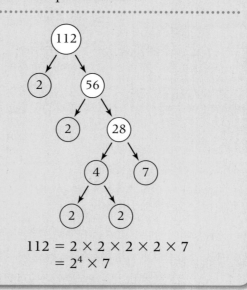

$84 = 2 \times 2 \times 3 \times 7$
$\quad = 2^2 \times 3 \times 7$

$112 = 2 \times 2 \times 2 \times 2 \times 7$
$\quad = 2^4 \times 7$

There may be more than one way of drawing the factor tree.

Example

Find the prime factor decomposition of 990.

$990 = 2 \times 3^2 \times 5 \times 11$

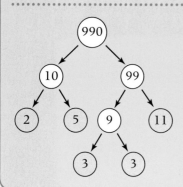

Examiner's tip
A factor tree helps you keep track of all the prime factors, even if you don't find them in ascending order.

1 Copy and complete these calculations to show the different ways that 24 can be written as a product of its factors.
   a 24 = □ × 2 **b** 24 = 3 × □   **c** 24 = 2 × 3 × □ **d** 24 = 4 × □

2 Each of these numbers has just two prime factors, which are not repeated. Write each number as the product of its prime factors.
   a 77   **b** 51   **c** 65   **d** 91   **e** 119   **f** 221

3 Copy and complete the factor tree to find the prime factor decomposition of 18.

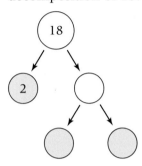

4 For each number, find its prime factors and write it as the product of powers of its prime factors.
   a 36   **b** 120   **c** 34   **d** 25
   e 48   **f** 90   **g** 27   **h** 60

5 Write the prime factor decomposition for each of these numbers.
   a 1052   **b** 2560   **c** 630   **d** 825
   e 715   **f** 1001   **g** 219   **h** 289
   i 2840   **j** 2695   **k** 1729   **l** 3366
   m 9724   **n** 11 830   **o** 2852   **p** 10 179

A03 Problem

6 A cuboid is made from 210 small cubes.
The prime factor decomposition of 210 is 2 × 3 × 5 × 7.
One way of combining the prime factors is (2 × 3) × 5 × 7 = 6 × 5 × 7. The dimensions of the cuboid could be 6 × 5 × 7.

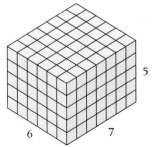

   a Find all the ways in which the prime factors of 210 can be combined to make 3 factors.
   b Using your answer to part **a**, list the dimensions of all the cuboids that could be made with 210 cubes.

Using prime factors: HCF and LCM

This spread will show you how to:

- Find the highest common factor and lowest common multiple of two numbers

Keywords

HCF (highest common factor)
LCM (least common multiple)
Multiple
Prime
Venn diagram

- The **highest common factor (HCF)** of two numbers is the largest number that is a factor of both of them.

    The HCF of 12 and 18 is 6.

- The **least common multiple (LCM)** of two numbers is the smallest number that is a **multiple** of both of them.

    The LCM of 12 and 18 is 36.

To find the HCF of two numbers:
- list all the factors of each number
- identify the largest factor that is in both lists.

    For example:
    Factors of 18 = {1, 2, 3, **6**, 9, 18}
    Factors of 24 = {1, 2, 3, 4, **6**, 8, 12, 24}
    HCF of 18 and 24 = 6

To find the LCM of two numbers:
- list the first few multiples of each number
- identify the smallest multiple that is in both lists.

    For example:
    Multiples of 18 = 18, 36, 54, **72**, 90, ...
    Multiples of 24 = 24, 48, **72**, 96, ...
    LCM of 18 and 24 = 72

You can find the HCF and LCM by writing the **prime** factor decomposition for each number in a **Venn diagram**.

Note that there are other methods for finding LCM and HCF.

Example

Find the HCF and LCM of 36 and 28.

Find the prime factor decomposition of each number: $36 = 2^2 \times 3^2$ and $28 = 2^2 \times 7$.

prime factors of 36   prime factors of 28

3   2   7
3   2

Write these prime factors in a Venn Diagram.

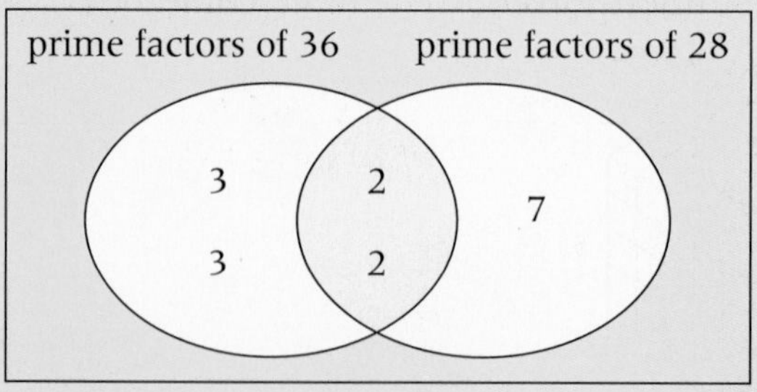

The common factors are in the middle (the 'intersection').

The HCF of 36 and 28 is the product of the numbers in the intersection: $2 \times 2 = 4$.

The LCM is the product of all of the numbers in the diagram: $2^2 \times 3^2 \times 7 = 4 \times 9 \times 7 = 252$

1 The diagram shows how a student listed all the factors of 36.
The lines show how the pairs of factors combine to make 36.

Factors of 36 = 1, 2, 3, 4, 6, 9, 12, 18, 36

  a Copy the diagram, and explain why the factor 6 has not been
    joined to another factor.

  b List the factors of 48 in the same way. Draw lines to show the
    factor pairs.

  c The highest common factor (HCF) of 36 and 48 is the largest
    number that is in both lists. Write down the HCF of 36 and 48.

2 Using the method from question **1**, find the HCF of these pairs
of numbers.

  a 7 and 8   **b** 4 and 5   **c** 6 and 9

  d 14 and 32   **e** 8 and 24   **f** 50 and 70

3 The diagram below shows how a student found the least common
multiple (LCM) of 8 and 6.

multiples of 8 = 8, 16, 24, 32, 40, 48 . . .
multiples of 6 = 6, 12, 18, 24, 30, 36 . . .

  a List the multiples of 12 and 9 in the same way.

  b The least common multiple (LCM) of 12 and 9 is the smallest
    number that is in both lists.
    Write the LCM of 12 and 9.

4 Using the method from question **3**, find the LCM of these pairs
of numbers:

  a 4 and 5   **b** 12 and 18   **c** 5 and 30

  d 12 and 30   **e** 14 and 35   **f** 8 and 20

5 Use the Venn diagram method to find the HCF of 24 and 80.

6 Use the Venn diagram method to work out the HCF and LCM
of these pairs of numbers.

  a 25 and 120 **b** 16 and 108   **c** 60 and 144

  d 42 and 56   **e** 15 and 35   **f** 20 and 110

DID YOU KNOW?

In 1880, John Venn,
an English logician,
introduced the logic
diagrams that now bear
his name.

Ordering fractions

This spread will show you how to:

● Compare and order fractions by rewriting them with a common denominator

Keywords

Ascending
Common
  denominator
Denominator
Descending
Least common
  multiple (LCM)
Numerator

Which fraction is bigger $\frac{2}{5}$ or $\frac{5}{12}$?

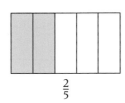

$$\frac{2}{5}$$

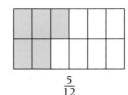

$$\frac{5}{12}$$

$$\frac{2}{5} \overset{\times 12}{\underset{\times 12}{=}} \frac{24}{60}$$

$$\frac{5}{12} \overset{\times 5}{\underset{\times 5}{=}} \frac{25}{60}$$

So $\frac{5}{12}$ is bigger than $\frac{2}{5}$.

● In the fraction $\frac{2}{5}$

- The top number, 2, is the **numerator**.
- The bottom number, 5, is the **denominator**.
- The **common denominator** of $\frac{2}{5}$ and $\frac{5}{12}$ is 60, the **LCM** of 5 and 12 (the original denominators).

The **least common multiple** (LCM) is the lowest number that two (or more) numbers will divide into exactly.

To compare two or more fractions:
● Find the common denominator.
● Work out the equivalent fractions.
● Compare the new numerators to write the fractions in ascending or descending order.

Ascending – going up
Descending – going down

Example

Write in ascending order $\frac{7}{8}$ $\frac{5}{6}$ $\frac{3}{4}$

....

$8 = 2^3$
$6 = 2 \times 3$
$4 = 2^2$

LCM of 8, 6 and 4 $= 2^3 \times 3$
$\qquad\qquad\qquad\quad = 24$

$\frac{7}{8} = \frac{21}{24}$ Multiply numerator and denominator by 3.

$\frac{5}{6} = \frac{20}{24}$ Multiply numerator and denominator by 4.

$\frac{3}{4} = \frac{18}{24}$ Multiply numerator and denominator by 6.

$\frac{18}{24} < \frac{20}{24} < \frac{21}{24}$

In ascending order the fractions are: $\frac{3}{4}, \frac{5}{6}, \frac{7}{8}$

1 Write the least common multiple of each pair of numbers.
   a 2 and 4   **b** 2 and 5   **c** 3 and 8   **d** 4 and 6
   e 4 and 10   **f** 7 and 5   **g** 15 and 20   **h** 20 and 30

2 The diagram shows that $\frac{1}{2} = \frac{2}{4}$
Draw diagrams to show that

   a $\frac{1}{2} = \frac{3}{6}$ **b** $\frac{2}{3} = \frac{4}{6}$

   c $\frac{3}{5} = \frac{9}{15}$ **d** $\frac{3}{4} = \frac{15}{20}$

3 Write each fraction as an equivalent fraction with a denominator of 60.
   a $\frac{1}{3}$ **b** $\frac{1}{4}$ **c** $\frac{2}{3}$ **d** $\frac{2}{5}$

4 Write each fraction as an equivalent fraction with a denominator of 24.
   a $\frac{3}{4}$ **b** $\frac{1}{3}$ **c** $\frac{3}{8}$ **d** $\frac{5}{12}$

5 Rewrite each fraction with the denominator shown.
   a $\frac{2}{3} = \frac{}{30}$ **b** $\frac{3}{7} = \frac{}{42}$ **c** $\frac{7}{9} = \frac{}{45}$ **d** $\frac{5}{8} = \frac{}{40}$

6 Write the fractions in each pair as equivalent fractions with a common denominator. Say which fraction in each pair is larger.
   a $\frac{1}{5}$ and $\frac{3}{10}$ **b** $\frac{2}{3}$ and $\frac{3}{4}$ **c** $\frac{2}{5}$ and $\frac{1}{3}$ **d** $\frac{7}{10}$ and $\frac{2}{3}$

7 Draw diagrams (like the ones in question **2**) to illustrate your answers to question 6.

8 Find the least common multiple of each set of numbers.
   a 2, 3 and 5   **b** 3, 4 and 6   **c** 2, 3 and 8   **d** 2, 4 and 7

9 Rewrite each set of fractions with a common denominator.
   a $\frac{1}{2}, \frac{2}{3}$ and $\frac{3}{4}$ **b** $\frac{1}{5}, \frac{3}{4}$ and $\frac{7}{20}$ **c** $\frac{1}{8}, \frac{7}{12}$ and $\frac{2}{3}$ **d** $\frac{2}{3}, \frac{3}{4}$ and $\frac{2}{7}$

10 Write each set of fractions in ascending order. Show your working.
   a $\frac{2}{3}, \frac{1}{5}$ and $\frac{2}{15}$ **b** $\frac{1}{4}, \frac{2}{5}$ and $\frac{7}{20}$ **c** $\frac{3}{7}, \frac{3}{8}$ and $\frac{5}{14}$ **d** $\frac{2}{3}, \frac{5}{6}$ and $\frac{2}{7}$

11 Write each set of fractions in descending order. Show your working.
   a $\frac{2}{5}, \frac{1}{2}, \frac{3}{10}$ and $\frac{1}{4}$ **b** $\frac{1}{4}, \frac{3}{20}, \frac{4}{5}$ and $\frac{1}{10}$

   c $\frac{2}{5}, \frac{3}{8}, \frac{3}{4}$ and $\frac{17}{40}$ **d** $\frac{5}{6}, \frac{11}{24}, \frac{7}{12}$ and $\frac{5}{8}$

Powers and indices

This spread will show you how to:

- Understand and use powers and index notation
- Find the highest common factor and lowest common multiple of two numbers

Keywords
HCF
Index
LCM
Power

- Repeated multiplications such as $2 \times 2 \times 2 \times 2$ can be written in **index** notation as 2^4.

You read 2^4 as 'two to the **power** 4'.
- You use powers when factorising, for example
$$24 = 2 \times 2 \times 2 \times 3 = 2^3 \times 3$$

But note that 2^2 is 'two squared', and 2^3 is 'two cubed'.

- When powers of the same number are multiplied together, you can find the answer by adding the indices.

For example
$$2^2 \times 2^3 = (2 \times 2) \times (2 \times 2 \times 2) = 2^5$$
$$\text{Similarly, } 3^5 \times 3^6 = 3^{(5 + 6)} = 3^{11}.$$

In 3^{11}, 11 is the index. The plural of index is indices.

Example

Find the value of **a** 2^3 **b** 3^2 **c** 5^3 **d** 10^4 **e** 2^8

....

a $2^3 = 2 \times 2 \times 2 = 8$ **b** $3^2 = 3 \times 3 = 9$ **c** $5^3 = 5 \times 5 \times 5 = 125$
d $10^4 = 10 \times 10 \times 10 \times 10 = 100 \times 100 = 10\,000$
e $2^8 = 2 \times 2 \times 2 \times 2 \times 2 \times 2 \times 2 \times 2 = 256$

Example

Write **a** 625 as a power of 5 **b** 100 000 as a power of 10
c 48 as a product of prime factors in index form.

....

a $625 = 5 \times 125 = 5 \times 5 \times 25 = 5 \times 5 \times 5 \times 5 = 5^4$
b $100\,000 = 10 \times 10 \times 10 \times 10 \times 10 = 10^5$
c $48 = 2 \times 24 = 2 \times 2 \times 12 = 2 \times 2 \times 2 \times 6 = 2 \times 2 \times 2 \times 2 \times 3 = 2^4 \times 3$

Example

Find, in index form, the values of
a $4^3 \times 4^7$ **b** $5^2 \times 5$ **c** $3^2 \times 3^4 \times 3^3$ **d** $2^2 \times 3^4 \times 2^5 \times 3^3$

....

a $4^3 \times 4^7 = 4^{(3 + 7)} = 4^{10}$ **b** $5^2 \times 5 = (5 \times 5) \times 5 = 5^3$
c $3^2 \times 3^4 \times 3^3 = 3^{(2 + 4 + 3)} = 3^9$
d $2^2 \times 3^4 \times 2^5 \times 3^3 = 2^{(2 + 5)} \times 3^{(4 + 3)} = 2^7 \times 3^7$

HCF is the highest number that is a factor of both 84 and 128.

Example

Find the **HCF** and **LCM** of 84 and 280.

....

$84 = 2 \times 2 \times 3 \times 7 = 2^2 \times 3 \times 7$
$280 = 2 \times 2 \times 2 \times 5 \times 7 = 2^3 \times 5 \times 7$
$\text{HCF} = 2^2 \times 7 = 28$
$\text{LCM} = 2^3 \times 3 \times 5 \times 7 = 840$

LCM is the lowest number that both 84 and 280 will divide into exactly.

1 Write these expressions in index form.

a 3×3 **b** $2 \times 2 \times 2$ **c** $3 \times 3 \times 3$ **d** $5 \times 5 \times 5 \times 5$

e $7 \times 7 \times 7$ **f** $10 \times 10 \times 10$ **g** $6 \times 6 \times 6 \times 6$ **h** $5 \times 5 \times 5$

2 Write these numbers in product form.
For example, $4^3 = 4 \times 4 \times 4$.

a 3^4 **b** 5^2 **c** 7^4 **d** 10^5 **e** 4^9 **f** 6^3 **g** 2^5 **h** 9^3

3 Find the value of each of these expressions.
For example, $5^3 = 5 \times 5 \times 5 = 125$.

a 4^2 **b** 4^3 **c** 2^5 **d** 10^2 **e** 10^3 **f** 3^3 **g** 2^3 **h** 3^2

4 Copy and complete the table to show the values of powers of 10.

Index form	Product	Value
10^6	$10 \times 10 \times 10 \times 10 \times 10 \times 10$	$1\,000\,000$
10^5		
10^4		
10^3		
10^2		
10^1		

5 Make a table, like the one in question **4**, to show the values of powers of 2 from 2^1 to 2^{10}.

6 Write

a 81 as a power of 9 **b** 125 as a power of 5

c 128 as a power of 2 **d** $100\,000$ as a power of 10

e 81 as a power of 3 **f** 343 as a power of 7

7 Write the answers to these multiplications in index form.

a $3^4 \times 3^2$ **b** $2^8 \times 2^1$ **c** $4^4 \times 4^4$

d $5^2 \times 5^3$ **e** $8^3 \times 8^5$ **f** $3^2 \times 3^5 \times 3^1$

g $2^3 \times 3^2 \times 2^4 \times 3^2$ **h** $5^2 \times 7^1 \times 5^2 \times 7^6$

8 Write each of these numbers as a prime number raised to a power.
For example, $49 = 7^2$.

a 16 **b** 121 **c** 27 **d** 125 **e** 169 **f** 625 **g** 243 **h** 256

9 Write these numbers as products of their prime factors, using index notation. For example
$$72 = 2 \times 36 = 2 \times 2 \times 18 = 2 \times 2 \times 2 \times 9 = 2 \times 2 \times 2 \times 3 \times 3$$
$$= 2^3 \times 3^2.$$

a 52 **b** 36 **c** 50 **d** 24 **e** 18 **f** 48 **g** 60 **h** 144

10 Find the HCF and LCM of

a 64 and 112 **b** 38 and 33

This spread will show you how to:
- Use index laws, including fractional and negative indices

There are rules that you can use when calculating with indices.

In these **index laws**, letters are used to represent numbers.

- Add **indices** when multiplying **powers** of the same number.
$$x^a \times x^b = x^{a+b}$$

- Subtract indices when dividing powers of the same number.
$$x^a \div x^b = x^{a-b}$$

$$5^4 \times 5^3 = 5^{4+3}$$
$$= 5^7$$

$$6^5 \div 6^2 = 6^{5-2}$$
$$= 6^3$$

Using the rule for multiplication, $7^3 \times 7^0 = 7^3$. Since multiplying by 7^0 leaves the 7^3 unchanged, 7^0 must be equal to 1.

- Any number (except 0) to the power 0 is 1: $x^0 = 1$ for any value of x, if $x \neq 0$.

You know that $3^2 \times 3 = (3 \times 3) \times 3 = 3^3$.
You can write this as $3^2 \times 3^1 = 3^{(2+1)} = 3^3$, so $3 = 3^1$.

- Any number to the power 1 is just the number itself: $x^1 = x$ for any value of x.

Example

Simplify these expressions, giving your answers in index form.
a $2^3 \times 2^2$ **b** $5^7 \div 5^3$ **c** $6^4 \times 6^2 \div 6^3$
d $7^2 \times 5^3 \times 7^3 \times 5^4$ **e** $(2^5 \times 3^4) \div (2^3 \times 3^2)$

. . . .

a $2^3 \times 2^2 = 2^{(3+2)}$ **b** $5^7 \div 5^3 = 5^{(7-3)}$
          $= 2^5$ $= 5^4$

c $6^4 \times 6^2 \div 6^3 = 6^{(4+2-3)}$ **d** $7^2 \times 5^3 \times 7^3 \times 5^4 = 7^{(2+3)} \times 5^{(3+4)}$
                $= 6^3$ $= 7^5 \times 5^7$

e $(2^5 \times 3^4) \div (2^3 \times 3^2) = 2^{(5-3)} \times 3^{(4-2)}$
                               $= 2^2 \times 3^2$

Example

Write the value of   **a** 19^0 **b** 8^1 **c** $(16^3 - 81 \times 17)^0$ **d** $(4.8)^1$

. . . .

a $x^0 = 1$ for any value of x, so $19^0 = 1$.
b $x^1 = x$ for any value of x, so $8^1 = 8$.
c There is no need to evaluate the expression in the bracket (except to note that it is none zero.)
The power of 0 means that $(16^3 - 81 \times 17)^0 = 1$.
d $x^1 = x$ for any value of x (including decimal numbers),
so $(4.8)^1 = 4.8$.

1   Write the answers to these multiplications in index form.
    a $7 \times 7 \times 7$ **b** 3×3^2 **c** 5×5^2 **d** $6 \times 6 \times 6^2$
    e $5^3 \div 5$ **f** $8^4 \times 8$ **g** $9^3 \times 9^2 \times 9$ **h** $8^7 \times 8$

2   Simplify these expressions, giving your answers in index form.
    a $6^2 \times 6^3$ **b** $4^5 \times 4^4$ **c** $2^6 \times 2^7$ **d** $11^5 \times 11^2$
    e $1^{17} \times 1^{13}$ **f** $7^8 \times 7^4$ **g** $3^6 \times 3^6$ **h** $9^9 \times 9^1$

3   Simplify these expressions, giving your answers in index form where
    appropriate.
    a $7^8 \div 7^6$ **b** $8^6 \div 8^2$ **c** $3^3 \div 3^2$ **d** $9^{11} \div 9^8$
    e $4^7 \div 4^1$ **f** $2^9 \div 2^9$ **g** $12^8 \div 12^6$ **h** $6^{13} \div 6^{13}$

4   Simplify these expressions, giving your answers in index form.
    a $8^6 \times 8^2 \div 8^3$ **b** $5^7 \times 5^2 \div 5^4$ **c** $2^8 \times 2^3 \div 2^5$ **d** $9^6 \times 9^3 \div 9^7$
    e $8^5 \times 8^5 \div 8^2$ **f** $7^6 \times 7^5 \div 7^4$ **g** $4^6 \times 4^8 \div 4^4$ **h** $11^2 \times 11^2 \div 11^3$

5   Simplify these expressions, giving your answer in index form.
    a $3^4 \times 3^2 \div (3^3 \times 3^2)$ **b** $(5^6 \div 5^2) \times 5^4 \times 5^2$
    c $(4^5 \div 4^2) \div (4^6 \div 4^5)$ **d** $(7^9 \div 7^2) \div (7^2 \times 7^3)$
    e $(8^7 \div 8^4) \times 8^5 \times 8^3$ **f** $9^3 \times (9^5 \div 9^2) \times 9^4$

6   Simplify these expressions, giving your answers in index form.
    a $\dfrac{4^2 \times 4^2}{4^2}$ **b** $\dfrac{6^3 \times 6^4}{6^5}$ **c** $\dfrac{9^8}{9^2 \times 9^4}$ **d** $\dfrac{8^6 \div 8^3}{8^2}$
    e $\dfrac{5^9 \times 5^4}{5^3 \times 5^7}$ **f** $\dfrac{6^3 \times 6^4}{6^5 \div 6^3}$ **g** $\dfrac{8^9 \div 8^2}{8^7 \div 8^2}$ **h** $\dfrac{10^6 \div 10^2}{10^2 \times 10^2}$

7   Simplify these expressions as far as possible, giving your answers in
    index form.
    a $4^2 \times 3^3 \times 4^2$ **b** $8^5 \times 7^2 \div 8^2$ **c** $6^2 \times 5^3 \times 6^2 \times 5^3$
    d $4^5 \times 3^3 \times 3^3 \times 4^4$ **e** $5^4 \times 2^3 \div 5^2$ **f** $9^5 \times 7^2 \times 7^2 \times 9^2$
    g $8^2 \times 5^6 \times 8^3 \div 5^3$ **h** $3^4 \times 8^5 \times 3^4 \times 8^2$ **i** $9^3 \times 2^5 \div 2^3 \times 9^2$

8   Simplify these expressions, giving your answers in index form.
    a $\dfrac{5^2 \times 8^5}{8^2}$ **b** $\dfrac{6^5 \times 7^2}{6^3}$ **c** $\dfrac{6^4 \times 5^4}{6^2 \times 5^2}$ **d** $\dfrac{7^8 \times 5^6}{5^3 \times 7^2}$
    e $\dfrac{8^7 \times 3^5}{3^2 \times 8^5}$ **f** $4^3 \times \dfrac{4^5 \times 5^9}{4^3 \times 5^7}$ **g** $\dfrac{6^9 \times 7^5}{6^7 \times 7^3} \times 6^2$ **h** $4^3 \times \dfrac{7^6 \times 4^5}{4^4 \times 7^3} \times 7^2$

A03 Problem
9   Simplify these expressions, giving your answer in index form.
    a $x^2 \times x^5$ **b** $x^7 \div x^4$ **c** $x^{10} \times x \times x^4$
    d $x^4 \times x^{11} \div x^3$ **e** $x^5 \times x^7 \times y^3 \div y$ **f** $x^6 \times y^3 \times x^9 \times y^2$
    g $\dfrac{y^4 \times x^8}{x^0 \times y}$ **h** $\dfrac{x^3 \times y^7 \times z^4 \div y^2}{z^3 \times x \times y^3}$

This spread will show you how to:

- Use index laws, including fractional and negative indices

Keywords
Fractional
Index laws
Negative
Power
Reciprocal

Indices can be fractions as well as whole numbers.

$$5^{\frac{1}{2}} \times 5^{\frac{1}{2}} = 5^{(\frac{1}{2} + \frac{1}{2})} = 5^1 = 5$$

Since $\sqrt{5} \times \sqrt{5} = 5$, $5^{\frac{1}{2}}$ must represent the square root of 5.

- In general, $x^{\frac{1}{2}} = \sqrt{x}$ for any value of x.
  Similarly, $x^{\frac{1}{3}} = \sqrt[3]{x}$ (the cube root of x), and so on.

A **fractional** index means a root.

$$\frac{1}{5} = 1 \div 5 = 5^0 \div 5^1 = 5^{0-1} = 5^{-1}, \text{ so, } 5^{-1} \text{ means } \frac{1}{5}$$

- In general, x^{-1} is equal to $\frac{1}{x}$, for any value of x. This is the **reciprocal** of x.

$$\frac{1}{5^2} = 1 \div 5^2 = 5^0 \div 5^2 = 5^{0-2} = 5^{-2}, \text{ so, } 5^{-2} \text{ means } \frac{1}{5^2}$$

A **negative** index means a reciprocal.

- In general, x^{-n} means $\frac{1}{x^n}$. This is the reciprocal of x^n.

$$(5^2)^3 = 5^2 \times 5^2 \times 5^2 = 5^{(2+2+2)} = 5^6$$
The indices are multiplied: $(5^2)^3 = 5^{2 \times 3} = 5^6$

Note that zero has no reciprocal as 0^{-1} is not defined.

- In general, when finding a 'power of a power', multiply the indices: $(x^m)^n = x^{mn}$.

You can use these index laws to calculate quantities including powers.

Notice that $8 \times 8^{-1} = 1$. What about 6×6^{-1}? Or 50×50^{-1}?

Example

Evaluate

a $16^{\frac{1}{2}}$ **b** 8^{-1} **c** $(3^3)^2$

....

a $16^{\frac{1}{2}} = \sqrt{16} = 4$ **b** $8^{-1} = \frac{1}{8}$ **c** $(3^3)^2 = 27^2 = 729$

Example

Write these expressions as powers of the numbers indicated.

a 2 as a power of 4   **b** 0.125 as a power of 8   **c** $\frac{1}{16}$ as a power of 2

....

a $2 = \sqrt{4} = 4^{\frac{1}{2}}$ **b** $0.125 = \frac{125}{1000} = \frac{1}{8} = 8^{-1}$ **c** $\frac{1}{16} = \frac{1}{2^4} = 2^{-4}$

Example

Evaluate

a 10^{-4} **b** $2^{-\frac{1}{2}}$ **c** $100^{-\frac{1}{2}}$

....

a $10^{-4} = \frac{1}{10\,000} = 0.0001$ **b** $2^{-\frac{1}{2}} = \frac{1}{2^{\frac{1}{2}}} = \frac{1}{\sqrt{2}} = 0.707$

c $100^{-\frac{1}{2}} = \frac{1}{\sqrt{100}} = \frac{1}{10} = 0.1$ (by calculator)

1 Evaluate

　a 5^1　　　　**b** 6^1　　　　　**c** 6^0　　　　　**d** 7^0

　e $(4 + 88^2)^0$　**f** $(4^2 + 5^2)^1$　**g** $(92.5)^0$　　**h** 0^1

2 Evaluate

　a $100^{\frac{1}{2}}$　　**b** $16^{0.5}$　　**c** $49^{\frac{1}{2}}$　　**d** $4^{0.5}$

　e $64^{\frac{1}{2}}$　　**f** $9^{0.5}$　　**g** $121^{\frac{1}{2}}$　**h** $64^{0.5}$

　i $144^{\frac{1}{2}}$　　**j** $8^{\frac{1}{3}}$　　**k** $27^{\frac{1}{3}}$　　**l** $100^{0.5}$

3 Evaluate

　a $36^{\frac{1}{2}}$　　**b** 36^1　　**c** $81^{0.5}$　　**d** 81^1

4 Write these numbers in index form. For example, $\frac{1}{2} = 2^{-1}$.

　a $\frac{1}{3}$　　　**b** $\frac{1}{5}$　　　**c** $\frac{1}{7}$　　　**d** $\frac{1}{11}$

　e 0.5　　**f** 0.2　　**g** 0.1　　**h** $0.\dot{3}$

> Remember that x^{-1} is the reciprocal of x, and vice versa.

5 Evaluate

　a 9^{-1}　　**b** 9^0　　　**c** $9^{0.5}$　　**d** 9^1

　e 9^2　　**f** 9^3　　**g** 9^5　　**h** 9^{-3}

6 Write these expressions in index form. For example, $\frac{1}{5^2} = 5^{-2}$.

　a $\frac{1}{7^2}$　　**b** $\frac{1}{9^2}$　　**c** $\frac{1}{2^2}$　　**d** $\frac{1}{2^3}$

　e $\frac{1}{2^5}$　　**f** $\frac{1}{3^4}$　　**g** $\frac{1}{5^3}$　　**h** $\frac{1}{6^4}$

7 Write these expressions in fraction form. For example, $7^{-3} = \frac{1}{7^3}$.

　a 8^{-2}　　**b** 7^{-3}　　**c** 5^{-2}　　**d** 9^{-4}

　e 3^{-2}　　**f** 9^{-3}　　**g** 4^{-5}　　**h** 6^{-6}

8 Evaluate

　a 4^{-2}　　**b** 4^{-1}　　**c** 4^0　　　**d** $4^{0.5}$

　e 4^1　　**f** 4^2　　**g** 4^3　　**h** $4^{-0.5}$

9 Write these expressions in index form. For example, $\frac{1}{\sqrt{2}} = \frac{1}{2^{\frac{1}{2}}} = 2^{-\frac{1}{2}}$.

　a $\frac{1}{\sqrt{3}}$　　**b** $\frac{1}{\sqrt{5}}$　　**c** $\frac{1}{\sqrt{7}}$　　**d** $\frac{1}{\sqrt{11}}$

10 Simplify these expressions. For example, $(5^2)^3 = 5^{2 \times 3} = 5^6$.

　a $(2^2)^2$　　**b** $(2^3)^2$　　**c** $(3^2)^3$　　**d** $(4^{0.5})^2$

　e $(5^2)^4$　　**f** $(4^{-2})^3$　　**g** $(7^2)^6$　　**h** $(5^2)^{-2}$

11 Evaluate these expressions, giving your answers in index form.

　a $2^2 \div 2^4$　　**b** $3^5 \div 3^6$　　**c** $4^3 \div 4^9$

　d $(3^4 \times 5^5) \div (3^5 \times 5^6)$　　**e** $[(5^4 \times 7^3) \div (5^2 \times 7^5)]^2$

This spread will show you how to:

● Understand and use standard form in calculations with large and small numbers

You can use **standard form** to represent large and small numbers.

● In standard form, a number is written as $A \times 10^n$.
  - A is a number between 1 and 10 (but not including 10). Using algebra, $1 \leqslant A < 10$.
  - The value of n is an integer.

  For example, $856 = 8.56 \times 10^2$ and $0.00312 = 3.12 \times 10^{-3}$.

13×10^5 is *not* in standard form, because 13 is larger than 10.
0.75×10^4 is *not* in standard form, because 0.75 is less than 1.

The correct version is 1.3×10^6.

The correct version is 7.5×10^3.

Example

Write these numbers in standard form.
a 235   **b** 0.23×10^6 **c** 0.45   **d** 0.000 000 416

....

a $235 = 2.35 \times 10^2$ **b** $0.23 \times 10^6 = 2.3 \times 10^5$
c 4.5×10^{-1} **d** 4.16×10^{-7}

You can calculate with numbers in standard form.
● Multiplication works like this:
  $(3 \times 10^5) \times (4 \times 10^3) = (3 \times 4) \times 10^{(5+3)} = 12 \times 10^8 = 1.2 \times 10^9$
  $(2.5 \times 10^{-1}) \times (5 \times 10^4) = (2.5 \times 5) \times 10^{-1+4}$
  $= 12.5 \times 10^3 = 1.25 \times 10^4$
● Division works like this:
  $(1.4 \times 10^8) \div (7 \times 10^5) = (1.4 \div 7) \times 10^{(8-5)} = 0.2 \times 10^3 = 2 \times 10^2$
  $(6 \times 10^{-2}) \div (1.2 \times 10^{-4}) = (6 \div 1.2) \times 10^{-2+4} = 5 \times 10^2$

Multiplication – add the indices.

Division – subtract the indices.

Example

Calculate
a $(4.2 \times 10^3) \times (2 \times 10^2)$ **b** $(3.6 \times 10^5) \div (1.2 \times 10^3)$
c $(5.4 \times 10^4) \times (2 \times 10^{-3})$ **d** $(4.8 \times 10^{-6}) \div (1.2 \times 10^{-6})$

....

a $(4.2 \times 10^3) \times (2 \times 10^2) = (4.2 \times 2) \times (10^3 \times 10^2)$
$= 8.4 \times 10^{(3+2)} = 8.4 \times 10^5$

b $(3.6 \times 10^5) \div (1.2 \times 10^3) = (3.6 \div 1.2) \times (10^5 \div 10^3)$
$= 3 \times 10^{(5-3)} = 3 \times 10^2$

c $(5.4 \times 10^4) \times (2 \times 10^{-3}) = (5.4 \times 2) \times (10^4 \times 10^{-3})$
$= 10.8 \times 10^{4-3} = 1.08 \times 10^2$

d $(4.8 \times 10^{-6}) \div (1.2 \times 10^{-6}) = (4.8 \div 1.2) \times (10^{-6} \div 10^{-6})$
$= 4 \times 10^{-6+6} = 4$

1 Write these numbers in standard form.
  a 200   **b** 800   **c** 9000   **d** 650
  e 6500   **f** 952   **g** 23.58   **h** 255.85
  i 0.3   **j** 0.0047 **k** 0.000 078 **l** 0.4485

2 These numbers are in standard form. Write each of them as an 'ordinary' number.
  a 5×10^2 **b** 3×10^3 **c** 1×10^5 **d** 2.5×10^2
  e 4.9×10^3 **f** 3.8×10^6 **g** 7.5×10^{11} **h** 8.1×10^{18}

3 Write these measurements using standard form.
  a One hundredth of a kilometre   **b** Two thousandths of a gram
  c Five millionths of a metre   **d** 11 thousandths of a litre

4 Although they are written as multiples of powers of 10, these numbers are not in standard form. Rewrite each of them correctly in standard form.
  a 60×10^1 **b** 45×10^3 **c** 0.65×10^1 **d** 0.05×10^8
  e 28×10^{-2} **f** 0.4×10^{-1} **g** 13.5×10^{-4} **h** 12×10^{-8}

5 Evaluate these calculations, giving your answers in standard form. Do not use a calculator.
  a $(2 \times 10^2) \times (2 \times 10^3)$ **b** $(3 \times 10^4) \times (3 \times 10^3)$
  c $(5 \times 10^3) \times (5 \times 10^4)$ **d** $(8 \times 10^7) \times (3 \times 10^5)$
  e $(2.5 \times 10^{-3}) \times (2 \times 10^2)$ **f** $(4.6 \times 10^{-6}) \times (2 \times 10^{-2})$
  g $(4 \times 10^4) \div (2 \times 10^6)$ **h** $(8.4 \times 10^{-2}) \div (2 \times 10^6)$

6 Evaluate these calculations, showing your working. Do not use a calculator; give your answers in standard form.
  a $(4 \times 10^4) \div (2 \times 10^2)$ **b** $(8.4 \times 10^9) \div (4.2 \times 10^5)$
  c $(2 \times 10^6) \div (4 \times 10^4)$ **d** $(3 \times 10^5) \div (4 \times 10^2)$

7 Work out these calculations without using a calculator, giving your answers in standard form.
  a $(5 \times 10^{-1}) + (2 \times 10^{-2})$ **b** $(4 \times 10^{-2}) + (6 \times 10^{-3})$
  c $(2 \times 10^{-2}) + (9 \times 10^{-4})$ **d** $(1.5 \times 10^{-2}) - (2 \times 10^{-3})$

You may find it easier to convert the numbers to 'ordinary' numbers first, and then convert the answers back to standard form.

8 The speed of light is approximately 3×10^8 metres per second. Copy and complete the table to show the time taken for light from the Sun to reach the various planets.

Planet	Mean distance from Sun (m)	Light travel time
Mercury	5.79×10^{10}	
Earth	1.50×10^{11}	
Mars	2.28×10^{11}	
Jupiter	7.78×10^{11}	
Pluto	5.90×10^{12}	

Summary

Check out
You should now be able to:

- Identify factors, multiples and prime numbers from a list of numbers
- Find the prime factor decomposition of positive integers
- Find the Highest Common Factor and Least Common Multiple of two or three numbers
- Compare, order and simplify fractions
- Understand and use index notation and index laws, including integer, fractional and negative powers
- Understand and find reciprocals
- Use standard form to represent large and small numbers
- Calculate with numbers written in standard index form

Worked exam question

a Write as a power of 5

   i $5^{10} \div 5^2$

   ii $\dfrac{5^4 \times 5^3}{5}$

   (3)

b Write down the reciprocal of 5   (1)

a

  i $5^{10} \div 5^2 = 5^8$

  ii $5^4 \times 5^3 = 5^7$

    $\dfrac{5^4 \times 5^3}{5} = \dfrac{5^7}{5}$

    $5^7 \div 5 = 5^6$

> Subtract the indices: $10 - 2 = 8$

> Add the indices: $4 + 3 = 7$

> The answers must be a power of 5

b

  $\dfrac{1}{5}$

> The reciprocal of a number is 1 divided by that number.
>
> Either $\frac{1}{5}$ or 0.2 are possible answers.

Exam questions

1 The number 40 can be written as $2^m \times n$, where m and n are prime numbers. Find the value of m and the value of n. (2)

(Edexcel Limited 2005)

2 **a** Express the following numbers as products of their prime factors.
   i 60
   ii 96 (4)
   b Find the Highest Common Factor of 60 and 96. (1)
   c Work out the Lowest Common Multiple of 60 and 96. (2)

(Edexcel Limited 2003)

3 **a** Write 126 as a product of its prime factors. (2)
   b Find the Highest Common Factor (HCF) of 84 and 126 (2)

(Edexcel Limited 2007)

4 Find the Lowest Common Multiple (LCM) of 32 and 48 (2)

5 **a** Find the Highest Common Factor (HCF) of 24, 60 and 108 (2)
   b Work out the Least Common Multiple (LCM) of 8, 16 and 24 (2)

6 **a** Write these five fractions in order of size.
   Start with the smallest fraction.
   $$\frac{3}{4} \qquad \frac{1}{2} \qquad \frac{3}{8} \qquad \frac{2}{3} \qquad \frac{1}{6}$$
   (2)
   b Write these numbers in order of size.
   Start with the smallest number.
   $$65\% \qquad \frac{3}{4} \qquad 0.72 \qquad \frac{2}{3} \qquad \frac{3}{5}$$
   (2)

(Edexcel Limited 2004)

7 Work out the value of
   a $(2^2)^3$ **b** $(\sqrt{3})^2$ **c** $\sqrt{2^4 \times 9}$ (4)

(Edexcel Limited 2003)

8 Find the value of
   a 4^0 **b** $4^{\frac{1}{2}}$ **c** 4^{-1} (4)

9 Write in standard form
   a 456 000 (1)
   b 0.000 34 (1)
   c 16×10^7 (1)

(Edexcel Limited 2006)

10 **a** Write the number 28 000 in standard form. (1)
   b Write 5.42×10^{-4} as an ordinary number. (1)

11 Work out $(3.2 \times 10^5) \times (4.5 \times 10^4)$
   Give your answer in standard form correct to 2 significant figures. (2)

(Edexcel Limited 2005)

Averages and frequency graphs

Over 30% of the numbers in everyday use begin with the digit 1 whilst less that 5% begin with a 9. 'Benford's law', as it is called, makes it possible to detect when a list of numbers has been falsified. This is particularly useful in fraud investigations for detecting 'made-up' entries on claim forms and expense accounts.

What's the point?

One of the tasks statisticians work on is finding ways to display and characterise different data sets using statistical 'fingerprints'. This is so that they can identify and measure any possible differences to decide if they are significant or not.

You should be able to

■ **find quarters of an integer**

1 Calculate.

a $\frac{1}{2}$ of 124   b $\frac{1}{2}$ of 140   c $\frac{1}{4}$ of 240

d $\frac{1}{4}$ of 180   e $\frac{3}{4}$ of 136   f $\frac{3}{4}$ of 144

■ **read and interpret a line graph**

Use this graph showing the cost per day to hire a power tool for questions **2** and **3**.

2 How much does it cost to hire the power tool for

a 3 days   b 5 days?

3 Mike has £40.
What is the maximum number of days he can hire the power tool?

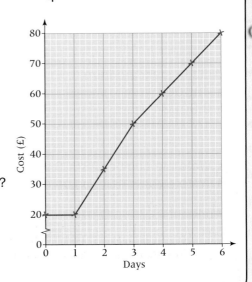

What I need to know	What I will learn	What this leads to
KS3 Plot points and draw graphs	■ Calculate measures of average and spread.	**A-level** Maths, Biology, Economics, Geography
D1 Calculate averages from frequency tables **D2** Calculate quartiles and draw box plots	■ Draw and interpret cumulative frequency diagrams and box plots ■ Compare sets of data	Commerce

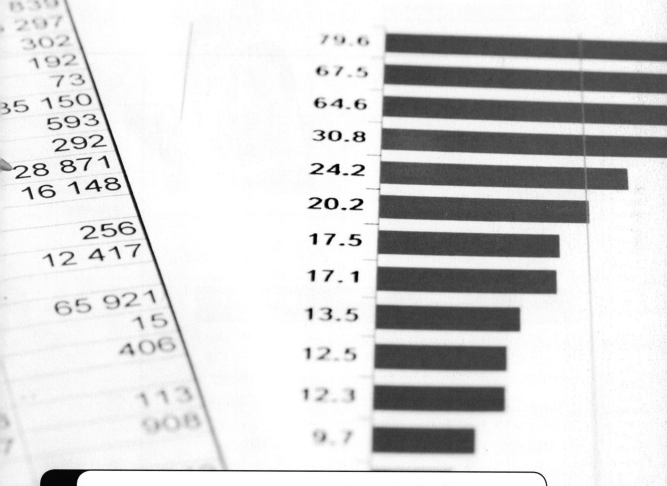

What is the most likely time during a football game for a team to score a goal?

It is frequently stated by football commentators that teams are most likely to concede a goal within five minutes of scoring a goal themselves. Is this true? Investigate and write a report on your results

Averages of grouped data

This spread will show you how to:

• Use grouped frequency tables

Keywords
Estimate
Grouped
  frequency table
Mean
Median
Modal class

p.82

• You can put large amounts of continuous data into a **grouped frequency table**.

A grouped frequency table does not tell you the actual data values so you can only find estimates of the averages.

• You use **estimates** of averages to summarise the data.

The table shows the time taken, to the nearest minute, by a group of students to solve a crossword puzzle.

Time, t, minutes	Frequency
$5 < t \leqslant 10$	2
$10 < t \leqslant 15$	14
$15 < t \leqslant 20$	13
$20 < t \leqslant 25$	6
$25 < t \leqslant 30$	1

For these data, work out

a the **modal class**
b the class containing the **median**
c an estimate for the **mean**

The modal class is the class with the greatest frequency.

a Modal class is $10 < t \leqslant 15$
b Class containing median is $15 < t \leqslant 20$
c

Time t minutes	Frequency	Midpoint	Word length × frequency
$5 < t \leqslant 10$	2	7.5	$7.5 \times 2 = 15$
$10 < t \leqslant 15$	14	12.5	$12.5 \times 14 = 175$
$15 < t \leqslant 20$	13	17.5	$17.5 \times 13 = 227.5$
$20 < t \leqslant 25$	6	22.5	$22.5 \times 6 = 135$
$25 < t \leqslant 30$	1	27.5	$27.5 \times 1 = 27.5$
Total	**36**		**580**

Put two extra columns in the table.

Total number of students   Total time

Find the totals of the Frequency and Word length × frequency columns.

$$\text{Estimated mean} = \frac{\text{Estimated total time}}{\text{Total nubmber of students}}$$

$$\text{Estimated mean} = \frac{580}{36} = 16.1$$

1 The grouped frequency tables give information about the time taken
to solve four different crosswords.
For each table, copy the table, add extra working columns and find
  i the modal class   **ii** the class containing the median
iii an estimate of the mean.

a

Time, t, minutes	Frequency
$5 < t \leqslant 10$	2
$10 < t \leqslant 15$	14
$15 < t \leqslant 20$	13
$20 < t \leqslant 25$	6
$25 < t \leqslant 30$	1

b

Time, t, minutes	Frequency
$0 < t \leqslant 10$	3
$10 < t \leqslant 20$	6
$20 < t \leqslant 30$	4
$30 < t \leqslant 40$	5
$40 < t \leqslant 50$	2

c

Time, t, minutes	Frequency
$5 < t \leqslant 10$	8
$10 < t \leqslant 15$	5
$15 < t \leqslant 20$	7
$20 < t \leqslant 25$	4
$25 < t \leqslant 30$	0
$25 < t \leqslant 35$	1

d

Time, t, minutes	Frequency
$5 < t \leqslant 15$	3
$15 < t \leqslant 25$	9
$25 < t \leqslant 35$	7
$35 < t \leqslant 45$	8
$45 < t \leqslant 55$	2
$45 < t \leqslant 65$	1

2 Alfie kept a record of his monthly mobile phone bills for one year.

Phone bill, b, pounds	Frequency
$10 < b \leqslant 20$	6
$20 < b \leqslant 30$	2
$30 < b \leqslant 40$	3
$40 < b \leqslant 50$	1

a Write the class interval that contains the median.
b Calculate an estimate for the mean cost of Alfie's mobile phone bill.

3 The heights of 50 Year 10 students were measured. The results are
shown in the table.

Height, h, cm	Number of students
$150 \leqslant h < 155$	3
$155 \leqslant h < 160$	5
$160 \leqslant h < 165$	15
$165 \leqslant h < 170$	25
$170 \leqslant h < 175$	2

a What is the modal group?
b Estimate the mean height.
c Which class interval contains the median?

This spread will show you how to:
- Draw frequency polygons
- Use frequency polygons to compare two data sets

Keywords
Class interval
Frequency
  polygon
Midpoint
Modal

- You can represent grouped data in a **frequency polygon**.

To draw a frequency polygon for continuous data you plot the **midpoint** of each **class interval** against the frequency.

Example

The tables show the ages of people attending concerts to see the bands Badness and Cloudplay.

Badness

Age, a, years	Frequency
$20 < a \leqslant 30$	1600
$30 < a \leqslant 40$	4300
$40 < a \leqslant 50$	2100
$50 < a \leqslant 60$	1000

Cloudplay

Age, a, years	Frequency
$10 < a \leqslant 20$	2800
$20 < a \leqslant 30$	4600
$30 < a \leqslant 40$	3300
$40 < a \leqslant 50$	1200

a Draw frequency polygons for these data.
b Make comparisons, with reasons, between the ages of people attending these concerts.

a

Midpoint	25	35	45	55
Frequency	1600	4300	2100	1000

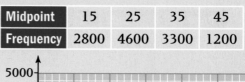

Midpoint	15	25	35	45
Frequency	2800	4600	3300	1200

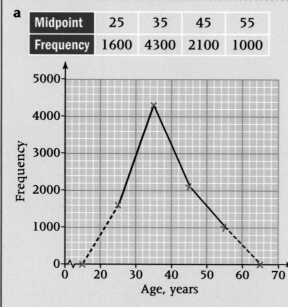

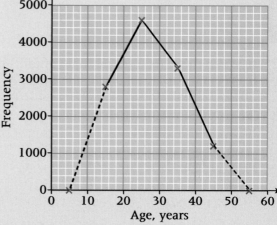

b The **modal** age is greater at Badness concerts than Cloudplay concerts. The highest frequency for Badness is in the class interval 30–40 years old, whereas the highest frequency for Cloudplay is in the interval 20–30 years old.

1 The tables show the ages of the first 100 people to visit a garden centre on a weekday and a Sunday.

Weekday

Age, *a*, years	Frequency
$0 < a \leqslant 20$	16
$20 < a \leqslant 40$	28
$40 < a \leqslant 60$	32
$60 < a \leqslant 80$	24

Sunday

Age, *a*, years	Frequency
$0 < a \leqslant 20$	22
$20 < a \leqslant 40$	45
$40 < a \leqslant 60$	18
$60 < a \leqslant 80$	15

a Draw frequency polygons for these data.
b Find, for each data set, the class which contains the modal age.
c Compare the ages of people at the garden centre on a weekday and a Sunday.

2 Jayne kept a daily record of the number of miles she travelled in her car during two months.

December

Miles travelled, *m*	Frequency
$0 < m \leqslant 20$	3
$20 < m \leqslant 40$	8
$40 < m \leqslant 60$	10
$60 < m \leqslant 80$	6
$80 < m \leqslant 100$	4

January

Miles travelled, *m*	Frequency
$0 < m \leqslant 20$	0
$20 < m \leqslant 40$	5
$40 < m \leqslant 60$	12
$60 < m \leqslant 80$	8
$80 < m \leqslant 100$	6

a Draw frequency polygons for these data.
b Find, for each month, the class which contains the modal number of miles
c Compare the number of miles Jayne travelled in December and January.

3 David carried out a survey to find the time taken by 120 teachers and 120 office workers to travel home from work.

Teachers

Time taken, *t*, minutes	Frequency
$0 < t \leqslant 10$	12
$10 < t \leqslant 20$	33
$20 < t \leqslant 30$	48
$30 < t \leqslant 40$	20
$40 < t \leqslant 50$	7

Office workers

Time taken, *t*, minutes	Frequency
$10 < t \leqslant 20$	2
$20 < t \leqslant 30$	21
$30 < t \leqslant 40$	51
$40 < t \leqslant 50$	28
$50 < t \leqslant 60$	18

a Draw frequency polygons for these data.
b Work out for each data set the class which contains the modal time taken.
c Make comparisons between the time taken by the teachers and office workers to travel home from work.

This spread will show you how to:
- Draw cumulative frequency polygons for grouped data
- Find the modal class of a data set

Keywords
Cumulative frequency
Grouped frequency
Modal class
Upper bound

If you have a large amount of data, you can group the data in a **grouped frequency** table.

- You can represent grouped data on a **cumulative frequency** diagram.

Example

The heights of 120 boys are given in the table.

Height, h, cm	$145 \leqslant h < 150$	$150 \leqslant h < 155$	$155 \leqslant h < 160$	$160 \leqslant h < 165$	$165 \leqslant h < 170$
Frequency	8	27	48	31	6

a Draw a cumulative frequency table for these data.
b Use the table to draw a cumulative frequency diagram.
c Write the **modal class** interval.

a

Height, h, cm	< 150	< 155	< 160	< 165	< 170
Cumulative frequency	8	35	83	114	120

Upper bound of each class.

b

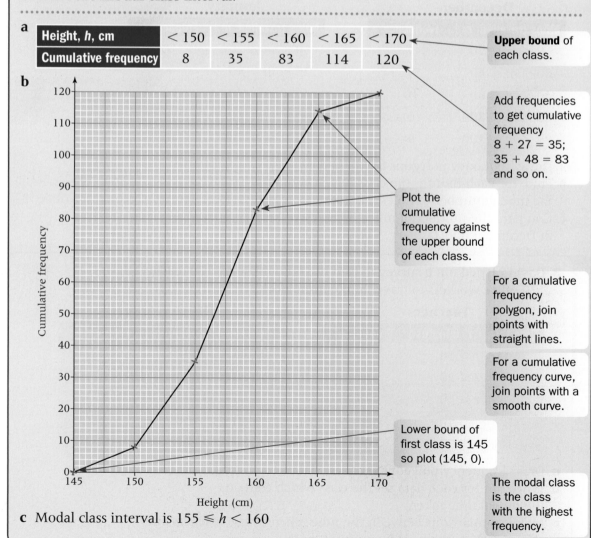

Add frequencies to get cumulative frequency
8 + 27 = 35;
35 + 48 = 83
and so on.

Plot the cumulative frequency against the upper bound of each class.

For a cumulative frequency polygon, join points with straight lines.

For a cumulative frequency curve, join points with a smooth curve.

Lower bound of first class is 145 so plot (145, 0).

The modal class is the class with the highest frequency.

c Modal class interval is $155 \leqslant h < 160$

For each of these data sets
a draw a cumulative frequency table
b draw a cumulative frequency diagram
c write the modal class interval.

> You will need the cumulative frequency diagrams from this exercise in **D5.4** and **D5.6**.

1 The heights of 100 girls.

Height, h, cm	$145 \leqslant h < 150$	$150 \leqslant h < 155$	$155 \leqslant h < 160$	$160 \leqslant h < 165$	$165 \leqslant h < 170$
Frequency	7	25	46	17	5

2 The ages of teachers in a school.

Age, A, years	$20 \leqslant A < 30$	$30 \leqslant A < 40$	$40 \leqslant A < 50$	$50 \leqslant A < 60$	$60 \leqslant A < 70$
Frequency	18	37	51	28	16

3 The times taken to complete a crossword puzzle.

Time, t, minutes	$0 \leqslant t < 10$	$10 \leqslant t < 20$	$20 \leqslant t < 30$	$30 \leqslant t < 40$	$40 \leqslant t < 50$	$50 \leqslant t < 60$
Frequency	4	11	29	37	27	12

4 The weights of a sample of cats and kittens.

Weight, w, grams	$1500 \leqslant w < 2000$	$2000 \leqslant w < 2500$	$2500 \leqslant w < 3000$	$3000 \leqslant w < 3500$	$3500 \leqslant w < 4000$
Frequency	9	22	37	20	12

5 The heights of sunflowers growing in one field.

Height, h, cm	$40 \leqslant h < 60$	$60 \leqslant h < 80$	$80 \leqslant h < 100$	$100 \leqslant h < 120$	$120 \leqslant h < 140$	$140 \leqslant h < 160$
Frequency	2	17	28	39	24	10

6 The total spent by 100 shoppers at Tesbury's superstore.

Amount, p, £	$0 \leqslant p < 10$	$10 \leqslant p < 20$	$20 \leqslant p < 30$	$30 \leqslant p < 50$	$50 \leqslant p < 70$	$70 \leqslant p < 100$
Frequency	16	14	23	17	15	15

Further cumulative frequency diagrams

This spread will show you how to:

- Estimate the median and the upper and lower quartiles from a cumulative frequency diagram

Keywords
Cumulative frequency
Interquartile range
Lower quartile
Median
Upper quartile

- You can estimate the **median** and the **upper** and **lower quartiles** from grouped data.
- You use a cumulative frequency diagram to estimate measures.

Example

The heights of 120 boys are summarised in the cumulative frequency graph.
Use the graph to estimate

a the median
b the **interquartile range**
c the number of boys with height
    i less than 153 cm   **ii** greater than 163 cm.

In grouped data you do not know the individual values so you can only make estimates.

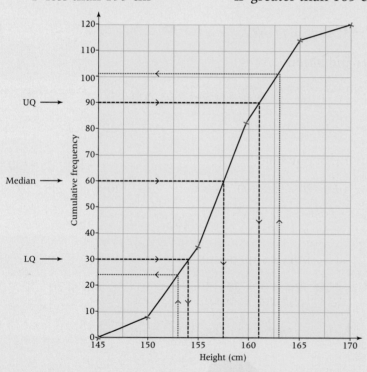

To estimate the measures, draw a line from the known values across the graph, then down to the horizontal axis.

Read estimates on the horizontal axis.

a Median = 60th value
           $= 157\frac{1}{2}$ cm

Total 120: $\frac{1}{2}$ of 120 = 60th value.

b UQ = 90th value
       = 161 cm
   LQ = 30th value
       = 154 cm
  IQR = 161 − 154 = 7 cm

$\frac{3}{4}$ of 120 = 90th value.

$\frac{1}{4}$ of 120 = 30th value.

IQR = UQ − LQ.

When the data set is large, you can use $\frac{1}{2}n$, $\frac{1}{4}n$ and $\frac{3}{4}n$ to find the median and quartiles.

c **i** 24 boys are less than 153 cm
    ii 19 boys are greater than 163 cm

Read up from 153 to the graph and across to the vertical axis.
Read up from 163 and across, then subtract 19 from the total 120.

1 Use the table and graph you have drawn in Exercise **D5.3** question **1** to estimate
　a the median
　b the interquartile range
　c the number of girls with height
　　i less than 152 cm　　　　**ii** greater than 163 cm.

You will need some of your answers from this exercise in exercise **D5.6**.

2 Use the table and graph you have drawn in Exercise **D5.3** question **2** to estimate
　a the median
　b the interquartile range
　c the number of teachers who are aged
　　i less than 35　　　　**ii** greater than 55.

3 Use the table and graph you have drawn in Exercise **D5.3** question **3** to estimate
　a the median
　b the interquartile range
　c the number of people who took
　　i less than 25 minutes
　　ii more than 45 minutes to complete the puzzle.

4 Use the table and graph you have drawn in Exercise **D5.3** question **4** to estimate
　a the median
　b the interquartile range
　c the number of cats and kittens that weighed
　　i less than 2200 g　　　　**ii** more than 3600 g.

5 Use the table and graph you have drawn in Exercise **D5.3** question **5** to estimate
　a the median
　b the interquartile range
　c the number of sunflowers that were
　　i less than 130 cm　　　　**ii** greater than 90 cm.

6 Use the table and graph you have drawn in Exercise **D5.3** question **6** to estimate
　a the median
　b the interquartile range
　c the number of shoppers who spent
　　i less than £20
　　ii more than £80.

Comparing data sets

This spread will show you how to:

● Use cumulative frequency diagrams to compare two data sets

Keywords
Cumulative
  frequency
Interquartile
  range
Median

● You can compare two data sets using information from **cumulative frequency** diagrams.

● You can compare data using a measure of average, such as the **median** and a measure of spread, such as the **interquartile range**.

Example

These cumulative frequency graphs summarise the weights of a sample of 100 men and 100 women.
Make three comparisons between the weights of the men and women.

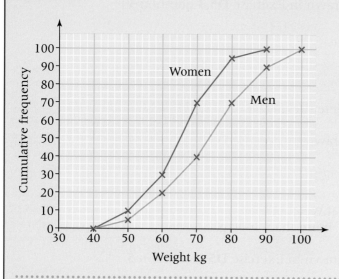

1 Median weight of women = 65 kg
Median weight of men = 73 kg
On average, the women are lighter than the men.

2 Range of women's weights = 90 − 40
= 50 kg

Range of men's weights = 100 − 40
= 60 kg

3 For the women, upper quartile = 71 kg
lower quartile = 57 kg
IQR = 14 kg

For the men, upper quartile = 82 kg
lower quartile = 62 kg
IQR = 20 kg

The middle half of the women's weights varies less than the middle half of the men's weights.

1 Write three comparisons between the test results of a group of girls and a group of boys.

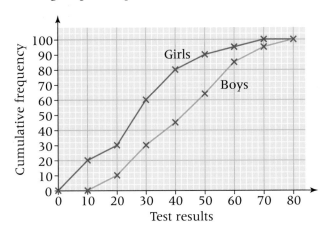

2 Write three comparisons between the heights of samples of sunflowers grown by two farmers.

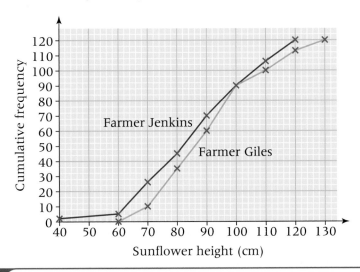

3 Write three comparisons between the mobile phone bills paid by samples of boys and girls.

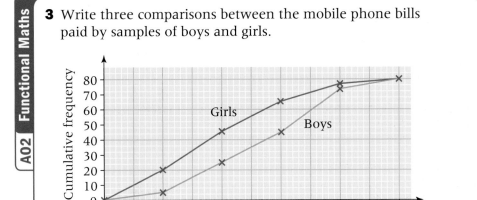

This spread will show you how to:

- Draw box plots

Keywords

Box plot
Cumulative frequency
Interquartile range (IQR)
Lower quartile (LQ)
Median
Range
Upper quartile (UQ)

p.150

- You use a **box plot** to show the **range**, the **median** and the **IQR** of a set of data.

Example

The heights of 120 boys are summarised in this **cumulative frequency** graph.
Use the graph to draw a box plot.

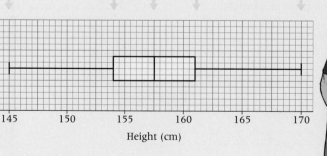

Box plots are also called box and whisker diagrams.

Use the lower bound of the first class and the upper bound of the last class as the lowest value and highest value.

From the graph:
Median = 60th value = 157.6 cm
UQ = 90th value = 161 cm
LQ = 30th value = 154 cm

The box shows the IQR.
The whiskers show the range.

1 Use the table, graph and values you found from question **1** in Exercises **D5.3** and **D5.4** to draw a box plot.

2 Use the table, graph and values you found from question **2** in Exercises **D5.3** and **D5.4** to draw a box plot.

3 Use the table, graph and values you found from question **3** in Exercises **D5.3** and **D5.4** to draw a box plot.

4 Use the table, graph and values you found from question **4** in Exercises **D5.3** and **D5.4** to draw a box plot.

5 Use the table, graph and values you found from question **5** in Exercises **D5.3** and **D5.4** to draw a box plot.

6 Use the table, graph and values you found from question **6** in Exercises **D5.3** and **D5.4** to draw a box plot.

7 Use these cumulative frequency diagrams to draw two plots.
  a Ages of students in a school of 400 students.
  b Waiting times of 80 patients in a doctor's surgery.

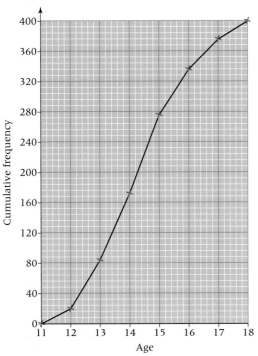

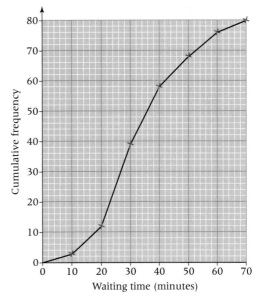

This spread will show you how to:

- Compare two sets of data using box plots

Keywords

Box plot
Interquartile
  range (IQR)
Median
Skewed
Spread
Symmetrical

- You can compare two or more data sets by using information found in **box plots**.

Example

These box plots summarise the heights of samples of
13- and 14-year-old boys and girls.

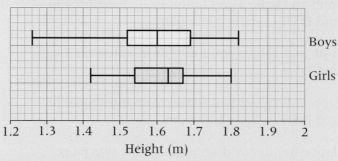

Boys

Girls

Write four comparisons between the heights of the boys and the girls.

Compare like
measures such
as the medians,
IQR, etc.

1 **Median** height of girls = 1.63 m   Median height of boys = 1.60 m
On average, the girls are taller than the boys.

2 Range of girls' heights = 1.8 − 1.42   Range of boys' heights = 1.82 − 1.26
                         = 0.38 m   = 0.56 m

The **spread** of boys' heights is greater than the spread of the girls' heights.

3 **IQR** for girls = 1.67 − 1.54   IQR for boys = 1.69 − 1.51
              = 0.13 m   = 0.18 m

The IQR for the boys is greater than for the girls so the middle half
of the heights is more varied for the boys.

4 The boys' median height is near the centre of the box.
The boys' heights are symmetrical.
The girls' median height is nearer to the UQ than the LQ so the girls'
heights are negatively **skewed**.

In **negatively skewed** data, there
are more values at the upper end
of the range.

In **positively skewed** data there
are more values at the lower end
of the range.

In **symmetrical** data there are
about the same number of values
at each end of the range.

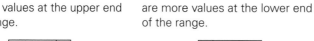

1 The box plots summarise the waiting times, to the nearest minute, of a group of patients at the doctor and the dentist.

Write four comparisons between the waiting times.

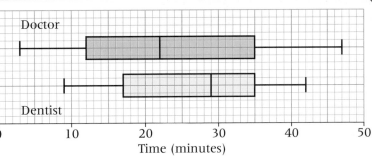

2 The box plots summarise the reaction time, to the nearest tenth of a second, of a group of boys and girls.

Write four comparisons between the reaction times.

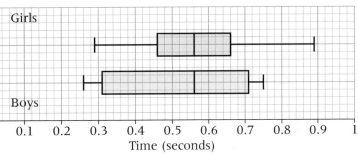

3 The box plots summarise the French and English test results of a group of students.

Write four comparisons between the test results.

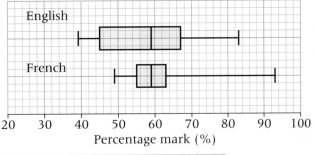

4 The box plots summarise the average length of a phone call, to the nearest minute, made by two groups of girls aged 13 and 17.

Write four comparisons between the average times.

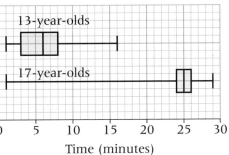

Summary

Check out

You should now be able to:

- Draw and interpret cumulative frequency diagrams
- Use cumulative frequency graphs to find median, quartiles and interquartile range
- Draw and interpret box plots for large sets of data
- Compare distributions using box plots and cumulative frequency graphs
- Calculate the modal class, class containing the median and estimated mean for large sets of grouped data
- Use frequency polygons to compare two sets of data

Worked exam question

The table gives information about the times, in minutes, that 106 shoppers spent in a supermarket.

Time (t minutes)	Frequency
$0 < t \leq 10$	20
$10 < t \leq 20$	17
$20 < t \leq 30$	12
$30 < t \leq 40$	32
$40 < t \leq 50$	25

a Find the class interval that contains the median. (1)

b Calculate an estimate for the mean time that the shoppers spent in the supermarket.

Give your answer correct to 3 significant figures. (4)

(Edexcel Limited 2007)

Time (t minutes)	Frequency	Mid-value	F × Mv
$0 < t \leq 10$	20	5	100
$10 < t \leq 20$	17	15	255
$20 < t \leq 30$	12	25	300
$30 < t \leq 40$	32	35	1120
$40 < t \leq 50$	25	45	1125
			2900

Frequency × mid-value

Use consistent mid-values.

a $30 < t \leq 40$

b An estimate of the mean = 2900 ÷ 106 = 27.4 minutes

Show this division calculation.

Exam questions

1 60 students take a science test.
The test is marked out of 50.
The table shows information about the students' marks.

Science mark	0–10	10–20	20–30	30–40	40–50
Frequency	4	13	17	19	7

On a copy of the grid, draw a frequency polygon to show this information.

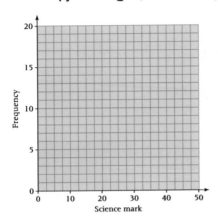

(2)

(Edexcel Limited 2008)

2 Here is the cumulative frequency curve of weights of 120 girls
at Mayfield Secondary School.

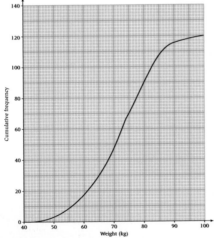

Use the cumulative frequency curve to find an estimate for the
a median weight,
b interquartile range of the weights.

(3)

(Edexcel Limited 2007)

Functional Maths 6: Holiday

Mathematics can help you to plan and budget for a holiday, as well as to understand currency, temperature and other units of measure at your destination.

holiday

Paris

LOUISE'S family are planning to go on holiday. Her parents will pay for the trip, but she must raise her own spending money.

Screen 13 Cinemaland
VALID FOR DATE OF PERFORMANCE ONLY
MANAGEMENT RESERVES THE RIGHT TO REFUSE ADMISSION
29/11/09
17:50
A GOOD NIGHT   £4.50
D
YOU WERE SERVED BY DS AT TERMINAL 4. PAID BY: Cash

Hire charge £1.99 per film!   **SIGN UP NOW!!**
4 easy steps to online

1 How much money could Louise save in the three months from March to May if she hired a DVD once a week instead of going to the cinema?

2 A neighbour offers to pay Louise £10 per week if she takes her dog for a 30-minute walk every weekday before school.

What hourly rate of pay does this represent?

How much would Louise earn if she walked the dog every weekday throughout March, April and May?

3 Louise's brother sold 20 of his CDs and 11 of his DVDs to raise his holiday money. In the first month he sold 9 CDs and 4 DVDs for a total of £36.50. In the second month he sold 5 CDs and 5 DVDs for a total of £30.

How much money does he raise from selling all of the CDs and DVDs?

How could you raise money towards a holiday fund or to buy a new item? How long would it take you to reach your target amount?

If you are going on holiday outside of the UK, then you will need to convert your money from £ Sterling to the local currency of your destination. Many European countries now use the Euro, €.

Suppose that you are charged £70 (with no commission) to buy 91.7EUR.

What is the exchange rate? Give your answer as a ratio £ Sterling : Euro.

Some companies charge a commission fee to exchange currency. With an added charge of 1%, how many Euros would you now receive (at the same exchange rate) for £70?

In 2000 Tim went on a holiday to France and Germany. To prepare for his trip he bought 700 French Francs (FRF) and 100 German Deutsche Marks (DEM) for £99.86. He was then given 200 FRF and 40 DEM as a present from his parents, who paid £32.23 for this currency. Calculate the unit per £ Sterling (GBP) rates for the FRF and the GEM (assuming that the rates were the same for both Tim and his parents).

Use the Internet to research the conversion rates used to convert the national currencies to the Euro in 2002. How did this affect the strength of the currencies of the Eurozone compared to the GBP? Compare this to today's unit per GBP exchange rate.

a $y = 3x + 2$

x	1	2	3
y			

b $2x + y = 12$

x	-1	0	
y			0

Deciding on your method of transport is an important part of planning a holiday.

Some travel options between Oxford and Paris are shown.

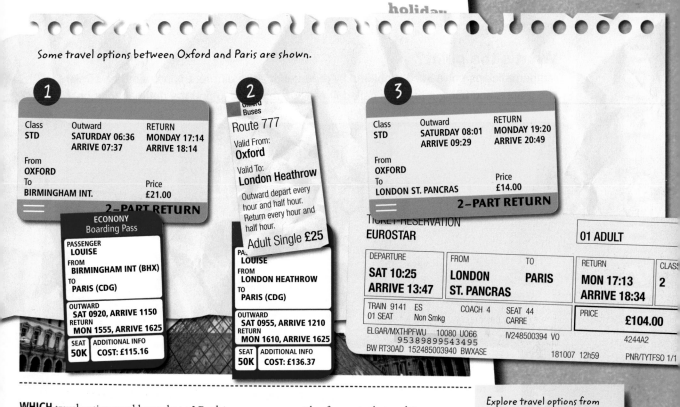

1

Class STD	Outward SATURDAY 06:36 ARRIVE 07:37	RETURN MONDAY 17:14 ARRIVE 18:14
From OXFORD		
To BIRMINGHAM INT.	Price £21.00	

2 – PART RETURN

ECONOMY
Boarding Pass

PASSENGER LOUISE	
FROM BIRMINGHAM INT (BHX)	
TO PARIS (CDG)	
OUTWARD SAT 0920, ARRIVE 1150 RETURN MON 1555, ARRIVE 1625	
SEAT 50K	ADDITIONAL INFO COST: £115.16

2

Oxford Buses
Route 777

Valid From: **Oxford**

Valid To: **London Heathrow**

Outward depart every hour and half hour. Return every hour and half hour.

Adult Single £25

PASSENGER LOUISE	
FROM LONDON HEATHROW	
TO PARIS (CDG)	
OUTWARD SAT 0955, ARRIVE 1210 RETURN MON 1610, ARRIVE 1625	
SEAT 50K	ADDITIONAL INFO COST: £136.37

3

Class STD	Outward SATURDAY 08:01 ARRIVE 09:29	RETURN MONDAY 19:20 ARRIVE 20:49
From OXFORD		
To LONDON ST. PANCRAS	Price £14.00	

2 – PART RETURN

TICKET-RESERVATION
EUROSTAR

01 ADULT

DEPARTURE SAT 10:25 ARRIVE 13:47	FROM LONDON ST. PANCRAS	TO PARIS	RETURN MON 17:13 ARRIVE 18:34	CLASS 2
TRAIN 9141 ES 01 SEAT Non Smkg	COACH 4	SEAT 44 CARRE	PRICE £104.00	

ELGAR/MXTHPFWU 10080 U066 IV248500394 V0 4244A2
95389899543495
BW RT30AD 152485003940 BWXASE 181007 12h59 PNR/TYTFSO 1/1

WHICH travel option would you choose? Explain your response with reference to the travel times and costs. All times given are local. Paris is in the time zone GMT + 1 hour.

The foreign travel legs of the same journey options can be paid for in Euros for the following prices:

Return flight BHX to Paris CDG 151.49€; return Eurostar journey 130€; return flight London Heathrow to Paris CDG 162.82€.

How does each of the prices in Euros compare with the corresponding price in GBP?

Explore travel options from your hometown to different destinations. Be careful, there are some times hidden costs such as additional taxes and fees.

■ Different countries often use different units of measure for quantities such as temperature.

An internet site states that the maximum and minimum temperatures in Rome on a particular day are 34°C (93°F) and 22°C (63°F) respectively. The formula used to convert temperatures is of the form °C = aF° + b where a and b are constants.

Use the information to set up two simultaneous equations involving a and b.

Hence find the values of a and b and derive the formula used by the website.

What are the maximum and minimum temperatures in your home town today? Use the formula in the example to convert the temperatures you have found from °C to °F.

STREET MAP
Paris
1:13,000 and 1:8,600

Formulae and simultaneous equations

In the business world, people try to increase profits by maximising their productivity and minimising their costs. However, there are many factors (called constraints) such as the number of workers, the capacity of their factories, cost of materials, *etc.*, which they must take into account.

What's the point?
Mathematicians solve such problems by representing the different constraints as straight line graphs and using a process called linear programming to find the optimal solution.

Check in

You should be able to

■ **recognise formulae**

1   You will have met some formulae before.
    Match each formula with the information that it is designed to find.

Formula:
$A = lw$
$A = \pi r^2$
$a^2 + b^2 = c^2$
$V = lwh$
$A = \frac{1}{2}bh$

To find:
Area of triangle
Volume of a cuboid
Area of a rectangle
Area of a circle
Length of sides in a right angle triangle

■ **rearrange and solve equations**

2   Solve to find the value of x.

   a $3x - 2 = 15$ **b** $4(4x - 5) = 20$ **c** $3x^2 = 75$

   d $5\sqrt{x} = 20$ **e** $x^3 + 1 = 9$ **f** $10 - x = 8$

■ **plot straight line graphs**

4   Copy and completing the table of coordinates for the given line and use it to draw a graph of the line.

   a $y = 3x + 2$ **b** $2x + y = 12$

x	1	2	3
y			

x	-1	0	
y			0

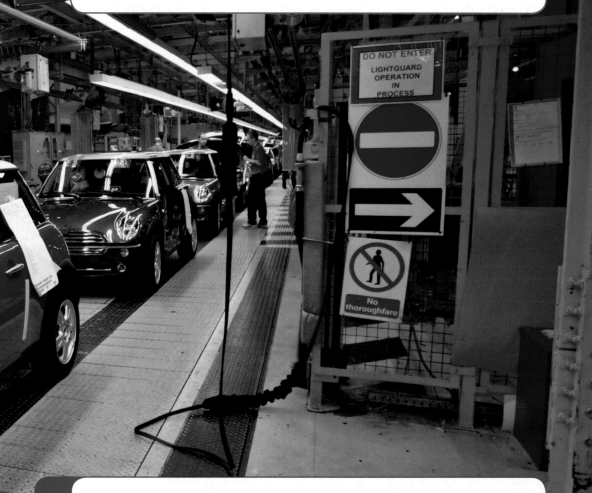

Orientation

What I need to know	What I will learn	What this leads to
A1 Plot straight line graphs	■ Manipulate mathematical formulae	**A6** Further simultaneous equations
A2 Collect like terms and expand brackets	■ Solve simultaneous, linear equations	
A4 Write formulae Use inverse operations and solve equations	■ Solve and write linear inequalities	Economics, Engineering, Operational research

DO NOT ENTER

LIGHTGUARD
OPERATION
IN
PROCESS

No
thoroughfare

Rich task

In a store there are a range of different mobile phone packages available.

Package 1   Pay as you go 10p per min.

Package 2   £5 per month and then all calls at 5p per min.

Package 3   £12 per month, with a 100 mins of free calls, and then all calls at 3p per min.

Package 4   £25 per month, with 600 mins of free calls, and then all calls at 2p per min.

Which package is the best value for money?

Rearranging formulae

This spread will show you how to:

• Rearrange a formula in order to change its subject

Keywords
Rearrange
Subject

p.134

• The **subject** of a formula is the variable before the equals sign.

For example, in $A = \pi r^2$, A is the subject of the formula.

• You can **rearrange** a formula in order to change its subject.

Example

p.232

Rearrange the formula $A = \pi r^2$ to make r the subject.

$A = \pi r^2$	Divide both sides by π.
$\dfrac{A}{\pi} = r^2$	Square root both sides.
$\sqrt{\dfrac{A}{\pi}} = r$	r is now the subject.
$r = \sqrt{\dfrac{A}{\pi}}$	Put the subject on the left-hand side.

Start by writing
the formula.
Use inverse
operations to get
r on its own.

Example

Make h the subject of the formula $A = \frac{1}{2}bh$.

$A = \frac{1}{2}bh$	Clear this fraction by multiplying both sides by 2.
$2A = bh$	Divide both sides by b.
$\dfrac{2A}{b} = h \rightarrow h = \dfrac{2A}{b}$	

Example

Make x the subject of these formulae

a $V = u + bx$ **b** $M = axy - c^2$

....

a $V = u + bx$ Subtract u from both sides.

     $V - u = bx$ Divide both sides by b.

     $\dfrac{V - u}{b} = x$

        $x = \dfrac{V - u}{b}$

b $M = axy - c^2$ Add c^2 to both sides.

     $M + c^2 = axy$

     $\dfrac{M + c^2}{ay} = x$ Divide both sides by ay.

        $x = \dfrac{M + c^2}{ay}$

1 Make x the subject of each formula.

a $C = ax + b$ **b** $M = x - b - c$ **c** $K = \dfrac{x}{t} - q$ **d** $W = t + xy$

e $H = \dfrac{x + z}{p}$ **f** $D = p(x - q)$ **g** $AB = x - ct$ **h** $Y = mx + c$

2 James and Sebastian are rearranging the formula $C = a(x - b)$ in order to make x the subject. They both come up with solutions that look different but are, in fact, correct. Can you explain why?

> James' solution
>
> $$\dfrac{C}{a} + b = x$$

> Sebastian's solution
>
> $$\dfrac{C + ab}{a} = x$$

3 Make y the subject of each formula.

a $c = y^2$ **b** $k = \dfrac{1}{4}y - 2$ **c** $M = xyz + t$ **d** $2x = \sqrt{y}$

e $p = y^3 + 2$ **f** $T = ky^2$ **g** $R = \dfrac{1}{3}ayz$ **h** $\sqrt[3]{y} = p$

4 Richard has made a mistake with his rearranging whilst trying to make p the subject of this formula. Copy his working and explain where he has gone wrong.

> $K = mp^3$
>
> $\sqrt[3]{k} = mp$
>
> $\dfrac{\sqrt[3]{k}}{m} = p$

5 These are the stages in changing the subject

of the formula $c = \dfrac{8(D + k)}{ab}$.

Put them in order.

$D + k = \dfrac{1}{8}abc$ $\dfrac{8(D + k)}{ab} = c$ $D = \dfrac{1}{8}abc - k$ $8(D + k) = abc$

6 A formula to change from degrees Celsius to degrees Fahrenheit is

$$F = \dfrac{9(C + 40)}{5} - 40$$

a Use this to change $30°C$ into $°F$.
b Rearrange to make C the subject of the formula.
c Use your new formula to find the Celsius equivalent of $-32°F$.

Rearranging harder formulae

This spread will show you how to:

- Rearrange a formula in order to change its subject

Keywords

Rearrange
Subject

Some formulae can be difficult to **rearrange**.
This is the case when

The new **subject** is subtracted in the original formula.
For example to make x the subject of the formula

$$p - x = k$$ Start by adding x to both sides.
$$p = k + x$$ Now subtract k from both sides.
$$p - k = x$$

This removes the 'subtracted from'.

The new subject is in the denominator in the original formula.
For example to make x the subject of the formula

$$\frac{p}{x} = k$$ Start by multiplying both sides by x.

$$p = kx$$ Now divide both sides by k.

$$\frac{p}{k} = x$$

Remember that x is the reciprocal of $\frac{1}{x}$.

You multiply by the reciprocal to remove the subject from the denominator.

Example

a Make x the subject of the formula $t(p - ax) = y$.

b Make y the subject of the formula $\frac{p}{y} + k = w$.

. . . .

a $t(p - ax) = y$

$$p - ax = \frac{y}{t}$$ Divide both sides by t.

$$p = \frac{y}{t} + ax$$ Add ax to both sides.

$$p - \frac{y}{t} = ax$$ Subtract $\frac{y}{t}$ from both sides.

$$\frac{p - \frac{y}{t}}{a} = x$$ Divide both sides by a.

You can go one step further and tidy up the numerator.

$$\frac{p - \frac{y}{t}}{a} = x \implies \frac{\frac{pt - y}{t}}{a} = x \implies \frac{pt - y}{at} = x$$

b $\frac{p}{y} + k = w$

$$\frac{p}{y} = w - k$$ Subtract k from both sides.

$$p = y(w - k)$$ Multiply both sides by y.

$$\frac{p}{w-k} = y$$ Divide both sides by $(w - k)$.

1 Make p the subject of each formula.

a $m = px - q$ **b** $p^2 - r = w$ **c** $\sqrt[3]{p} + h = m$ **d** $\frac{p}{t} - g = h$

e $\frac{1}{2}p + r = q$ **f** $bp^2 = k$ **g** $apw = z$ **h** $2x + y = \sqrt{p}$

2 Make x the subject of each formula.

a $k - x = w$ **b** $t - ax = p$ **c** $y = b - tx$ **d** $m = n(a-x)$

e $\frac{k}{x} = w$ **f** $m = \frac{t}{x}$ **g** $\frac{h}{x} + p = g$ **h** $\frac{p}{x^2} = k$

3 The formula $S = \frac{d}{t}$ connects speed, distance and time.

a Use the formula to find S when a distance of 27 miles is travelled in $\frac{3}{4}$ hour.

b Rearrange the formula to make t the subject.

c Use the formula to find the time taken to travel 60 km at 42 km/h.

4 This formula contains two difficult operations. Make k the subject.

$$p - \frac{t}{k} = q$$

5 Put these cards in order to give the steps in changing the subject of the formula $x = \frac{2(p-y)}{ab}$.

a $abx = 2(p - y)$

b $y + \frac{1}{2}abx = p$

c $y = p - \frac{1}{2}abx$

d $x = \frac{2(p - y)}{ab}$

e $\frac{1}{2}abx = p - y$

6 The formula $T = 2\pi\sqrt{\frac{L}{g}}$ is used to find the time, T, that a pendulum of length L takes to swing freely under gravity, g.

a Rearrange to make g the subject.

b Find a value of g (to 2 sf), given that a pendulum of length 0.4 m takes 1.27 seconds to complete its swing.

7 You can find the volume of a cylinder using the formula $V = \pi r^2 h$. Rearrange the formula to make the subject

a h **b** r

8 You can find the area of a trapezium using the formula $A = \frac{1}{2}(a + b)h$. Rearrange the formula to make b the subject.

Simultaneous equations

This spread will show you how to:
- Solve simultaneous equations by eliminating a variable

Keywords
Eliminate
Simultaneous
Solution

- Two equations that have the same **solution** are called **simultaneous** equations.
    For example,

$$x + y = 8$$
$$\text{and } x - y = 2$$

- You can solve simultaneous equations using algebra.

The solution is two numbers that add to 8 and with a difference of 2. They must be $x = 5$ and $y = 3$.

Example

Solve the simultaneous equations: $3x + 2y = 12$ and $3x + 8y = 30$.

....

$3x + 2y = 12$ (1)
$3x + 8y = 30$ (2)

Label the equations (1) and (2).
The x terms are identical so you can **eliminate** them.

$(3x + 8y) - (3x + 2y) = 30 - 12$
$\qquad\qquad 8y - 2y = 18$
$\qquad\qquad\quad 6y = 18$
$\qquad\qquad\quad\ y = 3$

Subtract equation (1) from equation (2).

$3x + 6 = 12$
$\quad 3x = 6$
$\quad\ x = 2$

Substitute 3 for y in equation (1).

Check: $3 \times 2 + 2 \times 3 = 12$ and
$3 \times 2 + 8 \times 3 = 30$

Write the equations one under the other so that you can compare them.

Subtract because the x terms have the same sign. (SSS)

Example

Solve $4x - 3y = 5$ (1)
$\qquad\ 8x + 3y = 1$ (2)

....

$(4x - 3y) + (8x + 3y) = 5 + 1$
$\qquad\qquad\qquad 12x = 6$
$\qquad\qquad\qquad\ \ x = \frac{1}{2}$
$8 \times \frac{1}{2} + 3y = 1$
$\qquad 4 + 3y = 1$
$\qquad\quad 3y = -3 \text{ so } y = -1$

In equation (1) 3y is positive, in equation (2) 3y is negative, so add the equations to eliminate the y-terms.

Substitute $\frac{1}{2}$ for x in equation (2).

Check in equation (1):
$4 \times \frac{1}{2} - 3 \times (-1) = 5$

Example

In a sweet shop, I spend £3.20 on three cans of soft drink and four bars of chocolate. The next day, I buy a can of soft drink and four bars of chocolate for £2. How much does each item cost?

....

Let c be the number of cans of drink I buy and b be the chocolate.
$3c + 4b = 320$ (1) $c + 4b = 200$ (2)
The differences between the two equations are 2 cans and 120p.
So, $2c = 120$ and $c = 60$.
$\qquad 180 + 4b = 320$ Substitute 60 for c in (1).
$\qquad\qquad 4b = 140, \text{ so } b = 35$
A can of soft drink costs 60 p and a bar of chocolate costs 35 p.

Choose letters for the variables.

Change from £ to pence to avoid working with decimals.

1 Which pairs of equations have the solution $x = 2$ and $y = 7$?

a
$$x + y = 9$$
$$x - y = -5$$

b
$$2x + y = 11$$
$$3x - y = -1$$

c
$$2x + 2y = 15$$
$$4x - y = 1$$

d
$$x + 2y = 16$$
$$x - 2y = 8$$

e
$$5x - y = 3$$
$$y - x = 5$$

2 Solve these pairs of simultaneous equations by subtracting one equation from the other.

a $3x + y = 15$
$x + y = 7$

b $6x + 2y = 6$
$4x + 2y = 2$

c $x + 5y = 19$
$x + 7y = 27$

d $5x + 2y = 16$
$x + 2y = 4$

e $m + 3n = 11$
$m + 2n = 9$

f $4x + 3y = -5$
$7x + 3y = -11$

3 Solve these pairs of simultaneous equations by adding one equation to the other.

a $3x + 2y = 19$
$8x - 2y = 58$

b $5x + 2y = 16$
$3x - 2y = 8$

c $7a - 3b = 24$
$2a + 3b = 3$

d $2x + 3y = 19$
$-2x + y = 1$

e $4x - 7y = 15$
$2x + 7y = 4\frac{1}{2}$

f $6p - 2q = -2$
$6p + 2q = 26$

4 Solve these pairs of simultaneous equations by either adding or subtracting in order to eliminate one variable.

a $x + y = 3$
$3x - y = 17$

b $5x - 2y = 4$
$3x + 2y = 12$

c $5a + b = -7$
$5a - 2b = -16$

d $3v + w = 14$
$3v - w = 10$

e $20p - 4q = 32$
$7p + 4q = 22$

f $3x - 2y = 11$
$3x + 4y = 23$

5 Solve these simultaneous equations.

$$a = 2b + 7$$
$$a + b - 1 = 10$$

$$6w = 38 - 2v$$
$$5w = 6 + 2v$$

6 Solve these problems by using simultaneous equations.

a How much does a lemon cost?

3 lemons
4 oranges
£1.27

4 oranges
5 lemons
£1.61

b The perimeter of this triangle is 30 cm. How long is the base?

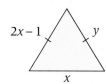

$2x - 1$ $\quad y$

x

Further simultaneous equations

This spread will show you how to:

- Solve simultaneous equations by eliminating a variable
- Use notation and symbols correctly and consistently within a given problem

Sometimes the **coefficients** of the x-terms (or the y-terms) in **simultaneous** equations are not the same.

You have to adjust the equations before you can **eliminate** the x-terms (or the y-terms).

$$3x + 5y = 4 \quad (1)$$
$$2x - y = 7 \quad (2)$$

If I multiply equation (2) by 5, I'll have $-5y$.

$$3x + 5y = 4$$
$$10x - 5y = 35$$

Now you can add the equations to eliminate the y-terms.

Example

Solve the simultaneous equations $\quad 3x + 5y = 4$
$$\qquad\qquad\qquad\qquad\qquad\qquad 2x - y = 7$$

$3x + 5y = 4 \qquad (1)$ Label the equations.
$\ 2x - y = 7 \qquad (2)$

$10x - 5y = 35 \qquad (3)$ Multiply equation (2) by 5.

To get $5y$ in both equations.

$(3x + 5y) + (10x - 5y) = 4 + 35$ Add equations (1) and (3) to
$\qquad\qquad\qquad 13x = 39$ eliminate the y-terms.
$\qquad\qquad\qquad\quad x = 3$

$9 + 5y = 4$ Substitute 3 for x in equation (1).
$\qquad y = -1$

Check the **solution** in equation (2)
$2 \times 3 - (-1) =$
$6 + 1 = 7$

Example

The difference between two numbers is 4. Treble the smaller number, subtract double the larger number is 1. What are the numbers?

Let the larger number be x and the smaller be y.
$\qquad x - y = 4 \qquad (1)$
$\ \ 3y - 2x = 1 \qquad (2)$
$-2x + 3y = 1 \qquad (2)$ Rearrange equation (2).

$3x - 3y = 12 \qquad (3)$ Multiply equation (1) by 3.

To get $3y$ in both equations.

$(-2x + 3y) + (3x - 3y) = 1 + 12$ Add equations (2) and (3),
$\qquad\qquad\qquad\qquad x = 13$ to eliminate the y-terms.

$13 - y = 4$ Substitute 13 for x in equation (1).
$\qquad y = 9$
The numbers are 9 and 13.

Check in equation (2)
$3 \times 9 - 2 \times 13$
$= 1$

1 Solve these simultaneous equations.

a $2x + y = 8$
  $5x + 3y = 12$

b $3x + 2y = 19$
  $4x - y = 29$

c $8a - 3b = 30$
  $3a + b = 7$

d $2v + 3w = 12$
  $5v + 4w = 23$

e $9p + 5q = 15$
  $3p - 2q = -6$

f $3x - 2y = 11$
  $2x - y = 8$

2 Make as many pairs of simultaneous equations as you can using these three cards. Solve your pairs.

| $2x + y = 12$ | $y - x = 15$ | $3x - 4y = 7$ |

3 Solve these simultaneous equations.
Remember to clear the fractions first.

a $\frac{x}{3} - \frac{y}{4} = \frac{3}{2}$
  $2x + y = 14$

b $\frac{a}{2} + 3b = 1$
  $5a - 7b = 47$

c $p - \frac{2q}{3} = \frac{26}{3}$
  $\frac{p}{4} + 3q + 1 = 0$

4 Write a pair of simultaneous equations to solve each problem.

  a Two numbers have a sum of 41 and a difference of 7. What numbers are they?

  b One number is 6 more than another. Their mean average is 20. What numbers are they?

  c 230 students and 29 staff are going on a school trip. They travel by large and small coaches. The large coaches seat 55 and the small coaches seat 39. If there are no spare seats and five coaches are to make the journey, how many of each coach are used?

  d In an isosceles triangle, the largest angle is 30° more than double the equal angles. What are the angles in the triangle?

  e Uncle Jack gave me a £25 book token at Christmas. At Firestone's Bookshop I can use the token to buy exactly 3 paperbacks and 1 hardback book or 1 paperback and 2 hardbacks.
  Find the cost of a paperback and a hardback.

Solving linear simultaneous equations graphically

This spread will show you how to:

- Use graphs to find the solutions or approximate solutions of two simultaneous equations

Keywords
Intersection
Simultaneous
Solution

p.38   You can solve **simultaneous** equations on a graph.

Example

Solve $3x - y = 2$ and $2x + y = 8$

Plot their graphs. (1) $3x - y = 2 \Rightarrow y = 3x - 2$

x	1	2	3
y	1	4	7

(2) $2x + y = 8 \Rightarrow y = 8 - 2x$

x	0	4	2
y	8	0	4

This line shows the solutions to $2x + y = 8$.

At the point of **intersection** of the two lines, (2, 4) both equations have a **solution**.

This line shows the solutions to $y = 3x - 2$.

The solution is $x = 2$, $y = 4$.

Check:
(1): $3 \times 2 - 4 = 2$
(2): $2 \times 2 + 4 = 8$

- The solution to simultaneous equations is where their graphs intersect.

Example

Three times one number plus twice another is 9. Twice the first number subtract the second is 13. Use a graphical method to find the two numbers.

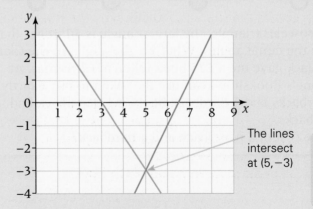

Let the numbers be x and y. Hence,
$3x + 2y = 9$
$2x - y = 13$

$3x - 2y = 9$

x	1	3	2
y	3	0	1.5

$2x - y = 13$

x	0	$6\frac{1}{2}$	3
y	−13	0	−7

The lines intersect at (5, −3)

Don't just give the coordinates as the solution. Say what each number is.

Hence, the solution is $x = 5$, $y = -3$.
The two numbers are 5 and −3.

1 Use the graph to solve these pairs of simultaneous equations.

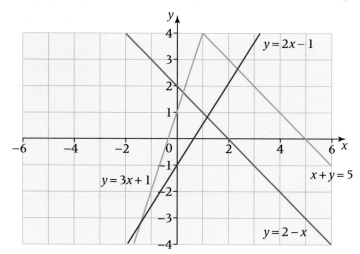

a i $y = 2x - 1$ and $x + y = 5$ **ii** $y = 3x + 1$ and $x + y = 5$
   iii $y = 2 - x$ and $y = 2x - 1$ **iv** $y = 3x + 1$ and $y = 2 - x$

b Using the graph, explain why the simultaneous equations $x + y = 5$ and $y = 2 - x$ have no solution.

2 Plot graphs to solve these simultaneous equations.
  a $y = 2x + 1$ **b** $y = 3x - 2$ **c** $2x + y = 5$
    $x + y = 10$ $x + y = 2$ $x - y = 4$

3 a Solve $3x + 2y = 4$ and $x + 4y = 3$ graphically.
  b Solve $3x + 2y = 4$ and $x + 4y = 3$ algebraically.
  c What should you notice about your answers to part **a** and part **b**?

4 Use a graphical method to solve these problems.
  a Twice one number plus three times another is 4.
     Their difference is 2. What are the numbers?
  b The sum of the ages of James and Isla is 4. The difference between twice Isla's age and treble James' age is 3. How old are they?

5 a By plotting graphs if necessary, explain why the simultaneous equations $y = 2x - 1$ and $y = 2x + 4$ have no solution.
  b Is it possible to have a pair of simultaneous equations with more than one solution?

6 Two lines intersect. One has gradient 4 and y-axis intercept 3. The other has gradient 6 and cuts the y-axis at $(0, 1)$.
  a Write the simultaneous equations they represent.
  b Solve the equations algebraically.
  c What is the point of intersection of the lines?

Problem

A03

This spread will show you how to:

• Solve simple inequalities, representing the solution on a number line

Keywords
Inequality
Number line
Reverse
Solve

• An **inequality** tells you about two quantities that are unequal.
• You can **solve** an inequality using inverse operations, for example

$3x + 2 > 5$ Subtract 2 from both sides.
$3x > 3$ Divide both sides by 3.
$x > 3$

$x < 3, 2x > 5,$
$x + 1 > 0$ are
inequalities.

• If you multiply or divide by a negative number, an inequality becomes false.
For example:

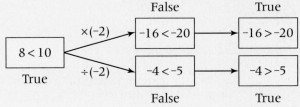

When you multiply or divide an inequality by a negative number you must **reverse** the inequality sign, for example

$7 - 2x \leq 3$ Subtract 7 from both sides.
$-2x \leq -4$ Multiply by -1 and **reverse** the inequality sign.
$2x \geq 4$ so $x \geq 2$

You can show the solution of an inequality on a **number line**, for example

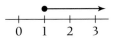

This shows that x can have a value equal to or greater than 1.

The full circle shows that x can have the value 1.

This shows that x can only have values greater than -2 and less than or equal to 1.

The open circle shows that x cannot have the value -2.

Example

Solve these inequalities and represent the solutions on a number line.
a $5x - 7 \geq 3x + 13$ **b** $16 > -4x$

a $5x - 7 \geq 3x + 13$
    $2x - 7 \geq 13$ Subtract $3x$ from both sides.
    $2x \geq 20$ Add 7 to both sides.
    $x \geq 10$

b $16 > -4x$ Divide both sides by 4.
    $4 > -x$ Multiply by -1 and reverse sign.
    $-4 < x,$ Or divide by -4.
    so $x > -4$

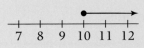

1　Use an inequality to represent each of these.
For example, I think of a number and it is more than 11: $x > 11$
  a I think of a number and it is 3 or below.
  b I think of a number and it is between 2 and 8 inclusive.
  c I think of a number and it is over −5 but below 12.

2　What inequalities are represented on these number lines?

a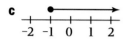
                                b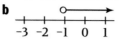

c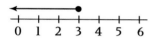
                                d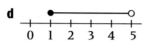

e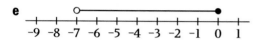

3　In question **2** which whole numbers satisfy the inequalities.

4　Is this statement true or false? The inequality $3 \geqslant x$ is represented on this number line.

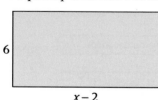

5　Solve these inequalities and represent the solutions on a number line.
  a $3x \leqslant 21$ **b** $2x - 5 > 17$ **c** $\dfrac{p}{2} + 6 \leqslant -2$
  d $28 < 7x + 49$ **e** $5y + 3 \leqslant 2y + 5$ **f** $-3y > 9$
  g $4(x + 2) \leqslant 16$ **h** $-6x < 30$ **i** $\dfrac{x}{-5} \geqslant -2$
  j $4p - 3 \leqslant 3(p - 2)$ **k** $3(x - 2) < 5(x + 6)$ **l** $6x - 4 \geqslant -2x$

6　**a** The area of this rectangle exceeds its perimeter. Write an inequality and solve it to find the range of values of x.

![rectangle with height 6 and width $x - 2$]

  b Given that x is an integer, find the smallest possible value that x can take.

7 Solve the inequality $-2 < 3x - 1 \leqslant 5$ and represent the solution on a number line.

Summary

Check out

You should now be able to:

- Change the subject of a formula
- Solve simultaneous equations in two unknowns by eliminating a variable
- Interpret a pair of simultaneous equations as a pair of straight lines and their solution as the point of intersection
- Solve simple inequalities and represent the solution set on a number line
- Set up and solve simple equations

Worked exam question

Solve $x + 2y = 4$
$3x - 4y = 7$ (3)

(Edexcel Limited 2005)

$$x + 2y = 4 \quad (1)$$
$$3x - 4y = 7 \quad (2)$$
$$3x + 6y = 12 \quad (3)$$
$$-4y - 6y = 7 - 12$$
$$10y = 5$$
$$y = 0.5$$
$$x + 2 \times 0.5 = 4$$
$$x + 1 = 4$$
$$x = 3$$
$$3 \times 3 - 4 \times 0.5 = 7$$
$$9 - 2 = 7$$

> Label the equations.

> $3 \times (1)$ labelled (3)
> $(3) - (2)$

> Substitute $y = 0.5$ in (1)

> Check $x = 3$ and $y = 0.5$ satisfy (2)

OR

$$x + 2y = 4 \quad (1)$$
$$3x - 4y = 7 \quad (2)$$
$$2x + 4y = 8 \quad (3)$$
$$3x + 2x = 7 + 8$$
$$5x = 15$$
$$x = 3$$
$$3 + 2y = 4$$
$$2y = 4 - 3$$
$$2y = 1$$
$$y = 0.5$$
$$3 \times 3 - 4 \times 0.5 = 7$$
$$9 - 2 = 7$$

> Label the equations.

> $2 \times (1)$ labelled (3)
> $(2) + (3)$

> Substitute $x = 3$ in (1)

> Check $x = 3$ and $y = 0.5$ satisfy (2)

Exam questions

1 The straight line $y + 2x = 5$ has been drawn on the grid.

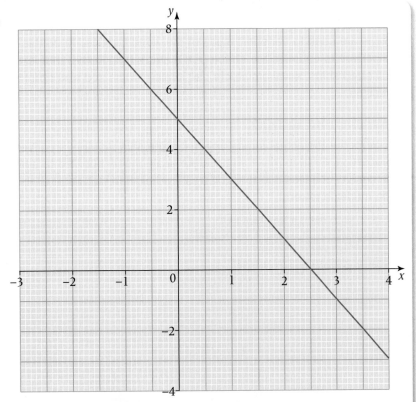

a Complete this table of values for $y = 2x - 1$

x	−1	0	1	2	3	4
y		−1		3	5	

(2)

b On a copy of the grid, draw the graph of $y = 2x - 1$ (2)

c Use your diagram to solve the simultaneous equations
$$y + 2x = 5$$
$$y = 2x - 1$$
(2)

(Edexcel Limited 2007)

2 a Make x the subject of the formula $y = 3x - 5$ (2)
  b Make a the subject of the formula $3(a - 3b) = 2a - 2$ (3)

3 Solve the inequality $2x - 1 < x + 9$ (1)

4 Solve $\quad 5x + 2y = 16$
$\qquad\quad x - 4y = 1$
(3)

(Edexcel Limited 2005)

Proportionality and accurate calculation

Ships are used to transport much of the world's goods. To do this efficiently it is important to optimise the shape of the ship's hull. Rather than experiment with costly, full sized ships, naval architects use scale models and water tanks to perfect their designs.

What's the point?

To work out the actual forces on a full sized ship requires an understanding of proportion. First, to make an accurate scale model but also to allow for effects such as density changes due to using fresh rather than salt water or the speed at which the model is towed.

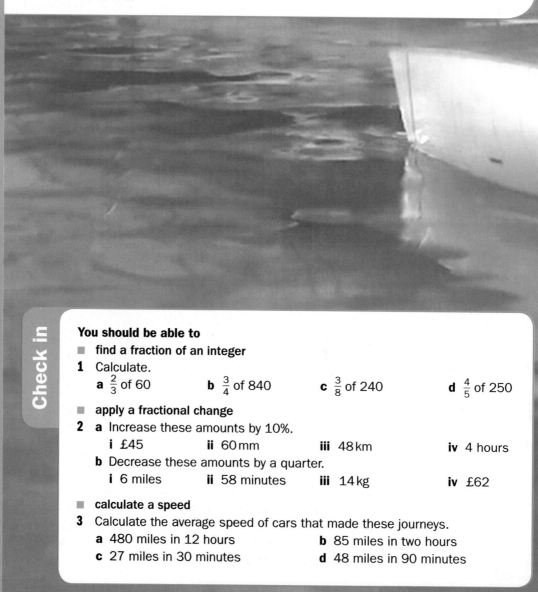

You should be able to

■ **find a fraction of an integer**

1 Calculate.

a $\frac{2}{3}$ of 60   **b** $\frac{3}{4}$ of 840   **c** $\frac{3}{8}$ of 240   **d** $\frac{4}{5}$ of 250

■ **apply a fractional change**

2 **a** Increase these amounts by 10%.

  i £45   **ii** 60 mm   **iii** 48 km   **iv** 4 hours

  b Decrease these amounts by a quarter.

  i 6 miles   **ii** 58 minutes   **iii** 14 kg   **iv** £62

■ **calculate a speed**

3 Calculate the average speed of cars that made these journeys.

  a 480 miles in 12 hours   **b** 85 miles in two hours

  c 27 miles in 30 minutes   **d** 48 miles in 90 minutes

Orientation

What I need to know	What I will learn	What this leads to
N1 Round numbers	■ Calculate proportional changes	**G5** Writting exact answers
N3 Use ratio and proportion Know how to calculate speed.	■ Use direct proportion ■ Use compound measures ■ Understand the effects of rounding	
N4 Do arithmetic with percentages	■ Use surds in exact calculation	**A-level** Maths, Chemistry, Physics

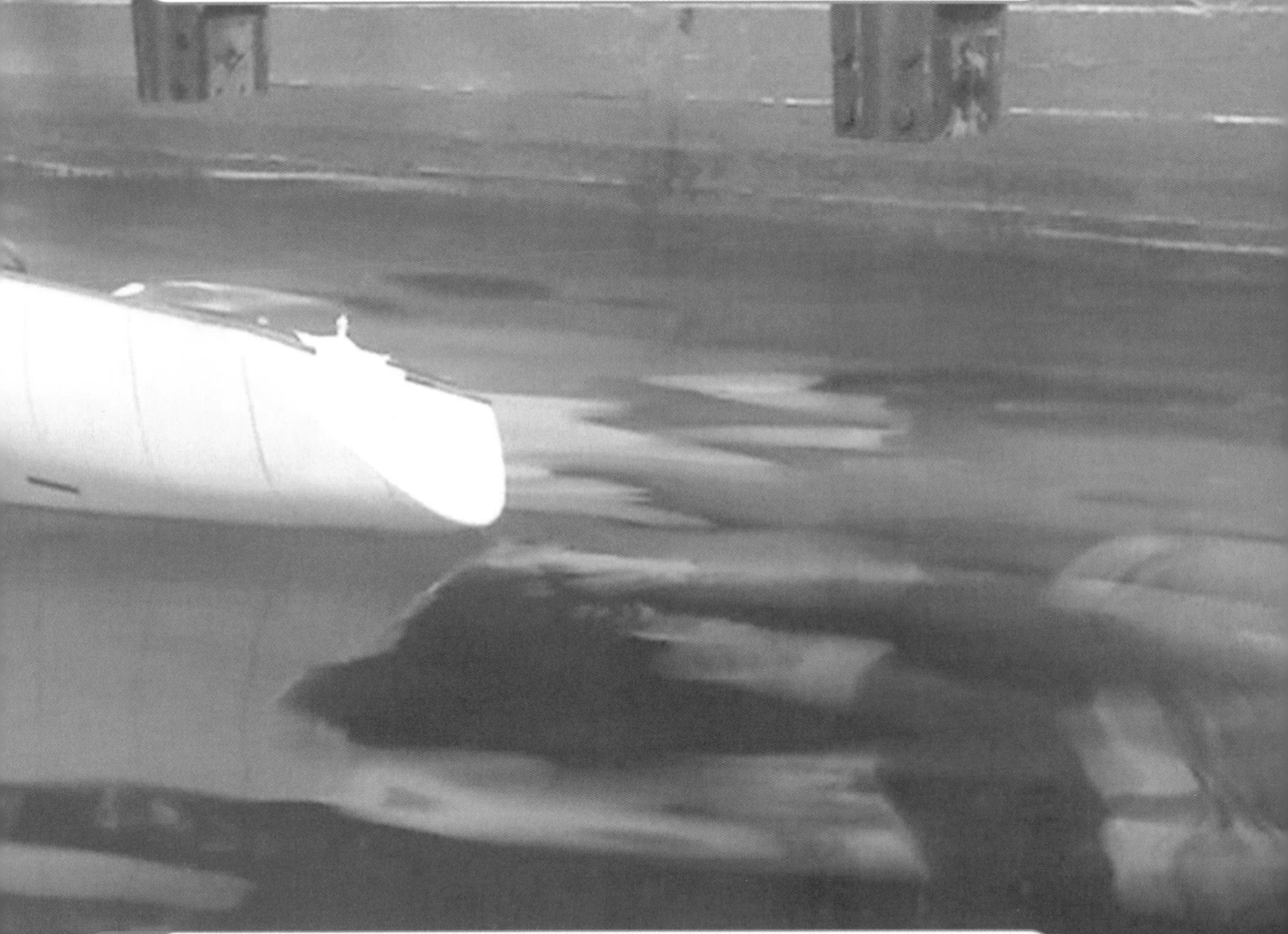

Rich task

A ship can sail at a steady speed of 30 km/h. There is enough fuel to last for 10 hours.
When the ship leaves port there is a strong current of 6 km/h which increases the speed of the ship to 36 km/h. Assuming the captain of the ship intends to use all the fuel, at what distance from port will he have to turn the ship around and head home? Remember on the return journey the strong current will be slowing down the speed of the ship.

This spread will show you how to:
- Understand direct proportion and ratio
- Solve problems involving proportion and proportional change

Keywords
Direct proportion
Ratio
Variable

Quantities which can change are called **variables**.

p.112

The amount of screenwash used is a variable.

- Two variables are in **direct proportion** if the **ratio** between them stays the same as the actual values vary.

When you multiply (or divide) one of the variables by a certain number, you have to multiply (or divide) the other variable by the same number.

Water (litres)	Capfuls
1 (× 4)	4
5 (× 4)	20

5 litres of water needs 20 capfuls of screenwash.

Example

A pipe 2.5 metres long weighs 35 kilograms. How much would 5.5 metres of the same pipe weigh?

Here are three different ways to solve this problem:

2.5 m weighs 35
⇓ ÷5   ⇓ ÷5
0.5 m weighs 7
⇓ × 11   ⇓ × 11
5.5 m weighs 77 kg

This is an informal scaling method.

2.5 m weighs 35
⇓ ÷2.5   ⇓ ÷2.5
1 m weighs 14 kg
⇓ × 5.5   ⇓ × 5.5
5.5 m weighs 77 kg

This is called the unitary method.

The formula $w = kx$ tells you the weight, w kg, of a pipe x metres long. The scale factor k is the weight of 1 metre of pipe, which is $35 \div 2.5 = 14$ kg, so the formula is: $w = 14x$.
Substituting $x = 5.5$ gives $w = 14 \times 5.5 = 77$ kg

This is an algebraic method.

5.5 metres of pipe weighs 77 kg.

Example

The cost of 12 pencils is £2.16. Work out the cost of 9 pencils.

12 pencils cost £2.16, so 3 pencils cost 54p (dividing by 4),
and 9 pencils cost £1.62 (multiplying by 3).

1 Ribbon costs £2.75 per metre. Find the cost of these lengths of ribbon.
   a 3 m **b** 4.5 m **c** 6.85 m **d** 27.55 m

2 A shop sells shelving at £3.45 per metre. Find the cost of these lengths of shelving.
   a 5 m **b** 3.45 m **c** 2.25 m **d** 4.85 m

3 A 2 m length of pipe weighs 8 kg. How much does a 3 m length of the same pipe weigh?

4 Four buckets of water weigh 60 kg. How much would 5 buckets of water weigh?

5 400 g of powder paint costs £2.40.
   a Find the cost of 100 g of the paint.
   b Use your answer to part **a** to find the cost of 300 g of the paint.

6 300 g of sherbet drops cost £1.20.
   a How much do 100 g of sherbet drops cost?
   b How much do 700 g of sherbet drops cost?

7 A pack of 250 tea bags contains 130 g of tea and costs £7.50. Calculate
   a the cost of one tea bag **b** the weight of tea. in one bag.

8 A shop sells five different types of luxury tea. Calculate the cost of 100 g of each brand, given that
   a 200 g of brand A costs £3.75 **b** 500 g of brand B costs £7.40
   c 300 g of brand C costs £5.20 **d** 250 g of brand D costs £5.10
   e 350 g of brand E costs £6.50.

9 Alan and Barry buy sand from a builders' merchant. Alan buys 35 kg of sand for £4.55. Barry buys 28 kg of the same sand. How much does he pay? Show your working.

A03 Problem

10 A shop sells drawing pins in two different packs.
    Pack A contains 120 drawing pins and costs £1.45.
    Pack B contains 200 of the same drawing pins, and costs £2.30.
    Calculate the cost of one drawing pin from each pack, and explain which pack is better value.

11 A store sells packs of paper in two sizes.

Regular	Super
150 sheets	500 sheets
Cost £1.05	Cost £3.85

Which of these two packs gives better value for money? You must show all of your working.

This spread will show you how to:

- Solve problems involving exchange rates

You can **convert** from one currency to another using an **exchange rate**.

Example

If €1 (1 euro) is worth 86p, find the value of
a €185 in pounds   **b** £260 in euros.

....

a The exchange rate is €1 = £0.86, so €185 = £0.86 × 185 = £159.10

b As an estimate, £250 ÷ 1.0 = €250,
£260 = 260 ÷ 0.86 = €302.33 to the nearest cent.

> To convert from one currency to another, simply multiply by the appropriate exchange rate.

> If you need to do the 'reverse' conversion, divide by the given exchange rate.

Example

Samantha travels from Ottawa to Buenos Aires, where she changes 375 Canadian dollars into Argentine pesos.

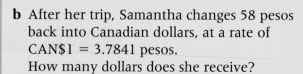

a How many pesos does she receive, if the exchange rate is CAN$1 = 3.5182 pesos?

b After her trip, Samantha changes 58 pesos back into Canadian dollars, at a rate of CAN$1 = 3.7841 pesos.
How many dollars does she receive?

....

a Estimate 400 × 4 = 1600   **b** Estimate 60 ÷ 4 = 15
$375 = 375 × 3.5182 = 1319.33 pesos   58 ÷ 3.7841 = CAN$15.33

Example

a Mary changed £900 into US dollars ($), when the rate of exchange was £1 = $1.65.

How many dollars did she receive?

b After her holiday Mary had $110 left. She changed them into pounds at an exchange rate of £1 = $1.71. How many pounds did she get?

....

a £900 = $1.65 × 900 = $1485.

b Divide by the conversion rate to convert dollars back to pounds.
So, $110 ÷ 1.71 = £64.33.

Example

Use the fact that 1 inch = 2.54 centimetres to convert
a 7.25 inches to centimetres   **b** 15 cm to inches.

....

a 7.25 inches = 7.25 × 2.54 cm = 18.4 cm

b 15 cm = 15 ÷ 2.54 inches = 5.91 inches

> Do these conversions in the same way as currency conversions.

1 The exchange rate between pounds and dollars is £1 = $1.65.
Convert these amounts from pounds to dollars.
   a £20   **b** £35   **c** £10.50   **d** £38.55

2 The exchange rate between pounds and euros is £1 = €1.16.
Convert these amounts from pounds to euros.
   a £5   **b** £30   **c** £59   **d** £264

3 One Australian dollar (A$1) is worth 42.7p.
Convert these amounts to pounds.
   a A$2.50   **b** A$45   **c** A$299   **d** A$715

4 £1 is worth 205 Japanese yen (£1 = ¥205).   See example 1**b**
Convert these amounts into pounds.
   a ¥410   **b** ¥2050   **c** ¥300   **d** ¥750
   e ¥6500   **f** ¥595

5 One US dollar is worth 0.8182 Canadian dollars (US$1 = CAN$0.8182).
   a Convert US$50 into Canadian dollars.
   b Convert CAN$50 into US dollars.

6 The table shows the exchange rates between
two currencies.

| £1 (pound) is worth €1.16 |
| $1 (dollar) is worth €0.70 |

   a Alan changes £300 into euros.
      How many euros does he receive?
   b Barbara changes €875 into dollars.
      How many dollars does she receive?

7 One pint is exactly 0.568261 litres. Find the number of pints in
one litre, giving your answer to 3 decimal places.

8 One pound weight (1 lb) is exactly 0.45359237 kg. Use this
information to convert the following weights to kilograms,
giving your answers to 3 decimal places.
   a 28 lb   **b** 2240 lb   **c** 375.5 lb   **d** 38.125 lb

9 One mile is exactly 1609.344 metres. Use this information to convert
   a 25 miles into kilometres   **b** 10 km into miles
   c 40 000 km into miles   **d** 4.5 miles into kilometres.

10 Use the approximate conversion rate of 1 gallon = 4.5 litres to convert
   a 35 gallons into litres   **b** 50 litres into gallons
   c 12.5 gallons into litres   **d** 38.8 litres into gallons.
   Give your answers to a suitable degree of accuracy.

This spread will show you how to:

- Solve problems involving proportion and proportional change

Keywords
Proportional change

Proportional change means increasing or decreasing a quantity by a certain percentage or fraction.

Daily Tabloid 40p

Town population down a quarter

Chairman's Report

'Revenue increased by 5%'

CEO Mr James Jameson's announcement in his Review of the company's Annual Accounts for 2005-2006.

Example

Find the new amount when £350 is increased by a quarter.

....

£350 ÷ 4 = £87.50   Work out the increase.
£350 + £87.50 = £437.50   Add it to original amount.

When a quantity is increased by 20% its new value is 120% of the original value, for example

£50 increased by 20% = 120% of £50

When a quantity is decreased by 20% its new value is 80% of the original value, for example

£50 decreased by 20% = 80% of £50

You can work out the new amounts in one step as in this example.

Example

When a new motorway was built between Utopia and Edenlandia the 6-hour journey time was decreased by 30%.

a Find the new journey time.

The motorway from Utopia to Edenlandia is being resurfaced resulting in a 15% increase in journey time.

b Find the new journey time due to resurfacing.

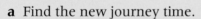

a New time is 70% of old time
New time = 70% × 6 hours $100\% - 30\% = 70\%$
       = 0.7 × 6 $70\% = 0.7$
       = 4.2 hours or 4 h 12 min
b New time due to resurfacing is 115% of old time.
New time = 115% × 4.2 hours $100\% + 15\% = 115\%$
       = 1.15 × 4.2 $115\% = 1.15$
       = 4.83 hours or 4 h 50 min

1 Find the result when these amounts are increased by one quarter.
a 20 g   b 160 cm   c 6 hours   d 500 kg

> You should be able to do questions 1 and 2 mentally.

2 Decrease these amounts by one third.
a 600 g   b 30 sec   c 45°   d 72 hours

3 Copy and complete the table to show the results of the proportional increases and decreases.

Original number	Proportional change	Result
42	Decrease by $\frac{1}{4}$	
110	Increase by $\frac{1}{5}$	
250	Increase by $\frac{1}{10}$	
450	Decrease by $\frac{2}{5}$	
965	Increase by $\frac{1}{10}$	

Functional Maths   AO2

4 A recipe for making 16 scones includes these ingredients
     60 g butter   3 teaspoons caster sugar   200 ml milk
Find the quantity of each ingredient needed to make 24 scones.

5 Increase these amounts by $\frac{1}{6}$.
a 240 g   b 300 g   c 200 g   d 750 g

6 Increase these amounts by 5%.
a £120   b £240   c £500   d £72

7 Calculate these percentage changes.
a £450 increased by 10%   b £600 decreased by 15%
c £900 increased by 6%   d £740 decreased by 11%

Functional Maths   AO2

8 Andrew earns £240 per week. He is awarded a pay rise of 4.5%.
Bella earns £260 per week. She is awarded a pay rise of 4%.
Whose weekly pay increases by the larger amount?
Show all your working.

Problem   AO3

9 Mr and Mrs Jones receive their electricity bill. The details are

Present meter reading	23 087	Service charge	£13.75
Previous meter reading	20 893	VAT	5%
Charge per unit	8.2 pence		

Find the total cost of the electricity including VAT. Show all your working.

This spread will show you how to:
- Use calculators effectively and efficiently, knowing when and when not to round the display
- Use calculators to calculate the upper and lower bounds of calculations

Keywords
Display
Effective
Efficient
Round

You need to know how to enter calculations into a calculator, and how to interpret the calculator **display**.

p.14

Calculator answers often need to be **rounded** to a suitable degree of accuracy.

Measurements (like lengths and weights) are correct to a certain degree of accuracy. You can use a calculator to find the upper and lower bounds of calculations involving measurements.

The length of a pipe given as 3.95 m is rounded to the nearest cm. The actual length is between 3.945 m and 3.955 m.

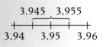

Example

Use a calculator to find the value of $(3.59 - 1.68) \div (2.4 \times 6.9)$. Write all of the digits on the calculator display, and round your answer to a suitable degree of accuracy.

A possible key sequence is:

The answer on a 10-digit display is

(3.59-1.68)÷■
0.115338164

Do not round before the end of a calculation. You should not give an answer with more significant figures than there were in the question.

Use 2 sf as a suitable degree of accuracy:
    the answer is 0.12 (2 sf).

Example

A model boat travels 3.9 metres in 7.3 seconds. Both measurements are correct to 1 dp. Find the upper and lower bounds of the speed of the boat in metres per second.

p.118

    Speed = distance ÷ time.

The distance 3.9 m is a rounded value in the range 3.85 m to 3.95 m, and 7.3 s is in the range 7.25 s to 7.35 s.

To get the maximum value of the speed, divide the largest possible distance by the shortest possible time:

    The upper bound of the calculation is
    $3.95 \div 7.25 = 0.544827 \ldots$ metres per second.

$Speed_{Max} = Distance_{Max} \div Time_{Min}$

Similarly,

$Speed_{Min} = Distance_{Min} \div Time_{Max}$

To get the minimum value of the speed, divide the smallest possible distance by the greatest possible time.

    The lower bound of the calculation is
    $3.85 \div 7.35 = 0.523809 \ldots$ metres per second.

1 Use a calculator to evaluate these, giving each answer to 2 sf.
  a $3.2 \times (2.8 - 1.05)$ **b** $2.8^2 \times (9.4 - 0.083)$
  c $16 \div (5.1^2 \times 7.2)$ **d** $(3.8 + 8.9) \times (2.2^2 - 7.6)$
  e $1.8^3 + 4.7^3$ **f** $52 \div (4.6 - 1.8^2)$

2 Work out these, giving your answers to a suitable degree of accuracy.
  a The total weight of two people weighing 68 kg and 73 kg.
  b The area of a rectangle with sides 2.2 m and 3.8 m.
  c The cost of 2.37 m of material at £5.75 per metre.
  d The time taken to travel 1 mile at a speed of 80 mph.

3 Write the upper and lower bounds of these measurements, which are
  all given to 3 significant figures.
  a 4.75 m   **b** 12.6 s   **c** 150 cm
  d 24.5 kg   **e** 8.07 g   **f** 4.33 s

You can draw a
number line to
help you.

4 A model car was rolled down a track. The length of the
  track was measured as 2.55 m, to the nearest centimetre.
  The time for the journey was 1.7 seconds, measured to the
  nearest tenth of a second.
  a Write the upper and lower bounds for the length of the track.
  b Write the upper and lower bounds for the time of the journey.
  c Use the relationship
      $$\text{Speed}_{Max} = \text{Distance}_{Max} \div \text{Time}_{Min}$$
    to find the maximum possible average speed of the car.
  d Use the relationship
      $$\text{Speed}_{Min} = \text{Distance}_{Min} \div \text{Time}_{Max}$$
    to find the minimum possible average speed of the car.

5 Find the maximum and minimum area of squares with side lengths
  given as
  a 8.00 m   **b** 6.40 cm   **c** 1.05 m
  d 3.00 mm   **e** 3.75 m   **f** 9.99 cm

Give your answers
to 3 sf.

6 The numbers in these calculations all relate to measurements.
  Find the upper and lower bound of each calculation.
  a $6.5 \times (1.2 + 4.6)$ **b** $1.2 - 0.8$

  c $(3.77 - 3.22) \times (2.43 - 1.75)$ **d** $\dfrac{(3.2 - 1.9)^2}{4.5 + 8.8}$

This spread will show you how to:

- Use surds and π in exact calculations

Keywords
Approximation
Surds
Surd form

- Numbers like $\sqrt{2}$ and $\sqrt{5}$ are called **surds**.

When calculating, first do the written calculation using surds.
Then use a calculator to find approximate values at the end, if required.

> You can carry out written calculations with surds without converting them to decimals.

- You can separate the square roots of a factorised number.

  For example:
  - $36 = 4 \times 9$, so $\sqrt{36} = \sqrt{(4 \times 9)} = \sqrt{4} \times \sqrt{9} = 2 \times 3 = 6$
  - $80 = 5 \times 16$, so $\sqrt{80} = \sqrt{(5 \times 16)} = \sqrt{5} \times \sqrt{16} = \sqrt{5} \times 4 = 4\sqrt{5}$

Decimal **approximations** are useful in practical contexts.

> If you wanted to mark out a square of area $5\,\text{m}^2$, you would give the side length as $2.24\,\text{m}$, not $\sqrt{5}\,\text{m}$.

Example

Evaluate   **a** $\sqrt{2}(1 + \sqrt{2})$ **b** $2\pi(4^2 + 4 \times 6)$

a Multiply each term in the bracket by $\sqrt{2}$.
$$\sqrt{2}(1 + \sqrt{2}) = \sqrt{2} \times 1 + \sqrt{2} \times \sqrt{2} = \sqrt{2} + 2$$
You can then find a decimal approximation:
$$\sqrt{2} = 1.414 \text{ to 3 dp so } 2 + \sqrt{2} \approx 3.414$$

b First simplify, leaving π in the expression.
$$2\pi(4^2 + 4 \times 6) = 2\pi(16 + 24) = 2\pi \times 40 = 80\pi$$
You can now find a decimal approximation:
$$\pi = 3.14 \text{ to 2 dp so } 80\pi \approx 251 \text{ to 3 sf.}$$

> This is normally written as $2 + \sqrt{2}$.

Leaving an answer in **surd form** is more accurate than a decimal approximation.

Example

Given that $\sqrt{3} = 1.73$ and that $\sqrt{7} = 2.65$ (both to 3 sf), find approximate values for   **a** $\sqrt{27}$ **b** $\sqrt{243}$ **c** $\sqrt{28}$ **d** $\sqrt{21}$,   without using the square root function on your calculator. Show your working, and give your answers to a suitable degree of accuracy.

a $\sqrt{27} = \sqrt{(9 \times 3)} = \sqrt{9} \times \sqrt{3} = 3\sqrt{3} = 3 \times 1.73 = 5.19$ to 3 sf

b $243 = 3 \times 81 = 3 \times 9 \times 9$
$\sqrt{243} = \sqrt{(3 \times 9 \times 9)} = \sqrt{3} \times \sqrt{9} \times \sqrt{9} = \sqrt{3} \times 3 \times 3 = 9\sqrt{3} = 15.6$ to 3 sf

c $\sqrt{28} = \sqrt{(4 \times 7)} = \sqrt{4} \times \sqrt{7} = 2\sqrt{7} = 5.30$ to 3 sf

d $\sqrt{21} = \sqrt{(3 \times 7)} = \sqrt{3} \times \sqrt{7} = 4.58$ to 3 sf

1 Simplify these expressions
   a $\sqrt{3} + \sqrt{3}$ **b** $\sqrt{5} + \sqrt{5}$ **c** $\sqrt{9} + \sqrt{4}$ **d** $\sqrt{7} + \sqrt{7} + \sqrt{7}$

2 Simplify these expressions
   a $\pi + \pi$ **b** $2\pi + \pi$ **c** $4 + \pi - 2 + \pi$
   d $\pi(4^2 - 6)$ **e** $2\pi(4^2 - 7)$ **f** $4\pi(2 + \sqrt{4})$

3 Simplify these expressions
   a $\sqrt{16} + \sqrt{3}$ **b** $\sqrt{49} + \sqrt{2} - \sqrt{16}$
   c $3\sqrt{7} + \sqrt{49} - \sqrt{7}$ **d** $17 + \sqrt{17} - \sqrt{9}$

4 Simplify these expressions
   a $\pi(6^2 - 4)$ **b** $\sqrt{49} + \pi(7 - 5)$ **c** $4(7 + \sqrt{2})$ **d** $\pi(8^2 - \sqrt{4})$

5 Simplify these expressions, giving your answers in surd form where necessary.
   a $\sqrt{2} \times \sqrt{2}$ **b** $\sqrt{5} \times \sqrt{5}$ **c** $\sqrt{3}(\sqrt{3} + 3)$ **d** $\sqrt{4}(\sqrt{3} + 4)$

6 Use a calculator to find an approximate decimal value for each of these expressions. Give your answers to 2 decimal places.
   a $4\sqrt{2}$ **b** $\sqrt{5} + 1$ **c** $2 + \sqrt{5}$ **d** $36\pi - 7$

7 Simplify these expressions, then use a calculator to find approximate decimal values for each one.
   Give your answers correct to 3 significant figures.
   a $4(3 + \sqrt{5})$ **b** $\sqrt{5}(5^2 - 5)$ **c** $\sqrt{7}(2 + \sqrt{7})$ **d** $\pi(2^3 + \sqrt{5})$

8 Simplify these expressions
   a $\sqrt{20}$ **b** $\sqrt{125}$ **c** $\sqrt{18}$ **d** $\sqrt{98}$

9 You are told that $\sqrt{2} \approx 1.414$, $\sqrt{3} \approx 1.732$ and $\sqrt{5} \approx 2.236$. Use this information to estimate the value of each of these, without using the square root button on your calculator. Show your working, and give your answers to 4 significant figures.
   a $\sqrt{2} + \sqrt{3}$ **b** $\sqrt{10}$ **c** $\sqrt{125}$ **d** $\sqrt{24}$

10 Use the square root key on your calculator to work out the answers to question **9**. Compare your answers, and explain which answers are more accurate.

11 Simplify these expressions by multiplying out the brackets.
   a $(3 + \sqrt{2})(4 + \sqrt{2})$ **b** $(4 + \sqrt{5})(3 + \sqrt{5})$
   c $(6 + \sqrt{3})(3 - \sqrt{3})$ **d** $(5 + \sqrt{5})(5 - \sqrt{5})$

Summary

Check out
You should now be able to:

- Describe and calculate proportions, using fractions, decimals or percentages
- Understand direct proportion
- Solve problems involving proportion and proportional change
- Use calculators effectively and efficiently for complex calculations
- Give answers to an appropriate degree of accuracy using upper and lower bounds
- Use surds and π in exact calculations

Worked exam question
The mass M grams of a cube with edges of length L cm and density D grams per cm^3 is given by the formula

$$M = DL^3$$

$D = 8$ correct to 1 significant figure.
$L = 6.4$ correct to 1 decimal place.

Calculate the upper bound of M.
Give your answer correct to 2 significant figures. (3)

(Edexcel Limited 2007)

$D = 8 \pm 0.5$

Upper bound for $D = 8.5$
Lower bound for $D = 7.5$

> Write down the upper bound for D.

$L = 6.4 \pm 0.05$

Upper bound for $L = 6.45$
Lower bound for $L = 6.35$

> Write down the upper bound for L.

$$M = DL^3$$
Upper bound for $M = 8.5 \times 6.45^3$
$= 2280.857$
$= 2300$

> Show this calculation.

> Round the answer to 2 significant figures.

Exam questions

1 Zac buys 29 identical text books.
The total cost is £248.24

Work out the cost of 41 of these text books (3)

A02

2 Tania went to Italy.
She changed £325 into euros (€).
The exchange rate was £1 = €1.68
a Change £325 into euros (€). (2)

When she came home she changed €117 into pounds.
The new exchange rate was £1 = €1.50
b Change €117 into pounds. (2)

(Edexcel Limited 2009)

3 Work out $\dfrac{4 \times 3.2 \times 10^{18} \times 2.5 \times 10^{18}}{3.2 \times 10^{18} - 2.5 \times 10^{18}}$

Give your answer correct to 3 significant figures. (3)

4 Kelly runs a distance of 100 metres in a time of 10.52 seconds.

The distance of 100 metres was measured to the nearest metre.
The time of 10.52 seconds was measured to the nearest
hundredth of a second.
a Write down the upper bound for the distance of 100 metres. (1)
b Write down the lower bound for the time of 10.52 seconds. (1)
c Calculate the upper bound for Kelly's average speed.
Write down all the figures on your calculator display. (2)
d Calculate the lower bound for Kelly's average speed.
Write down all the figures on your calculator display. (2)

(Edexcel Limited 2007)

Pythagoras' theorem and areas

The idea of distance is fundamental in geometry and Pythagoras' theorem gives a way of calculating distances in right-angled triangles. Coordinates based on two perpendicular axes give a means of specifying a point and naturally provide two sides of a right angled triangle. By putting these two ideas together a new algebraic approach to geometry is possible.

What's the point?

Mathematicians look for connections between apparently disconnected topics because of the insights they can give rise to. The new coordinate geometry eventually lead to a proof that it was impossible construct a square equal in area to a given circle using straightedge and compass.

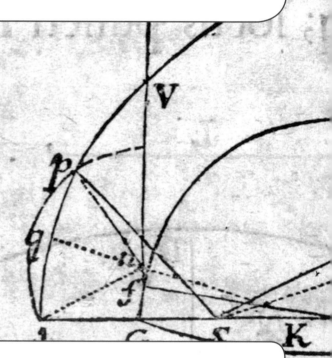

Check in

You should be able to

■ **do basic arithmetic**

1 Calculate

a 7^2 **b** $4^2 + 6^2$ **c** $3^2 + 5^2$ **d** $8^2 - 4^2$

e $7^2 - 2^2$ **f** $9^2 - 1^2$ **g** $\sqrt{13^2 - 12^2}$ **h** $\sqrt{25^2 - 24^2}$

■ **calculate areas and volumes of basic shapes**

2 Find the area of these shapes.

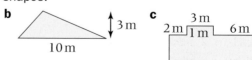

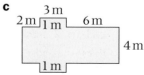

3 Find the volume and surface area of a cuboid measuring $3\,\text{m} \times 4\,\text{m} \times 5\,\text{m}$.

What I need to know	What I will learn	What this leads to
N5 Do decimal arithmetic **N7** Write exact answers using surds	■ Know and use Pythagoras' theorem ■ Find the area and circumference of circles ■ Find the volume and surface areas of prisms ■ Convert between units of measurement	**G7** Apply Pythagoras' theorem
G1 Calculate areas and volumes of basic shapes		**A-level** Maths

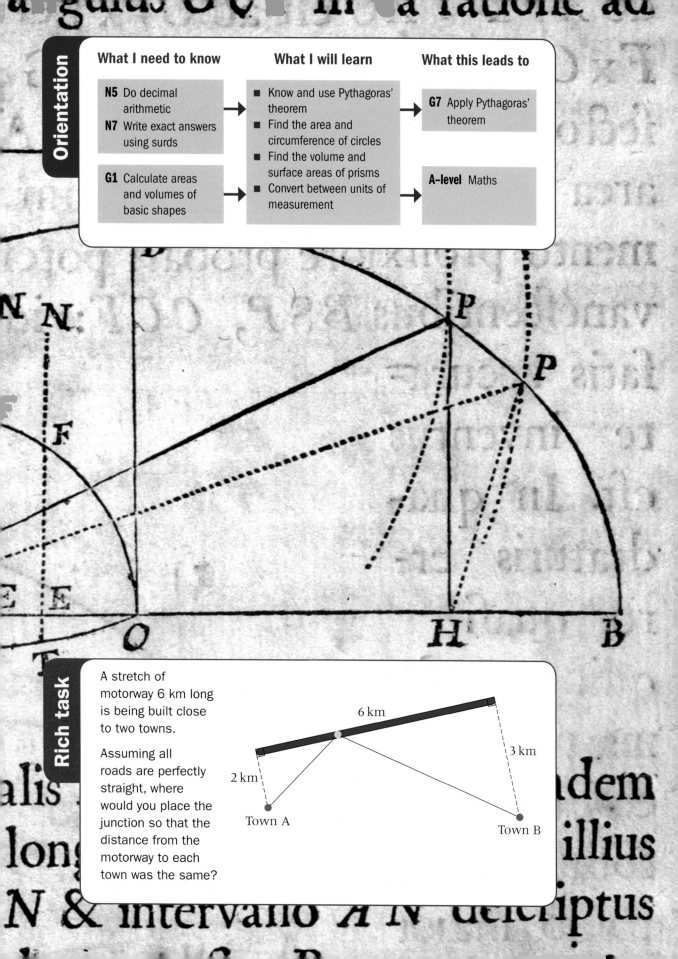

Rich task

A stretch of motorway 6 km long is being built close to two towns.

Assuming all roads are perfectly straight, where would you place the junction so that the distance from the motorway to each town was the same?

6 km

2 km

3 km

Town A

Town B

This spread will show you how to:

- Use angle properties of equilateral, isosceles and right-angled triangles
- Understand, recall and use Pythagoras' theorem in 2-D problems

Keywords
Equilateral
Hypotenuse
Isosceles
Pythagoras'
theorem
Right-angled
triangle
Scalene

You need to know the properties of these triangles.

Equilateral	**Isosceles**	**Scalene**	**Right-angled**
All sides equal All angles equal	Two sides equal Two angles equal	No sides equal No angles equal	One angle 90°

p.406
In a right-angled triangle the **hypotenuse** is the longest side. It is opposite the right-angle.

hypotenuse

In a right-angled triangle, the length of the hypotenuse squared equals the sum of the squares of the lengths of the other two sides.

- This is **Pythagoras' theorem**.

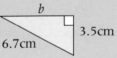

$c^2 = a^2 + b^2$
c is the hypotenuse

Example

Work out the missing sides in these triangles.

a
7.6cm
4.2cm

b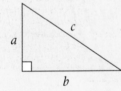
b
3.5cm
6.7cm

c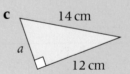
14 cm
12 cm
a

To find the hypotenuse: 'square, add and square root'.

a $c^2 = a^2 + b^2$
$\quad = 4.2^2 + 7.6^2$
$\quad = 17.64 + 57.76$
$\quad = 75.4$
$c = \sqrt{75.4}$
$\quad = 8.7\,cm$

b $c^2 = a^2 + b^2$
$b^2 = c^2 - a^2$
$\quad = 6.7^2 - 3.5^2$
$\quad = 44.89 - 12.25$
$\quad = 32.64$
$b = \sqrt{32.64}$
$\quad = 5.7\,cm$

c $c^2 = a^2 + b^2$
$a^2 = c^2 - b^2$
$\quad = 14^2 - 12^2$
$\quad = 196 - 144$
$\quad = 52$
$a = \sqrt{52}$
$\quad = 7.2\,cm$

To find a shorter side: 'square, subtract and square root'.

1 Find the hypotenuse in each of these right-angled triangles.

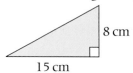 Give answers in this exercise to 1 dp where appropriate.

a
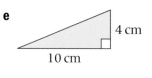
3 cm
4 cm

b
8 cm
15 cm

c
5 cm
12 cm

d
5 cm
9 cm

e
4 cm
10 cm

f
7 cm
7 cm

g
6.2 cm
10.9 cm

h
3.5 cm
6.4 cm

2 In some of the triangles in question **1**, all three sides have integer (whole number) values.
Such sets of three numbers are called Pythagorean triples.
List the Pythagorean triples from question **1**.

3 Find the missing side in each of these right-angled triangles.

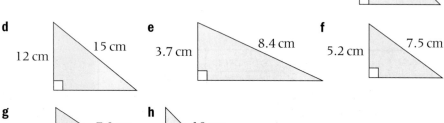

a
6 cm
10 cm

b
7 cm
21 cm

c
24 cm
26 cm

d
12 cm
15 cm

e
3.7 cm
8.4 cm

f
5.2 cm
7.5 cm

g
4.8 cm
7.3 cm

h
10 cm

4 Some of the triangles in question **3** are Pythagorean triples.
  a Write down the Pythagorean triples in question **3**.
  b Compare these with your answers to question **2**.
  c Comment on anything you notice.

5 Repeat question **1** parts **a** to **f** giving your answer as an exact expression written in surd form.

This spread will show you how to:

● Use Pythagoras' theorem in 2-D problems

Keywords
Hypotenuse
Pythagoras'
  theorem
Right-angled
  triangle

● In a **right-angled triangle**, the square on the **hypotenuse** equals the sum of the squares of the other two sides.

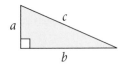

$$c^2 = a^2 + b^2$$
c is the hypotenuse

This is **Pythagoras' theorem.**
The hypotenuse is always opposite the right-angle.

To solve problems using Pythagoras' theorem

– sketch a diagram and label the right-angle

– label the unknown side

– round your answer to a suitable degree of accuracy.

    Unless the question tells you otherwise, round to 2 dp.

Example

A rectangle measures 5 cm by 8 cm.
Find the length of its diagonal.

....

Diagonal, d, is the hypotenuse of a right-angled triangle.
$$d^2 = 5^2 + 8^2$$
$$d^2 = 89$$
$$d = 9.43 \text{ cm}$$

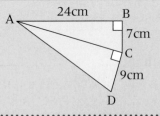

Mark all the facts you know on your diagram.

Example

Find the length AD.

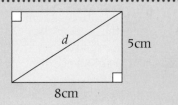

....

First find AC $AC^2 = 7^2 + 24^2$
                $AC^2 = 625$
                 $AC = 25 \text{ cm}$

AC is the hypotenuse of the right-angled triangle *ABC*.

In triangle ACD $AD^2 = 9^2 + 25^2$
                 $AD^2 = 706$
               $AD = 26.57 \text{ cm}$

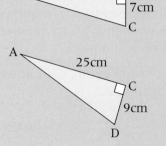

AD is the hypotenuse of the right-angled triangle *ACD*.

1 Find the length of the diagonal of each rectangle.

Give answers in this exercise to 2 dp where appropriate.

a
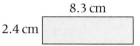
8.3 cm

2.4 cm

b
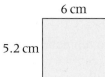
6 cm

5.2 cm

2 A rectangle has one side 4 cm and diagonal 10.4 cm.
Find the length of the other side.

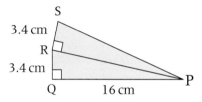
10.4 cm

4 cm

3 Find the length of the diagonal of a square with side length 8 cm.

4 Find the length of the side of a square with diagonal length 8 cm.

5 PQR and PRS are right-angled triangles.
Find the length PS.

S
3.4 cm
R
3.4 cm
Q   16 cm   P

6 ABC and ACD are right-angled triangles.
Find the length AB.

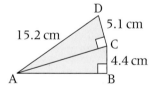
D
5.1 cm
15.2 cm
C
4.4 cm
A   B

7 A ladder of length 5.5 m leans against a wall.
The foot of the ladder is 1 m from the wall.
How far up the wall does the ladder reach?

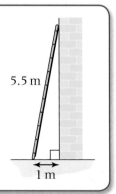
5.5 m

1 m

Pythagoras' theorem and coordinates

This spread will show you how to:

Keywords
Coordinates
Midpoint
Pythagoras'
theorem

- Find the coordinates of the midpoint of the line segment *AB*, and its length, given the points *A* and *B*
- Understand and use coordinates to represent points in three dimensions

Coordinates identify a point on a grid.

A (2, 1)   B (4, 4)

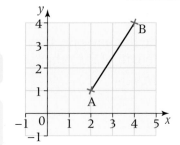

- The **midpoint** of two points is the mean of their coordinates.

- Midpoint of (x_a, y_a) and (x_b, y_b) is $\left(\dfrac{x_a + x_b}{2}, \dfrac{y_a + y_b}{2}\right)$

You can use **Pythagoras' theorem** to find the length of a line joining two points on a grid.

Example

Find the midpoint and length of the line joining A and B.

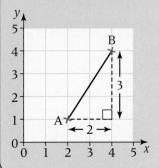

Midpoint

$\left(\dfrac{2 + 4}{2}, \dfrac{1 + 4}{2}\right) = (3, 2.5)$

Length

Draw a right-angled triangle and label the lengths of the two shorter sides.

$AB^2 = 2^2 + 3^2$
$AB^2 = 13$
$AB = 3.61$ (2 dp)

Using Pythagoras' theorem.

- You can use coordinates to identify a point in a 3-D grid.

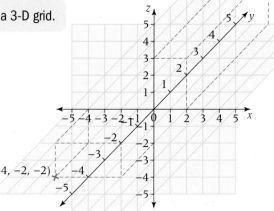

The point (2, 5, 3) is located
2 units along the *x*-axis,
5 units along the *y*-axis, and
3 units up the *z*-axis.

The point (−4, −2, −2) is located
4 units along the negative *x*-axis,
2 units along the negative *y*-axis,
and 2 units down the negative *z*-axis.

All three axes are perpendicular to each other.

1 Find the midpoints of these line segments.
 a A (6, 4) and B (2, 7)
 b C (−2, 5) and D (3, −3)
 c E (0, 6) and F (4, 0)
 d G (7, −1) and H (−4, 5)

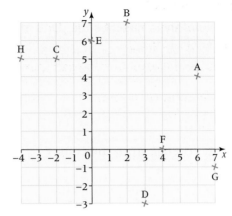

Use the diagram to check your answers.

Make a sketch to show the position of the points.

2 Use Pythagoras' theorem to find the length of the line segment joining each pair of points.
 a (2, 5) and (6, 8)
 b (7, 1) and (2, 8)
 c (0, 7) and (1, −3)
 d (−4, 5) and (4, −2)

3 Write the coordinates of these points.

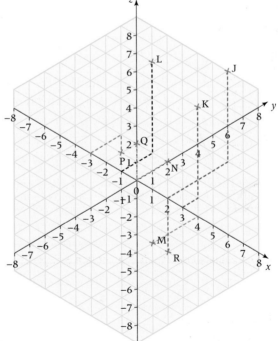

4 Here is a grid in three dimensions.
The points on the grid have coordinates of
A(1, 2, 5) B(1, 4, 2) C(2, 3, 1) D(5, 3, 4)
 a Find the midpoints of these line segments.
  i AB **ii** CD **iii** BC **iv** AD
 b Use Pythagoras' theorem to find the length of each line segment in part **ai** and **aii**.
 c Can you extend the use of Pythagoras' theorem to find the length of BC? You will need to use it twice.

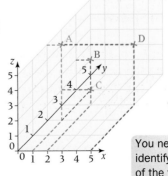

You need to identify the sides of the triangle.

Area and circumference of a circle

This spread will show you how to:

- Use the correct vocabulary to describe the parts of a circle
- Calculate the area and circumference of circles

In a **circle**:

- The **diameter**, d, is the distance across the circle through the centre.

- The **radius**, r, is the distance from the centre to the edge.

- The **circumference**, C, is the perimeter – the distance around the edge.

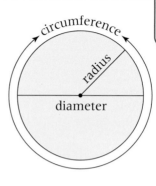

The circumference of a circle is in **proportion** to its diameter: $C \approx 3d$

The actual proportion is a decimal. You use a symbol, π (pi).

$d = 2 \times r$

π is about 3.142

- $C = \pi \times d$ or $C = 2 \times \pi \times r$

You can cut a circle into lots of sectors and lay them out side by side, to make a rectangle.

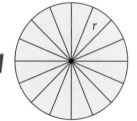

p.22

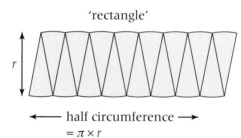

'rectangle'

r

$\longleftarrow$ half circumference $\longrightarrow$
$= \pi \times r$

Area of 'rectangle' = length $\times$ width = $(\pi \times r) \times r = \pi \times r^2$

- Area of a circle = $\pi \times r^2$ or $A = \pi \times r^2$

Example

Find **i** the circumference **ii** the area of each circle.

a

$r = 5$ cm

b

$d = 32$ mm

Use the π key on your calculator.

....

a i $C = 2 \times \pi \times r = 2 \times \pi \times 5$
$\quad$ = 31.415...
$\quad$ = 31.4 cm (to 3 sf)

ii $A = \pi \times r^2 = \pi \times 5^2$
$\quad$ = 78.539...
$\quad$ = 78.5 cm² (to 3 sf)

b i $C = \pi \times d = \pi \times 32$
$\quad$ = 100.530...
$\quad$ = 101 mm (to 3 sf)

ii $A = \pi \times r^2 = \pi \times 16^2$
$\quad$ = 804.247...
$\quad$ = 804 mm² (to 3 sf)

Round your answers to a sensible degree of accuracy. 3 sf is usually good practice.

1 Find the circumferences of these circles.

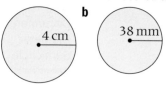
a

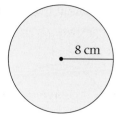
b

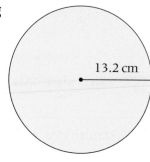
c 8 cm

d 7.5 cm

e 24 mm

f 42 mm

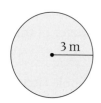
g 13.2 cm

Give your answers to 3 sf.

2 Find the circumferences of these circles.
   a radius = 12 mm   **b** radius = 23 cm   **c** diameter = 105 mm
   d diameter = 1.2 cm   **e** radius = 3.6 cm **f** diameter = 125 cm

3 Find the areas of the circles in question **1**.

4 Find the areas of the circles in question **2**.

5 A circular pond has radius 3 m.
   Work out the circumference of the pond.

3 m

6 A round hole has circumference of 44 cm.
   Work out the radius of the hole, to 1 decimal place.

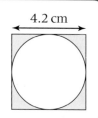
r

A03 **Problem**

7 Shamin is cutting out circles for an art project.
   She has squares of card that are 4.2 cm wide.
   a What is the area of the biggest circle she can
      cut out?
   b What area of card is left?

4.2 cm

This spread will show you how to:

- Calculate the length of an arc and the area of a semicircle

Keywords
Area
Circumference
Diameter
Semicircle

A **semicircle** is half a circle.

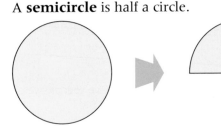

- Area of a semicircle $= \frac{1}{2} \times$ area of whole circle.

Example

Calculate the area of this semicircle.

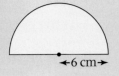
←6 cm→

Area $= \frac{1}{2} \times (\pi \times r^2)$
$= \frac{1}{2} \times (\pi \times 6^2)$
$= \frac{1}{2} \times (113.097 \ldots$
$= 56.548 \ldots \text{ cm}^2$

In terms of π
Area $= 18\pi$

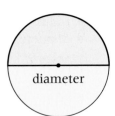
diameter

- Perimeter of a semicircle $= \frac{1}{2} \times$ circumference of whole circle + diameter.

Example

Calculate the perimeter of this semicircle.

←6 cm→

Circumference of whole circle $= 2 \times \pi \times r$
$= 37.699 \ldots \text{ cm}$
Perimeter of semicircle $= \frac{1}{2} \times$ circumference of whole circle + diameter
$= (\frac{1}{2} \times 37.699\ldots) + (2 \times 6)$
$= 18.849 \ldots + 12$
$= 30.8 \text{ cm (to 1 dp)}$

In terms of π
Perimeter

Example

A door is shaped in an arch, with a semicircle on top.
Calculate the perimeter of the door, giving your
answer to 1 decimal place.

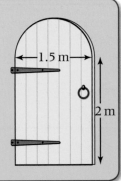
←1.5 m→
2 m

Perimeter of arch
$= 2\,\text{m} + 1.5\,\text{m} + 2\,\text{m} + \frac{1}{2} \times$ circumference of circle
Circumference $= 2 \times \pi \times r = 1.5 \times \pi = 4.712\ldots$
Perimeter of arch $= 2 + 1.5 + 2 + \frac{1}{2} \times 4.712$
$= 7.9\,\text{m}$ (to 1 dp)

In terms of π
Perimeter

1 Find the area of each semicircle.

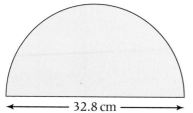

a

5 cm

b

7.2 cm

c

24 mm

Give your answers
to 1 dp.

d

←18 cm→

e

←— 32 mm —→

f

15.4 cm

g

←———— 32.8 cm ————→

h

18.1 cm

2 Find the area of each semicircle.
   a radius = 12 cm   **b** radius = 2.3 cm   **c** diameter = 12.9 m
   d diameter = 22.3 cm **e** radius = 9.5 mm   **f** diameter = 3.39 cm

3 Work out the perimeter of each semicircle in question **1**.

4 Work out the perimeter of each semicircle in question **2**.

A03 Problem

5 A flowerbed in the park is semicircular.
   It has a radius of 2 m.
   a Work out the area of the flowerbed.

   Percy the park keeper wants to plant
   flowers that each need an area of 0.3 m².
   b How many of these flowers can Percy plant in
   the flowerbed? What space does he have left?

6 Viaduct arches have straight sides 50 m high.
   The arch at the top is a semicircle with diameter 8 m.

   A spider crawls from ground level on one side,
   around the arch, and back down the other side.
   Work out how far it crawls.

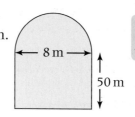

←— 8 m —→

50 m

Give your answer
in terms of π and
to 1 dp.

7 A bathroom window is a semicircle with an
   internal diameter of 80 cm.
   Work out the area of the glass in the window.

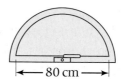

←— 80 cm —→

This spread will show you how to:

- Solve problems involving surface areas and volumes of prisms

Keywords
Cylinder
Pythagoras'
  theorem
Surface area
Volume

- **Surface area** is the total area of all the faces of a solid.

Example

A cube has surface area $150\,\text{cm}^2$.
Find the **volume** of the cube.

A cube has six faces.
Each face has the same area.
So area of one face is $150 \div 6 = 25\,\text{cm}^2$
Each face is a square, so length of each side $\sqrt{25} = 5\,\text{cm}$
Volume of cube = area of cross-section $\times$ length
$$= 25 \times 5$$
$$= 125\,\text{cm}^3$$

Each face is a
cross-section.

Example

This triangular prism has length $9\,\text{cm}$ and
its end faces are equilateral triangles with
side length $4\,\text{cm}$.
Work out its volume.

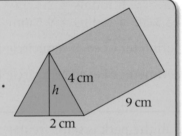

Using **Pythagoras' theorem** to find
the height, h, of the triangle.
$$h^2 = 4^2 - 2^2$$
$$h = \sqrt{12} = 3.464\ldots$$
Area of cross-section $= \frac{1}{2} \times 4 \times 3.464\ldots$
$$= 6.928\ldots$$
Volume of prism: $6.928\ldots \times 9 = 62.4\,\text{cm}^3$

To work out the
area of the cross-
section you need
the perpendicular
height of the
triangle.

Example

Calculate the surface area of a **cylinder** radius $= 3\,\text{cm}$
and length $= 7\,\text{cm}$.

Area of top circle
$$\pi \times 3^2 = 28.27$$
Area of bottom circle
$$\pi \times 3^2 = 28.27$$
Area of curved surface
$$2 \times \pi \times 3 \times 7 = 131.95$$
Surface area of cylinder
$$= 28.27 + 28.27 + 131.95$$
$$= 188.5\,\text{cm}^2 \text{ (to 1 dp)}$$

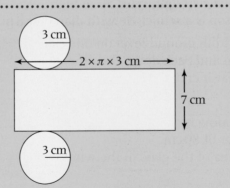

The curved surface
of a cylinder is a
rectangle.
Length of
rectangle =
circumference
of circle.
Width of rectangle
= height of
cylinder.

p.28
p.30

1 Find the volumes of the cubes with these surface areas.
   a Surface area 54 cm²   **b** Surface area 294 cm²
   c Surface area 96 cm²   **d** Surface area 1.5 m²

2 Find the volumes of these triangular prisms which have
   equilateral or isosceles triangles as cross-sections.

a

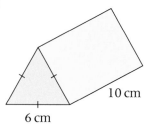

6 cm   10 cm

b

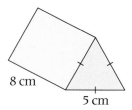

8 cm   5 cm

c

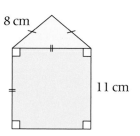

8 cm   11 cm

d

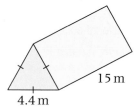

4.4 m   15 m

3 Find the surface areas of the cubes with these volumes.
   a Volume 512 cm³   **b** Volume 1000 cm³
   c Volume 216 cm³   **d** Volume 1 m³

4 Find the volume and total surface area of each cylinder.

a

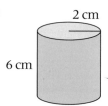

2 cm   6 cm

b

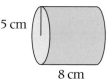

5 cm   8 cm

c

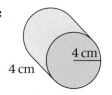

4 cm   4 cm

d

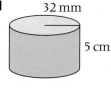

32 mm   5 cm

> Be careful with units in part **d**.

5 The volume of a cuboid is 80 cm³.
   The cuboid has square ends.
   The sides of the cuboid are
   whole numbers of centimetres.

   a Find the possible dimensions of the cuboid.
   b Find **i** the smallest possible surface area
           ii the largest possible surface area of this cuboid.

6 Repeat question **4** for cuboids with volumes
   a 24 cm³   **b** 64 cm³

Measures

This spread will show you how to:

- Convert between volume measures including cm^3 and m^3
- Understand the difference in dimensions of perimeter, area and volume

Keywords
Area
Dimension
Perimeter
Volume

- **Perimeter** is the distance around a shape.

  It is a length measured in mm, cm and m.

- **Area** is the space covered by a two-dimensional shape.

  It is the product of length × length, measured in mm^2, cm^2 and m^2.

- **Volume** is the space inside a three-dimensional solid.

  It is the product of length × length × length, measured in cubic mm^3, cm^3 and m^3.

Example

Change
a 3 700 000 cm to m
b 3 700 000 cm^2 to m^2
c 3 700 000 cm^3 to m^3

....

a 3 700 000 ÷ 100 = 37 000 m
b 3 700 000 ÷ (100 × 100) = 370 m^2
c 3 700 000 ÷ (100 × 100 × 100) = 3.7 m^3

Larger unit means smaller number, so divide.
Divide twice for squared units.
Divide three times for cubed units.

Example

Change
a 0.042 m to mm
b 0.038 m^2 to cm^2
c 7 cm^3 to mm^3

....

a 0.042 × 1000 = 42 mm
b 0.038 × (100 × 100) = 380 cm^2
c 7 × (10 × 10 × 10) = 7000 mm^3

Smaller unit means larger number, so multiply.
Multiply twice for squared units.
Multiply three times for cubed units.

1 Change

   a 40 000 cm to m   **b** 63 000 cm to m
   c 42 m to cm   **d** 1200 cm to m
   e 80 000 m to km   **f** 45 000 mm to m
   g 0.05 m to mm   **h** 0.0003 km to mm
   i 0.6 m to cm   **j** 0.007 km to mm
   k 0.00004 km to cm   **l** 18.05 km to m

2 Change

   a 2 600 000 mm^2 to m^2 **b** 700 000 000 cm^2 to m^2
   c 0.00045 m^2 to cm^2 **d** 0.12 m^2 to cm^2
   e 0.008 m^2 to cm^2 **f** 0.00045 m^2 to cm^2
   g 840 000 000 mm^2 to cm^2 **h** 3 000 000 cm^2 to m^2
   i 2 m^2 to cm^2 **j** 1 km^2 to m^2

3 Change

   a 3 cm^3 to mm^3 **b** 0.0002 m^3 to cm^3
   c 0.0048 m^3 to cm^3 **d** 3 000 000 mm^3 to cm^3
   e 10 000 000 cm^3 to m^3 **f** 50 000 000 cm^3 to m^3

4 The photo shows a window that
is rectangular with semi-circular top.
Write an expression for
   a the perimeter
   b the area.

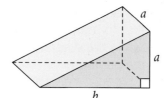

5 The diagram shows a door wedge.
Write an expression for
   a the surface area
   b the volume.

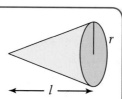

6 Jon said the volume of this shape is $\frac{5}{8}\pi rl$.
   a Explain why the expression cannot
     be correct.
   b What could the expression $\frac{5}{8}\pi rl$ represent?

G5

Summary

Check out

You should now be able to:

- Understand, recall and use Pythagoras' theorem in 2-D problems
- Find the coordinates of the midpoint of the line segment *AB*, and its length, given the points *A* and *B*
- Calculate the surface area and volume of right prisms
- Calculate the circumference and area of circles
- Convert between length measures, area measures and volume measures

Worked exam question

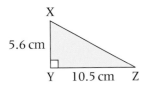

Diagram NOT accurately drawn.

In the triangle *XYZ*
XY = 5.6 cm
YZ = 10.5 cm
angle *XYZ* = 90°

a Work out the length of *XZ*. (3)

4 copies of the triangle are fitted together to make the shape shown in the diagram.

b Calculate the perimeter of the shape. (2)

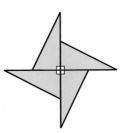

Diagram NOT accurately drawn.

(Edexcel Limited 2007)

....

a

$$5.6^2 + 10.5^2 = 31.36 + 110.25$$
$$= 141.61$$
$$XZ = \sqrt{141.61}$$
$$= 11.9 \text{ cm}$$

> Show the working to square, add and square root the numbers.

b

$$10.5 - 5.6 = 4.9 \text{ cm}$$
$$11.9 + 4.9 = 16.8 \text{ cm}$$
$$16.8 \times 4 \quad = 67.2 \text{ cm}$$

> Show the method to find the perimeter.

Exam questions

1

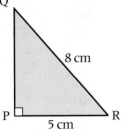

Diagram NOT accurately drawn.

PQR is a right-angled triangle.
PR = 5 cm   *QR* = 8 cm
Work out the length of *PQ*.
Give your answer correct to 3 significant figures. (3)

2

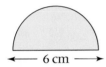

Diagram NOT accurately drawn.

Work out the perimeter of the semicircle.
Give your answer correct to 2 decimal places. (3)

3

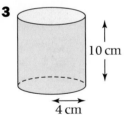

Diagram NOT accurately drawn.

A solid cylinder has a radius of 4 cm and a height of 10 cm.
The cylinder is made from wood.
The density of the wood is 0.6 grams per cm^3.
Work out the mass of the cylinder.
Give your answer correct to 3 significant figures. (4)

(Edexcel Limited 2008)

4 Change 2.5 m^2 to cm^2. (2)

(Edexcel Limited 2003)

Functional Maths 7: Radio maths

Mathematics can be used to explain how radio transmission works.

Radio transmitters use continuous sine waves to send and receive information such as music or speech.

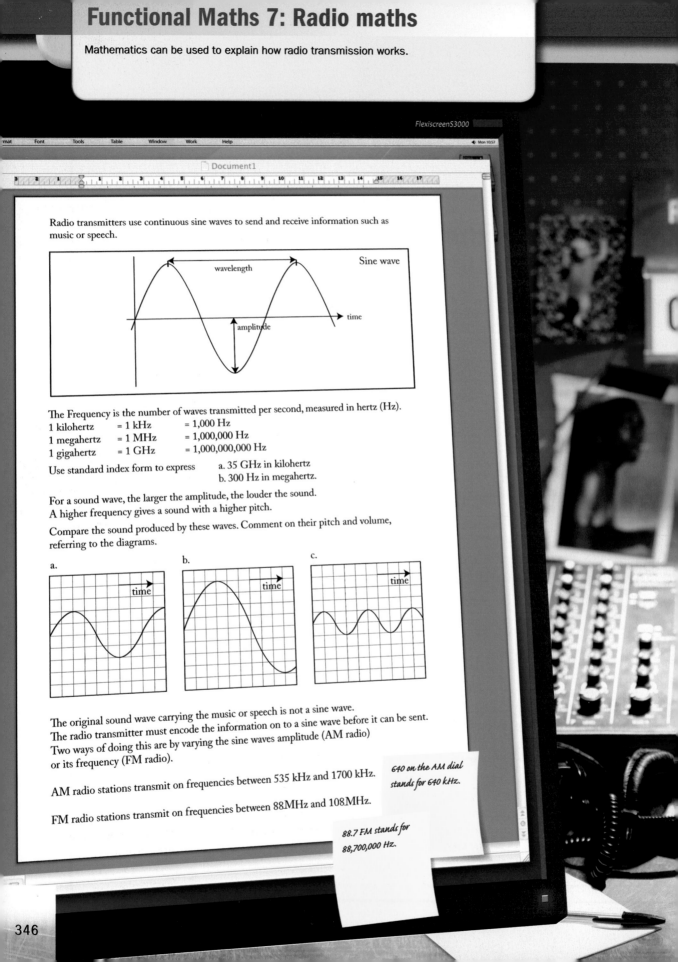

Sine wave

The Frequency is the number of waves transmitted per second, measured in hertz (Hz).

1 kilohertz	= 1 kHz	= 1,000 Hz
1 megahertz	= 1 MHz	= 1,000,000 Hz
1 gigahertz	= 1 GHz	= 1,000,000,000 Hz

Use standard index form to express
  a. 35 GHz in kilohertz
  b. 300 Hz in megahertz.

For a sound wave, the larger the amplitude, the louder the sound.
A higher frequency gives a sound with a higher pitch.

Compare the sound produced by these waves. Comment on their pitch and volume, referring to the diagrams.

The original sound wave carrying the music or speech is not a sine wave.
The radio transmitter must encode the information on to a sine wave before it can be sent.
Two ways of doing this are by varying the sine waves amplitude (AM radio)
or its frequency (FM radio).

AM radio stations transmit on frequencies between 535 kHz and 1700 kHz.

640 on the AM dial stands for 640 kHz.

FM radio stations transmit on frequencies between 88 MHz and 108 MHz.

88.7 FM stands for 88,700,000 Hz.

Wave speed (m/s) = frequency (Hz) × wavelength (m)

Maths FM transmits on the frequency 93.2 FM with a wavelength of 3.22m.
a. What is the frequency of the radio station in
    i.   Mhz   ii. Hz   iii. GHz?
    Give your answers in standard index form.

Maths AM transmits on a frequency of 930kHz. The wave speed is the same as for Maths FM.
b. What is the wavelength used by Maths AM?

The frequency (in kHz) of another radio station, Radio Alpha, is equal to its wavelength in metres.
c. Is this radio station on the AM or FM dial?
    Justify your answer and write down its AM or FM frequency.

Mathematics can also be applied to plan and produce radio programmes.

DJ Cool uses this wheel diagram to plan his hour-long show:

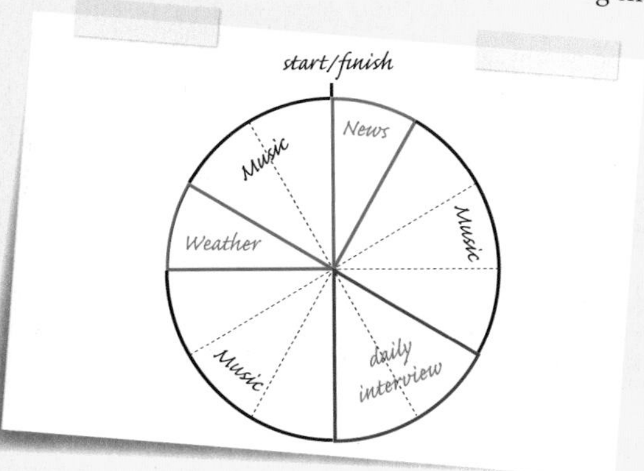

a. How many minutes of the show are taken up by
    i.   music   ii. speech-based material?

b. The radio station has a rule that at least 30% of every show should be made up of speech-based content. Does DJ Cool achieve this target? Explain your answer, referring to the information given in the diagram.

DJ Talk hosts an hour-long phone-in show from 4pm. The phone-in makes up 75% of the show. The news report is at 4pm and the weather forecast is at 4:45pm. 5-minute music sections are spread throughout the show.

c. Draw a wheel diagram to show how DJ Talk's show might look.

Investigate the frequency and wavelengths used by the radio stations that you and your friends and family listen to.

Consider some of the radio shows that you and your friends and family listen to. Do they use a format that could be shown on a wheel?

347

Further probability

How much will I be charged to insure my car? Insurance companies work out the likelihood of you having an accident in your car, based on age, type of vehicle, driving experience and other factors, and then calculate your risk. This determines the cost of your insurance premium.

What's the point?

Mathematicians need to work out the effect of combining the probabilities of events so that more complex probability questions, such as the risk of a car accident, can be calculated.

Check in

You should be able to

■ **do arithmetic with decimals**

1 Work out

a $1 - 0.45$	**b** $1 - 0.96$	**c** $1 - 0.28$	**d** $1 - 0.375$
e $0.2 + 0.4$	**f** $0.3 + 0.04$	**g** $0.65 + 0.25$	**h** $0.7 + 0.05$
i 0.5×0.36	**j** 0.25×0.68	**k** 0.64×0.3	**l** 0.16×0.75

■ **do arithmetic with fractions**

2 Work out

a $1 - \frac{5}{6}$	**b** $1 - \frac{1}{5}$	**c** $1 - \frac{7}{9}$	**d** $2 - 1\frac{1}{4}$
e $\frac{1}{5} + \frac{2}{3}$	**f** $\frac{3}{4} + \frac{1}{6}$	**g** $\frac{2}{3} \times \frac{5}{6}$	**h** $\frac{2}{9} \times \frac{4}{5}$

What I need to know	What I will learn	What this leads to
N1 Do arithmetic with decimals **N2** Do arithmetic with fractions	■ Calculate probabilities for mutually exclusive and independent events ■ Draw and use tree diagrams	**A level** Maths, Biology, Geography
D3 Calculate theoretical and experimental probabilities		Finance, Insurance, Meteorology, Quality assurance

A bag contains twice as many red balls as blue balls. What is the probability of drawing out a red and a blue ball on two successive draws?

Probability revision

This spread will show you how to:
- Understand and use the probability scale
- Find probabilities of mutually exclusive events
- Calculate theoretical probabilities

You write a probability as a decimal, fraction or percentage.

p.204–208

- Probability takes a value from 0 to 1.
- Probability of an **event** = $\dfrac{\text{number of favourable outcomes}}{\text{total number of outcomes}}$.

The total probability for all possible **outcomes** = 1.

- For an event A, P(not A) = 1 − P(A)

P(A) means the probability of event A.

Two or more events are **mutually exclusive** if they cannot happen at the same time.

- For two mutually excusive events A and B, P(A or B) = P(A) + P(B).
- **Expected number** = total number of trials × probability of event.

Sometimes called the OR rule.

Example

A spinner has 8 sides, of which 4 show squares, 3 show triangles, and 1 shows a circle.

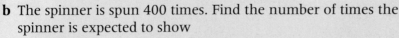

a The spinner is spun once. Find the probability that the spinner
  i shows a triangle
  ii does not show a triangle
  iii does not show a circle
  iv shows a square or a circle
  v shows a triangle or a square.

b The spinner is spun 400 times. Find the number of times the spinner is expected to show
  i a circle   **ii** a triangle.

a **i** P(triangle) = $\frac{3}{8}$

 ii P(not triangle)$=1-\frac{3}{8}=\frac{5}{8}$

 iii P(not circle) $=1-\frac{1}{8}=\frac{7}{8}$

 iv P(circle or square) = P(circle) + P(square) = $\frac{1}{8}+\frac{4}{8}=\frac{5}{8}$

 v P(triangle or square) = P(triangle) + P(square) = $\frac{3}{8}+\frac{4}{8}=\frac{7}{8}$

b **i** Expected number of circles = 400 P(circle) = 400 × $\frac{1}{8}$ = 50

 ii Expected number of triangles = 400 × P(triangle) = 400 × $\frac{3}{8}$ = 150

1 A spinner has ten equal sides, of which 4 show squares, 3 show
  pentagons, 2 show hexagons and 1 shows a circle. The spinner is
  spun once. Find the probability that the spinner
  a shows a square   **b** does not show a square
  c shows a square or pentagon   **d** shows a pentagon or circle
  e does not show a hexagon   **f** does not show a circle.

2 A fair coin is thrown 482 times.
  How many times would you expect the coin to land on tails?

3 An ordinary fair dice is rolled 612 times.
  How many times would you expect it to land on 6?

4 In a class of 28 students, 15 are girls and 13 are boys. 22 students
  wear glasses. One student is chosen at random.
  What is the probability that this student
  a is a boy   **b** is a girl
  c wears glasses   **d** does not wear glasses?

5 A spinner has 16 equal sides. The table shows the colours of the sides
  and the shapes drawn on them. The spinner is spun once.

	Circle	Triangle	Square
Red	3	5	1
Black	4	1	2

  a Find the probability that the spinner
    i shows red   **ii** shows a triangle
    iii shows a circle or a triangle   **iv** does not show a circle
    v shows a red triangle   **vi** shows a black square.
  b Copy and complete this sentence

  'The probability of the spinner landing on the red square is the same as the
  probability of the spinner landing on _____ _____'

6 A fair, eight-sided dice has equal sides. Two sides show 2s, three sides
  show 3s and three sides show 5s.
  a The dice is thrown once. Find the probability that the dice shows
    i an even number   **ii** an odd number
    iii a prime number   **iv** a multiple of 4.
  b The dice is thrown 256 times.
  How many times would you expect the dice to land on 2?

Independent events

This spread will show you how to:
- Draw tree diagrams for the outcomes of several events

Keywords
Event
Independent
Outcome

- Two or more events are **independent** if when one event occurs it has no effect on the other event(s) occuring.

When a dice is rolled and a coin is thrown, the number rolled on the dice has no effect on whether the coin lands on a head or a tail. The events are independent.

When two or more coins are thrown, a head or tail on the first coin has no effect on what shows on any of the other coins. All the events are independent.

You use the multiplication rule to find the probability of two or more independent events.

- For two independent events A and B, P(A and B) = P(A) × P(B)

Sometimes called the AND rule.

Example

A spinner has 12 equal sides: five green, four blue, two red and one white.

A fair coin is thrown and the spinner is spun.

a What is the probability of getting
  i a head and green   **ii** a tail and white
  iii a tail and red   **iv** a head and red?

b Why are the answers to **iii** and **iv** the same?

....

a **i** $P(H) = \frac{1}{2}$ $P(green) = \frac{5}{12}$

   $P(\text{head and green}) = P(H) \times P(green) = \frac{1}{2} \times \frac{5}{12} = \frac{5}{24}$

   ii $P(T) = \frac{1}{2}$ $P(white) = \frac{1}{12}$

   $P(\text{head and white}) = P(H) \times P(white) = \frac{1}{2} \times \frac{1}{12} = \frac{1}{24}$

   iii $P(T) = \frac{1}{2}$ $P(red) = \frac{2}{12}$

   $P(\text{head and red}) = P(H) \times P(red) = \frac{1}{2} \times \frac{2}{12} = \frac{2}{24}$

   iv $P(T) = \frac{1}{2}$ $P(red) = \frac{2}{12}$

   $P(\text{head and red}) = P(H) \times P(red) = \frac{1}{2} \times \frac{2}{12} = \frac{2}{24}$

$P(H) = P(T)$

b The answers to **iii** and **iv** are the same since the coin is fair.

Example

A dice is rolled twice.

Find the probability that on the first roll the dice shows a four and on the second roll the dice shows an odd number.

....

$P(4) = \frac{1}{6}$ $P(\text{odd number}) = \frac{3}{6}$ $P(4, \text{odd number}) = \frac{1}{6} \times \frac{3}{6} = \frac{3}{36}$

1 A fair coin is thrown and an ordinary dice is rolled.
  a Copy and complete the table to list all the outcomes.
  One has been done for you.

Coin/Dice	1	2	3	4	5	6
Head		H2				
Tail						

  b Find the probability of getting
  i a head and a 2   **ii** a tail and a 4   **iii** a tail and a 5.

2 A red dice and a blue dice are rolled.
  a Draw a table to show all the possible outcomes.
  b Find the probability of getting
    i 6 on the red dice and 6 on the blue dice
    ii 3 on the red dice and 5 on the blue dice
    iii 3 on the red dice and an odd number on the blue dice
    iv 5 or greater on the red dice and 1 on the blue dice.

3 An ordinary fair dice is rolled twice. Find the probability that the dice shows an even number on the first roll and a number greater than 4 on the second roll.

4 A spinner has ten equal sides: 4 show squares, 3 show pentagons, 2 show hexagons and 1 shows a circle. A fair coin is thrown and the spinner is spun. Find the probability of getting
  a a square and a head   **b** a circle and a head
  c a pentagon and a tail   **d** a hexagon and a tail
  e a circle and a tail.

5 A spinner has ten equal sides: 4 show squares, 3 show pentagons, 2 show hexagons and 1 shows a circle. The spinner is spun twice. Find the probability of getting
  a a circle on the first spin and a circle on the second spin
  b a circle on the first spin and a hexagon on the second spin
  c a triangle on the first spin and a square on the second spin
  d a square on the first spin and a circle on the second spin
  e a hexagon on first spin and a pentagon on second spin.

6 A 10 pence coin and a 2 pence coin are spun.
  a Draw a table to show all the possible outcomes.
  b Find the probability that the 10p shows tails and the 2p shows heads.

7 One coin is spun twice. Find the probability that both times the coin shows heads.

This spread will show you how to:

- Draw tree diagrams for the outcomes of several events

Keywords
Random
Replaced
Tree diagram

- You can use a **tree diagram** to show the possible outcomes of two events.

- In a tree diagram,
  - write the outcomes at the end of each branch
  - write the probability on each branch
  - the probabilities on each set of branches should add to 1.

Example

A bag contains 7 yellow and 3 blue marbles.

A marble is chosen at **random** from the bag, its colour is noted and then it is **replaced** in the bag.

The bag is shaken and then a second marble is chosen at random.

Draw a tree diagram to show all the possible outcomes.

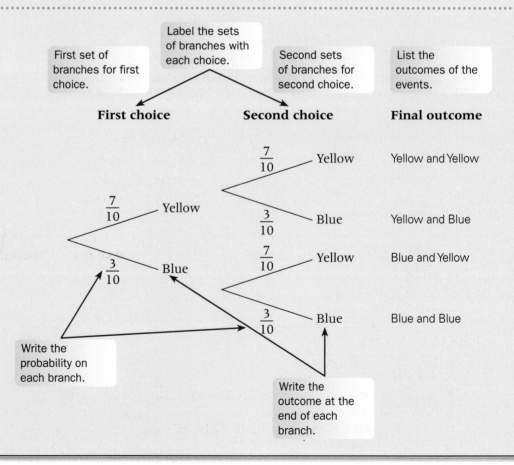

Label the sets of branches with each choice.

First set of branches for first choice.

Second sets of branches for second choice.

List the outcomes of the events.

First choice   **Second choice**   **Final outcome**

$\frac{7}{10}$ — Yellow   Yellow and Yellow

$\frac{7}{10}$ — Yellow

$\frac{3}{10}$ — Blue   Yellow and Blue

$\frac{3}{10}$ — Blue

$\frac{7}{10}$ — Yellow   Blue and Yellow

$\frac{3}{10}$ — Blue   Blue and Blue

Write the probability on each branch.

Write the outcome at the end of each branch.

1 A bag contains 6 purple counters and 11 orange counters. A counter is chosen at random from the bag, its colour noted and then it is replaced in the bag. The bag is shaken and a second counter is chosen at random. Copy and complete the tree diagram.

You will need the tree diagrams drawn for questions **2**, **3** and **4** for Exercise **D2.4**.

First choice   **Second choice**   **Final outcome**

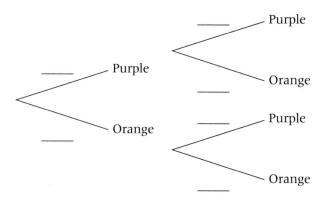

2 A bag contains 4 green and 5 red marbles. A marble is chosen at random from the bag, its colour noted and then it is replaced in the bag. The bag is shaken and a second marble is chosen at random. Draw a tree diagram to show all the possible outcomes.

3 A bag contains 3 white counters and 8 black counters. A counter is chosen at random from the bag, its colour noted and then it is replaced in the bag. The bag is shaken and then a second counter is chosen at random. Draw a tree diagram to show all the possible outcomes.

4 A 10 pence and a 2 pence coin are thrown. Draw a tree diagram to show all the possible outcomes.

5 A bag contains 2 white and 5 black counters. A counter is chosen from the bag and a fair coin is thrown. Draw a tree diagram to show all the possible outcomes.

A03 Problem

6 Dave owns 13 CDs. Three of the CDs are by his favourite group, Cloudplay. Dave chooses one of the CDs at random, notes whether it is a Cloudplay CD, and replaces it. He then chooses another one of the 13 CDs at random. Copy and complete the probability tree diagram that has been started.

First choice   **Second choice**   **Final outcome**

$\frac{3}{13}$ Cloudplay CD

Not Cloudplay CD

Using tree diagrams to find probability

This spread will show you how to:

- Draw and use tree diagrams for the outcomes of several events

Keywords

Tree diagram

- You can use a **tree diagram** to find the probability of an outcome of an event.

- To find the probability of an outcome, multiply the probabilities along the branches leading to that outcome.

Example

A bag contains 7 yellow and 3 blue marbles.

A marble is chosen at random from the bag, its colour noted and it is replaced in the bag.

The bag is shaken and then a second marble is chosen at random.

The tree diagram shows all the possible outcomes.

First choice	Second choice	Outcome

$\frac{7}{10}$ Yellow

$\frac{7}{10}$ Yellow $\qquad \frac{7}{10}$ Yellow $\qquad$ YY

$\qquad \frac{3}{10}$ Blue $\qquad$ YB

$\frac{3}{10}$ Blue $\qquad \frac{7}{10}$ Yellow $\qquad$ BY

$\qquad \frac{3}{10}$ Blue $\qquad$ BB

Use the tree diagram to find the probability of choosing

a two yellow marbles
b two blue marbles
c a yellow marble then a blue marble, in that order.

a $P(YY) = \frac{7}{10} \times \frac{7}{10}$

$= \frac{49}{100}$

b $P(BB) = \frac{3}{10} \times \frac{3}{10}$

$= \frac{9}{100}$

c $P(YB) = \frac{7}{10} \times \frac{3}{10}$

$= \frac{21}{100}$

1 A bag contains 3 green and 4 white marbles. A marble is chosen at random from the bag, its colour noted and then it is replaced in the bag. The bag is shaken and a second marble is chosen at random. The tree diagram shows all the possible outcomes. Use the tree diagram to find the probability of choosing

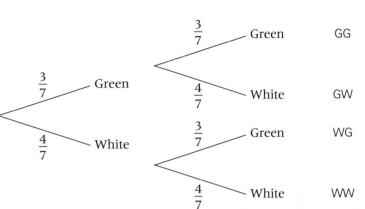

First choice   Second choice   Outcome

a two green marbles
b two white marbles
c a white marble then a green marble in that order.

2 A bag contains 4 green and 5 red marbles. A marble is chosen at random from the bag, its colour noted and then it is replaced in the bag. The bag is shaken and a second marble is chosen at random. Use the tree diagram from Exercise **D2.3**, question **2** to find the probability of choosing
a two green marbles   **b** two red marbles.

3 A bag contains 3 white counters and 8 black counters. A counter is chosen at random from the bag, its colour noted and then it is replaced in the bag. The bag is shaken and a second counter is chosen at random. Use the tree diagram from Exercise **D2.3**, question **3** to find the probability of choosing
a two white counters   **b** two black counters.

4 A 10 pence and a 2 pence coin are thrown. Use the tree diagram from Exercise **D2.3**, question **4** to find the probability of getting
a two heads   **b** two tails
c a head on the 10p and a tail on the 2p.

5 Bag A contains 5 red counters and 7 black counters. Bag B contains 2 yellow counters and 8 red counters. A counter is chosen from each bag.
a Copy and complete the tree diagram to show the possible outcomes.
b Find the probability of choosing
  i a black and a yellow counter
  ii two red counters.

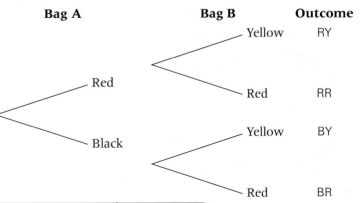

Bag A   Bag B   Outcome

6 Repeat question **3** but now suppose that the first counter is **not** replaced.

This spread will show you how to:
- Use a tree diagram to calculate the probability of an event that can happen in more than one way

Keywords
Event
Independent
Mutually exclusive
Outcome

You can use a tree diagram to find the probability of an **event** that can happen in more than one way.

- To find probabilities when an event can happen in different ways
  - multiply the probabilities along the branches
  - add the probabilities for the different ways of getting the chosen event.

Example

Pamela makes two pottery vases.

Each vase is made independently.

The probability that a vase cracks while it is in the kiln is 0.2.

a Find the probability that the vase does not crack while it is in the kiln.
b Draw a tree diagram to show all the possible outcomes for two vases.
c Find the probability that one of the vases will crack while it is in the kiln.

a P(does not crack) = 1 − P(cracks)
P(cracks) = 0.2
So, P(does not crack) = 1 − 0.2 = 0.8

The outcomes 'cracks' and 'does not crack' are **mutually exclusive**.

b

First Vase   Second Vase

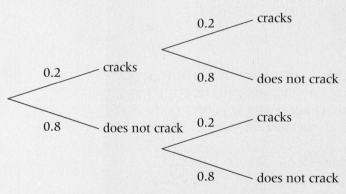

The state of the first vase has no effect on the state of the second vase – the events are **independent**.

c P(one will crack) = P(cracks, does not crack) + P(does not crack, cracks)
= (0.2 × 0.8) + (0.8 × 0.2)
= 0.16 + 0.16
= 0.32

Two ways of getting one cracked vase.

1 Batteries are placed in two toy cars, a red car and a blue car.
The probability that a battery lasts for more than 20 hours is 0.8.
  a Find the probability that a battery does not last for more than
  20 hours.
  b Draw a tree diagram to show the outcomes for the batteries in
  the two cars.
  c Find the probability that the batteries last more than 20 hours
    i in both cars   **ii** in only one car
    iii in at least one car.

2 A bag contains 6 white and 2 black counters. A counter is chosen
at random from the bag, its colour noted and then it is replaced.
A second counter is chosen at random.
  a Draw a tree diagram to show all the outcomes of choosing two
  counters.
  b Find the probability of choosing
    i two black counters   **ii** one counter of each colour
    iii at least one black counter.

3 Two delicate glasses are placed in a dishwasher. The probability that
a glass breaks in the dishwasher is $\frac{1}{20}$.
  a Find the probability that a glass does not break in the dishwasher.
  b Draw a tree diagram to show all the outcomes for the two glasses.
  c Find the probability that while in the dishwasher
    i neither glass breaks   **ii** only one glass breaks
    iii at least one glass breaks.

4 A spinner has five equal sectors. Three are coloured orange and two
are coloured black. The spinner is spun twice.
  a Draw a tree diagram to show all the outcomes of two spins on
  the spinner.
  b Find the probability that on two spins the spinner lands
    i on black both times   **ii** on black at least one time
    iii once on each colour.

5 Josh makes two model aeroplanes, a grey plane and an orange plane.
He flies both of them. The probability that one crashes is 0.1.
  a Find the probability that a model aeroplane does not crash.
  b Draw a tree diagram to show all the outcomes for the two model
  aeroplanes.
  c Find the probability that both model aeroplanes will crash.
  d Find the probability that only one of the aeroplanes will crash.

6 Repeat question **2** but now suppose that the first counter is
not replaced.

A03 Problem

359

Summary

Check out
You should now be able to:

- Understand and use the probability scale
- Calculate theoretical probabilities and expected frequencies
- Know that if two events, A and B, are mutually exclusive, then the probability of A or B occurring is P(A) + P(B)
- Know that if two events, A and B, are independent, then the probability of A and B occurring is P(A) × P(B)
- Show the possible outcomes of two or more events on a tree diagram
- Use a tree diagram to calculate the probability of combinations of independent events

Worked exam question
Fred did a survey of the time, in seconds, people spent in a queue at a supermarket.
Information about the times is shown in the table.

Time (t seconds)	Frequency
$0 < t \le 40$	8
$40 < t \le 80$	12
$80 < t \le 120$	14
$120 < t \le 160$	16
$160 < t \le 200$	10

A person is selected at random from the people in Fred's survey.

Work out an estimate for the probability that the person selected spent more than 120 seconds in the queue.　　　　　(2)

Edexcel Limited 2006

8
12
14
16
$\underline{10}$ +
60

$\dfrac{16}{60} + \dfrac{10}{60} = \dfrac{26}{60}$

OR

$16 + 10 = 26$

$\dfrac{26}{60}$

Show this addition sum.

Show this addition sum.

Exam questions

1 The probability that a biased dice will land on a four is 0.2
Pam is going to roll the dice 200 times.
a Work out an estimate for the number of times the dice will
land on a four. (2)

The probability that the biased dice will land on a six is 0.4
Ted rolls the biased dice once.
b Work out the probability that the dice will land on either
a four or a six. (2)

(Edexcel Limited 2004)

2 Tom and Sam each take a driving test.
The probability that Tom will pass the driving test is 0.8
The probability that Sam will pass the driving test is 0.6
a Complete a copy of the probability tree diagram.

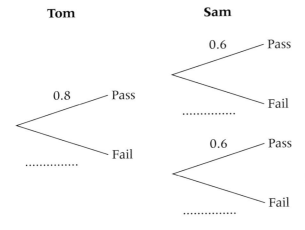

    Tom   **Sam**

(2)

b Work out the probability that both Tom and Sam will pass
the driving test. (2)
c Work out the probability that only one of them will pass the
driving test. (3)

(Edexcel Limited 2008)

Constructions and loci

When a company grow in size and opens a new branch in a different town, they need to find suitable location for their distribution warehouse. In an effort to save transport costs they try to choose a site which has good transport links and which is as close as possible to equidistant from each branch.

What's the point?

Mathematicians use constructions to find all the places on a map that are equidistant from the two towns. The resulting collection of possible places for the distribution warehouse is called the locus (a set of points which share the same mathematical property).

You should be able to

■ **measure lines and angles accurately**

1 Use a protractor to measure these angles.

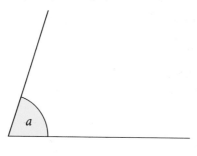
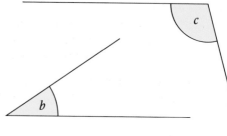

2 Use compasses to draw
  a a circle with radius 3 cm **b** an arc with radius 5 cm.

■ **do simple constructions**

3 **a** Draw a square *ABCD* with sides 4 cm.
   b Draw eight arcs, each with radius 3 cm, centred on each of the four corners of *ABCD*.
   c Join together the points where the arcs cut the sides of the square to form an octagon.

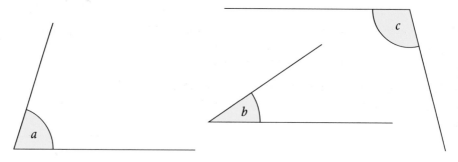

Orientation

What I need to know	What I will learn	What this leads to
KS3 Measure and draw accurately	■ Use bearings and scale drawings	**A-level** Geography
G2 Reason about angles	■ Do standard constructions ■ Find loci	Navigation, Engineering

Rich task

A rectangular field is 80 m long and 60 m wide. An electric fence goes from one corner to the opposite corner along the main diagonal.

A goat is tethered somewhere in the field. The goat can be tethered in different ways.

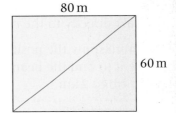

80 m

60 m

Fixed post   Between two fixed post   Running tether

Investigate where the goat should be positioned and how it should be tethered so that it can eat the most grass.

Bearings and scale drawings

This spread will show you how to:
- Use geometry to solve problems involving bearings
- Use and interpret maps and scale drawings

Keywords
Angle
Bearing

- A **bearing** is an **angle** measured in a clockwise direction from north.

p.408

To find the bearing of A from B

- Imagine you are standing at B, facing north.
- Turn clockwise until you face A.

The angle you have turned through is the bearing of A from B.

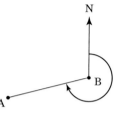

You always write bearings with 3 digits, for example 070°, 190°, 230°

The bearing of A from B is 256°.

Example

The bearing of G from B is 028°. Find the bearing of B from G.

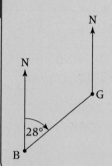

Bearing of B from G is the angle at G measured clockwise from north to B.

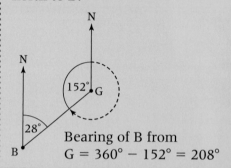

Bearing of B from G = 360° − 152° = 208°

Interior angles are supplementary.

Or using alternate angles, bearing of B from G is 180° + 28° = 208°

p.88

Example

A church, C, is 10 km due west of a school, S.
Joe is 6 km from the school on a bearing of 320°.
He wants to walk directly to the church.

Draw a diagram to show the positions of Joe, the church and the school, and use it to find the bearing Joe should take.
Use a scale of 1 cm to 2 km.

Label point S.
Draw C 5 cm west of S.
Draw the north line at S.
Measure and draw the 320° bearing from S and, 3 cm from S, mark a point, J, to show Joe's position.

Draw the line JC.
Draw the north line at J.
Measure the clockwise angle between the north line and JC.
The bearing Joe needs is 233°.

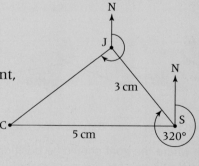

Scale 1 cm to 2 km, so represent 10 km by a line 5 cm long.

Represent 6 km by a line 3 cm long.

Draw and measure lengths and angles carefully or your answer will be inaccurate.

1 These diagrams are drawn accurately.
Measure the bearing of T from S in each.

a N

T

S

b T

N

S

2 P and Q are points 2 cm apart. Draw diagrams to show the position of points P and Q where the bearing of Q from P is
   a 070°   **b** 155°   **c** 340°   **d** 260°

3 These diagrams have not been drawn accurately.
Find the bearing of X from Y in each case.

a N

136°

X

Y

b N

95°

X   Y

c N

X

248°

Y

Draw a sketch to help you.

4 Find these bearings.
   a The bearing of A from B is 104°. Work out the bearing of B from A.
   b The bearing of E from F is 083°. Work out the bearing of F from E.
   c The bearing of J from K is 297°. Work out the bearing of K from J.

5 A youth club (Y) is 4 km due east of a school (S).
Hazel leaves school and walks 5 km on a bearing of 042° to her house (H).
   a Make a scale drawing to show the position of Y, S and H.
   Use a scale of 1 cm to 1 km.
   b Hazel walks directly from her house to the youth club.
   What bearing does she take?

A03 Problem

6 A lighthouse, L, is 6 km on a bearing of 160° from a point H at the harbour.
A boat, B, is 3 km from L on a bearing of 125°.
   a i Make a scale drawing to show the positions of L, H and B.
   ii What bearing should B travel to go directly to H?
   The boat moves 4 km due west.
   b i Mark on your drawing the new position of B.
   ii What bearing should B now travel to go directly to H?

Constructing triangles

This spread will show you how to:
* Use a ruler and compasses to draw standard constructions

Keywords
Arc
Construct
Radius
Triangle

You can **construct** a unique **triangle** when you know

| Two sides and the angle between them (SAS) | or | Two angles and a side (ASA) | or | Right angle, the hypotenuse and a side (RHS) | or | Three sides (SSS) |

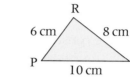

Any two triangles constructed using any one of these four sets of information will be congruent

You will need a ruler and a protractor for SAS, ASA and RHS triangles.

You will need a ruler and compasses for SSS triangles.

Example

a Construct this equilateral triangle ABC with side length 4 cm.

b Construct another equilateral triangle with base AB and side length 4 cm.

c What special quadrilateral have you drawn?

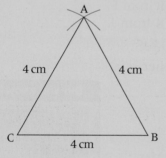

a

Draw BC 4 cm long.

Draw arcs of **radius** 4 cm from B and C to intersect at A.

Join AC and BC.

The construction **arcs** show your method, so do not erase them.

You can use this method to construct an angle of 60° (angles in an equilateral triangle = 60°).

b

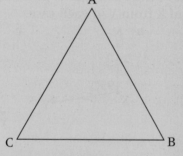

Draw arcs of radius 4 cm from A and B to intersect at D. Join AD and DB.

Four equal sides, two pairs of parallel sides, opposite angles equal.

c A rhombus

1 Use a straight edge and compasses or a protractor to construct these triangles.
  a Sides 8 cm, 4 cm, 7 cm   (SSS) **b** 3 cm, 30°, 4 cm (SAS)
  c Sides 10 cm, 7.5 cm, 6 cm (SSS) **d** 8 cm, 2 cm, 90° (RHS)
  e Sides 6 cm, 9 cm, 5 cm   (SSS) **f** 45°, 4 cm, 45° (ASA)

> It helps to draw a rough sketch first.

2 a Explain why you cannot construct a triangle with sides 9 cm, 4 cm, 3 cm.
  b Explain what happens when you try and construct a triangle with sides 9 cm, 4 cm, 5 cm.

> Try to construct the triangles to see what happens.

3 Without drawing the construction, write whether these sets of three sides will make a triangle.
  a Sides 5 cm, 5 cm, 9 cm   **b** Sides 2 cm, 2 cm, 2 cm
  c Sides 29 cm, 26 cm, 4 cm   **d** Sides 22 cm, 12 cm, 10 cm
  e Sides 20 cm, 7 cm, 9 cm   **f** Sides 14 cm, 8 cm, 6 cm
  g Sides 15 cm, 60 mm, 100 mm   **h** Sides 120 mm, 8 cm, 9 cm

4 Construct isosceles triangles with sides
  a 7 cm, 7 cm, 5 cm   **b** 5 cm, 5 cm, 7 cm

5 a Construct an equilateral triangle ABC with sides 5 cm.
  b Construct a second equilateral triangle with base AB and sides 5 cm to get a rhombus.
  c Follow the steps in **a** and **b** to construct a rhombus with sides 3.5 cm.

6 a Construct a triangle with sides 5 cm, 12 cm, 13 cm.
  b What type of triangle is this?

7 a Construct a triangle ABC with sides AB = 3 cm, BC = 4 cm, CA = 5 cm.
  b Construct triangle ADC with side CA from the triangle in part **a**, and side AD = 4 cm and side DC = 3 cm.
  c What special type of quadrilateral is this?

A03 Problem

8 a Construct a triangle ABC such that BC = 12 cm, AC = 7 cm and ∠B = 30°.
  b Now try to draw a second, **different** triangle with the same measurements. (Move the position of A.)
  c Are SSA triangles unique?

> This triangle is called SSA because you have two sides and the non-included angle.

Constructing bisectors

This spread will show you how to:

- Use a ruler and compasses to draw standard constructions

Keywords

Bisect
Equidistant
Perpendicular

- **Bisect** means cut into two equal parts.

You can use a straightedge and compasses to construct an angle bisector.

- To bisect angle ABC

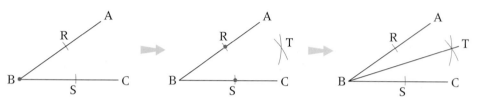

Use the same compass radius throughout the construction. Start at the red dots.

BRTS is a rhombus.

- All points on the angle bisector are **equidistant** from the arms of the angle.

Equidistant means equal distance from.

- The **perpendicular** bisector of a line bisects the line at right angles.

- To construct the perpendicular bisector of line AB

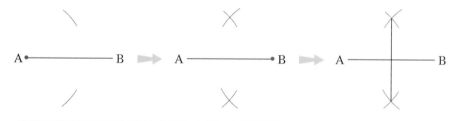

Use the same compass radius throughout the construction.

Start at the red dots.

- All points on the perpendicular bisector of AB are equidistant from A and B.

Example

Construct an angle of 45°.

....

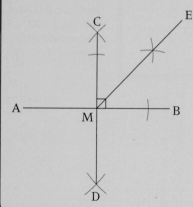

Draw a line AB
Construct the perpendicular bisector CD.
Construct the angle bisector of ∠BMC.
∠BME = 45°

45° is $\frac{1}{2}$ of 90°. Construct a perpendicular bisector to AB (90°) and then bisect the angle.

1 Trace these angles.
Construct the angle bisector of each angle, using ruler and
compasses.

a 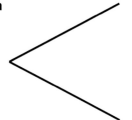 **b** 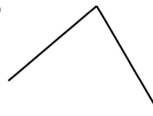 **c**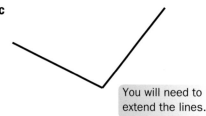

You will need to
extend the lines.

2 **a** Construct an equilateral triangle with sides 5 cm.
b Construct the angle bisector of each angle of the triangle.
c What do you notice about the three angle bisectors?

See the example
in G6.2 for help
with constructing
an equilateral
triangle.

3 Follow these steps to construct an angle of 30°.
a Construct an equilateral triangle with sides 4 cm.
b Bisect one of the base angles.

Angles in an
equilateral
triangle = 60°.

4 Draw lines AB for these lengths and construct their perpendicular
bisectors.
a 6 cm   **b** 9 cm   **c** 5.6 cm   **d** 10 cm   **e** 11.2 cm
Check by measuring that each bisector intersects the line AB at its
midpoint.

5 **a** Construct an equilateral triangle with sides 5 cm.
b Construct the perpendicular bisectors of each side of the triangle.
c Compare your diagram with the one for question **2**. Write down
what you notice.

6 **a** Construct two triangles with sides 8 cm, 5 cm, 7 cm. Label them
triangle A and triangle L.
b On triangle A construct the angle bisector of each internal angle of
the triangle.
c On triangle L construct the perpendicular bisectors of each side of
the triangle.
d Compare and comment on your answers to **b** and **c**.

7 **a** Draw a line AB, 8 cm long, and construct its perpendicular bisector.
b Construct an angle of 45° where the perpendicular bisector
intersects AB.
c What other angles have you created in this construction?

Further constructions

This spread will show you how to:

Keywords
Perpendicular

• Use a ruler and compasses to draw standard constructions

You can construct a **perpendicular** from a point to a line or from a point on a line.

• To construct the perpendicular from a point X to a line YZ

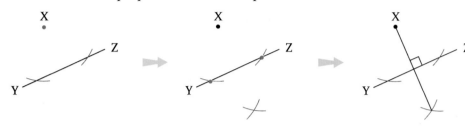

Start at the red dots.

Keep the same compass radius throughout the construction.

• To construct the perpendicular from a point E on a line DF

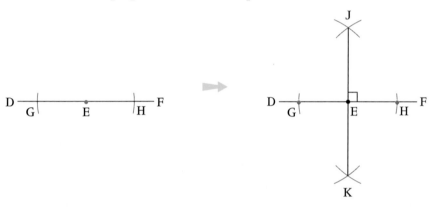

Start at the red dots.

Change your radius for the second part of the construction to a larger one.

• The shortest distance from a point to a line is the perpendicular distance.

Example

Construct a right-angled triangle with sides 3 cm, 4 cm and 5 cm.

Draw a line longer than 4 cm.
Mark the points A and B, 4 cm apart.

Construct a line perpendicular to A.
Mark point C at 3 cm above A on this line.

Draw the third side of the triangle.

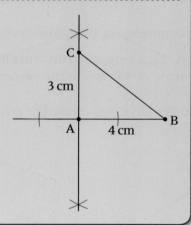

Start by drawing one of the shorter sides.

In a right-angled triangle, the hypotenuse is the longest side. So the other two sides (3 cm and 4 cm here) meet at right angles.

1 Trace these lines and the points marked X.
For each, use ruler and compasses to construct
a perpendicular from the point X to the line.
Check your constructions using a protractor.

Leave in the
construction lines
and arcs.

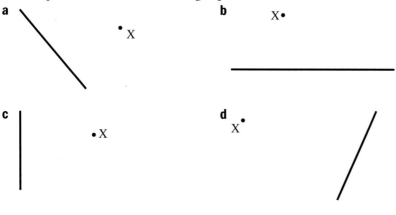

2 Trace these lines and mark the point X.
For each, use ruler and compasses to construct the perpendicular
from the point X on the line. Show all construction lines.
Check your constructions using a protractor.

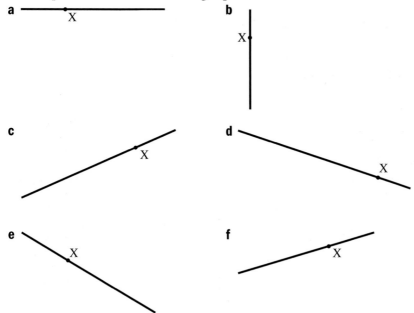

This spread will show you how to:

- Find loci by reasoning and using diagrams

Keywords
Bisector
Equidistant
Locus
Perpendicular

A **locus** is the path traced out by a moving point.

- The locus of a point which is a constant distance from another point is a circle.

- The locus of a point at a constant distance from a fixed line is a parallel line.

Loci is the plural of locus.

- The locus of a point that is **equidistant** from two other fixed points is the **perpendicular bisector** of the line joining the fixed points.

- The locus of a point equidistant from two intersecting lines is the angle bisector of the lines.

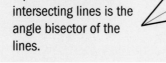

Example

P and Q are two points 2.5 cm apart on a line.

   P   Q
   •   •

Shade in the region that satisfies all these conditions:

- Right of the perpendicular to the line PQ at point P.
- Closer to P than to Q.
- More than 1 cm from P.

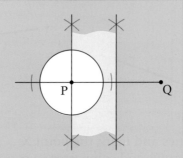

Construct the perpendicular to the line at point P.

Construct the perpendicular bisector of PQ. Points to the left are nearer to P than Q.

Draw a circle radius 2 cm, centre P. Points outside are more than 2. cm from P.

Example

ABCD is the plan of a garden.

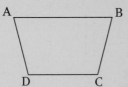

A tree is to be planted in the garden so that it is

- nearer to BC than to BA
- nearer to AD than AB.

Shade the region where the tree may be planted.

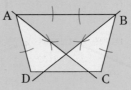

Construct the angle bisector of ∠ABC. Shade the region between this line and BC.

Construct the angle bisector of ∠BAD. Shade the region between this line and AD.

The tree can be planted in the region where the shadings overlap.

1 **a** Draw points A and B, 6 cm apart.
 b Shade in the region that satisfies both these conditions.
  i Closer to A than to B
  ii Less than 4 cm from B

2 **a** Draw points J and K, 5 cm apart.
 b Shade the region that satisfies both these conditions.
  i More than 4 cm from J
  ii More than 3 cm from K

3 **a** Trace the points X, Y and Z.
 b Shade the region that satisfies all three
 of these conditions.
  i Closer to X than to Y
  ii Closer to the line XZ than to the line XY
  iii More than 1 cm from X

X•

Z•   Y•

4 **a** Draw a rectangle PQRS where PQ = 5 cm
 and QR = 3 cm.
 b Shade the region of the rectangle that is within
 4 cm of P and within 2.5 cm of R.

5 **a** Construct a right-angled triangle ABC where angle ABC = 90°,
 AB = 6 cm, BC = 4.5 cm.
 b Shade the region that satisfies all three of these conditions.
  i Closer to A than to B
  ii Less than 4 cm from A
  iii Less than 4 cm from C

6 The diagram shows the rectangular
 garden of a house.
 There are two trees, *T*, in the garden.
 A radio mast is to be placed in the
 garden.
 It must be more than 5 m from the
 rear of the house.
 It must be more than 3 m from a tree.
 Using a scale of 1 cm : 2 m, draw a
 scale diagram and shade the possible
 site for the radio mast.

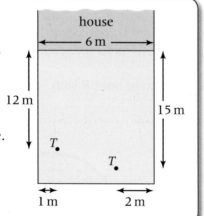

G6

Summary

Check out

You should now be able to:

- Use geometry to solve problems involving bearings
- Use and interpret maps and scale drawings
- Use a ruler and compasses to draw standard constructions
- Construct triangles and other 2-D shapes using ruler, protractor and compasses
- Construct loci

Worked exam question

The diagram shows the position of two boats, P and Q.

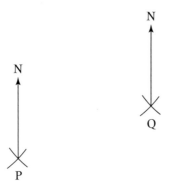

The bearing of a boat R from boat P is 060°
The bearing of boat R from boat Q is 310°
In the space above, draw an accurate diagram to show the position of boat R.
Mark the position of boat R with a cross (×). Label it R.   (3)

(Edexcel Limited 2009)

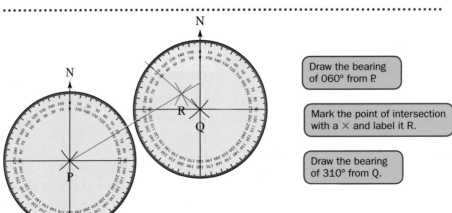

Draw the bearing of 060° from P.

Mark the point of intersection with a × and label it R.

Draw the bearing of 310° from Q.

Exam questions

1 Use ruler and compasses to construct an angle of 30°.
You must show all your construction lines.

(3)

2 Use ruler and compasses to construct the perpendicular to the line segment *AB*
that passes through the point *P*.
You must show all construction lines.

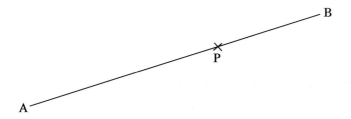

(2)
(Edexcel Limited 2004)

A02

3 The map shows part of a lake.
In a competition for radio controlled boats, a competitor has to
steer a boat so that
its path between *AB* and *CD* is a straight line
this path is always the same distance from *A* as from *B*
a On a copy of the map, draw the path the boat should take.

Scale: 1 cm represents 10 metres

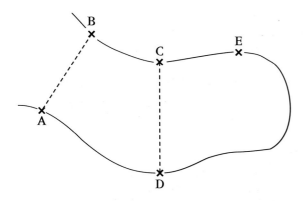

(2)

There is a practice region for competitors.
The practice region is that part of the lake that is less than
30 metres from point *E*.
The scale of the map is 1 cm represents 10 metres.
b Shade this practice region on your copy of the map.

(2)
(Edexcel Limited 2007)

375

Equations and graphs

The world is currently trying to respond to the effects on the environment of global warming. These effects are being predicted by the use of complex mathematical models. The models use mathematical functions along with current data on a wide range of variables.

What's the point?

The graphs of mathematical functions show how one quantity changes in relation to another. A mathematical model uses this idea to represent the real world by using a number of mathematical functions to represent a range of changing quantities and factors.

Check in

You should be able to

■ evaluate formulae

1 Evaluate these expressions for   **i** $x = 3$ **ii** $x = -2$.
   a x^2 **b** x^3 **c** $2x^2$ **d** $x^3 + x$
   e $3x^2 - x$ **f** $2x^3 + 2x$ **g** $x^3 + 4x^2$ **h** $2x^3 - 4x^2 - x$

■ use compound measures

2 Give the speed in miles per hour for each journey.
   a 20 miles is travelled in 1 hour.   **b** 40 miles is travelled in $\frac{1}{2}$ hour.
   c 30 miles is travelled in 20 minutes. **d** 120 miles is travelled in 5 hours.

■ interpet straight line graphs

3 For each line give its
      i gradient   **ii** y-axis intercept   **iii** direction.
   a $y = 3x + 4$ **b** $y = 10 - 4x$ **c** $2y = 8x + 10$
   d $2y - 4x = 15$ **e** $y = 7$ **f** $x = 2y - 4$

■ find the equation of a straight line

4 Give the gradient of each line segment.

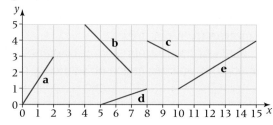

A1 Plot graphs and find equations of straight lines

A5 Solve simultaneous equations

N3 Use compound measures

- Draw and interpret real-life graphs
- Solve equations by trial and improvement and graphically
- Plot more complex graphs

A-level
Maths, Sciences, Economics

Engineering

How bouncy is a ball?
Find ways to compare the bounciness of different balls used in sports such as football, tennis, golf, squash, table tennis, *etc.*
Investigate and write a report on your results.

Distance–time graphs

This spread will show you how to:
- Draw and interpret distance–time graphs
- Understand and use compound measures, including speed

- You can represent a journey on a distance–time graph.
- **Time** is always plotted on the horizontal axis.
- **Distance** is plotted on the vertical axis.

This graph shows Dan's journey on his bike.

p.118

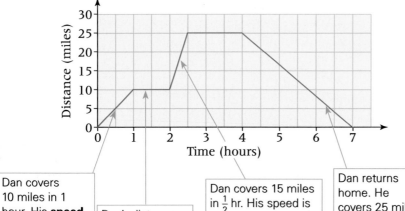

Dan covers 10 miles in 1 hour. His **speed** is 10 mph.

Dan's distance does not change. He has stopped for 1 hour.

Dan covers 15 miles in $\frac{1}{2}$ hr. His speed is 30 mph. He cycles much faster than before (maybe downhill).

Dan returns home. He covers 25 miles in 3 hours.

Example

Janine leaves home at 1 p.m. and cycles to her friend's house, 30 km away, at a speed of 20 km/h. She stays for 2 hours, then cycles home, arriving at 6 o'clock.

Draw a distance-time graph to represent the journey and determine her speed on the way home and her average speed for the entire journey.

Janine returning home, is shown by the graph going down, back to the x-axis.

On the journey home, Janine covers the 30 km in $1\frac{1}{2}$ hours. This means that she has covered 10 km in each half hour and, hence, 20 km in each hour. Her speed is 20 km/h.

Since she covers 60 km in 5 hours, her average speed for the whole journey, including the stop, is 12 km/h.

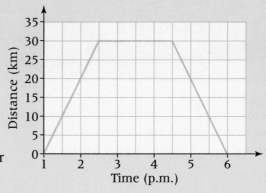

average speed = $\dfrac{\text{total distance}}{\text{total time}}$

1 The distance–time graph shows the journey of a car between Birmingham and Stoke-on-Trent.

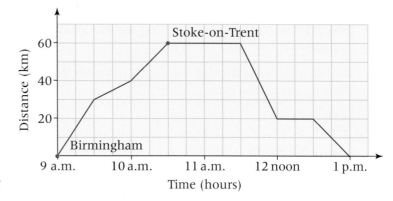

  a How far is it from Birmingham to Stoke-on-Trent?

  b For how long did the car stop?

  c What was the speed of the car for the first part of the journey?

  d Between which two times was the car travelling fastest?

  e What was the average speed of the car for the whole journey?

2 Three students have drawn distance–time graphs. Two have made mistakes. Which two students have made a mistake and what mistake is it?

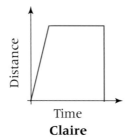

Claire

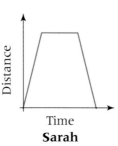

Sarah

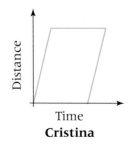

Cristina

3 Construct a distance–time graph to show each journey.

  a A car travels between Bristol and London. On the outward journey, it travels the 120 miles to London in $2\frac{1}{4}$ hours. The driver remains in London for $1\frac{1}{2}$ hours. The car travels half way back to Bristol at 40 miles per hour, as the motorway is busy, then the remaining distance at 80 miles per hour.

  b Two brothers both went to see each other on the same day. Henry left his home at 2 p.m. to go and see Leo, who lives 5 miles away. Henry walked at an average speed of 4 miles per hour, but he stopped half way for a 15 minute rest. At 2:30 p.m., Leo set out on his bicycle from his home in order to go and visit Henry. He cycled straight there in $\frac{1}{2}$ hour.

> Draw the two journeys on one graph.

A03 Problem

4 The following graph represents the journey of a car. Construct a graph of speed (km/h) against time (h) for this journey.

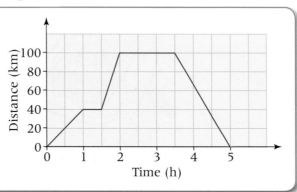

Other real-life graphs

This spread will show you how to:
- Draw and interpret graphs modelling real-life situations

Keywords
Model

You can use graphs to **model** the depth of water flowing in or out of a container at a constant rate.

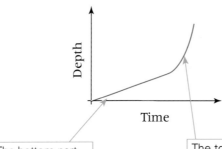

Imagine water filling this container.

The bottom part of the container has the same diameter, so fills at a steady rate

The top part starts wide then narrows. As the container gets narrower, it fills faster.

- In a real-life graph involving time, time is usually represented by the x-axis.

Example

Sketch a graph to show what happens as

A Sam fills a bath with both taps running

B He realises it is too hot so turns off the hot tap

C He turns off the cold tap

D He gets in

E Has a long soak

F Gets out

G Pulls out the plug.

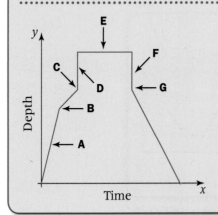

Label the axes with the quantities they represent. A scale is not needed for a sketch graph.

1 Match the four sketch graphs with the containers.

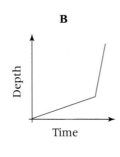   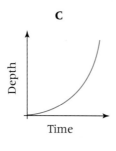   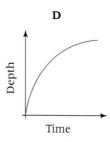

A   B   C   D

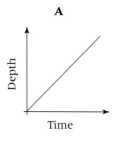

1   2   3   4

2 The sketch graph shows Andrew's height from 3 to 23 years of age. Explain what the graph shows at each stage and explain why this might be.

3 Sketch a graph of depth against time as this container is filled.

A02 Functional Maths

4 Construct sketch graphs to represent these situations.
  a A woman is pregnant and puts on weight. When her baby arrives, she finds it difficult to lose any of the weight for six months, but then joins an exercise class and eats healthily. She is back to her natural weight a year and a half after becoming pregnant. (Graph is weight against time.)
  b A frozen chicken is taken out of the freezer and left to defrost. Two hours later it is put in the microwave briefly to speed up and finish off the process. The chicken is then put in the oven to roast for Sunday lunch. (Graph is temperature against time.)

A03 Problem

5 A skateboarder at a skate park uses a ramp, as shown.
For one go on the ramp, construct a sketch graph of
  a speed against time
  b acceleration against time.

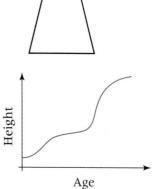

This spread will show you how to:

● Use trial and improvement to find approximate solutions of equations

Keywords
Approximate
Cubic
Systematic
Trial and
  improvement

Some equations don't have exact solutions and can't be solved by an algebraic method. This includes most **cubic** equations.

$x^3 + x = 145$
and $10 - x^3 = 7x$
are cubic equations.

No!

Can you factorise them?

Neither can I!

● You can find an **approximate** solution, correct to several decimal places if necessary, using **trial and improvement**.

Example

Solve the equation $x^3 + x = 145$.
Give your answer correct to 1 dp.

x	x^3	$x^3 + x$	
5	125	130	Too small
6	216	222	Too big
5.5	166.375	171.875	Too big
5.1	132.651	137.751	Too small
5.2	140.608	145.808	Too big
5.15	136.59088	141.74088	Too small

Put your trials in a table. Be **systematic**.

Start with whole numbers then find a better approximation.

The solution is between $x = 5.1$ and $x = 5.2$.

Test the x–value half way between 5.1 and 5.2.

The solution is between $x = 5.15$ and $x = 5.2$, so $x = 5.2$ correct to 1 dp.

You need to work to one more dp than is required in the answer.

Example

p.30

The volume of this cuboid is $20\,\text{cm}^3$.
Find its dimensions, correct to 1 dp.

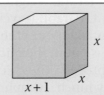

x

$x+1$ x

Volume $= x \times x \times (x + 1) = x^2(x + 1) = x^3 + x$
So $x^3 + x = 20$.

The solution is between $x = 2.55$ and
$x = 2.6$, so $x = 2.6$ to 1 dp.
The dimensions of the cuboid are 3.6 cm,
2.6 cm and 2.6 cm.

x	x^3	$x^3 + x$	
2	8	10	Too small
3	27	30	Too big
2.5	15.625	18.125	Too small
2.6	17.576	20.176	Too big
2.55	16.581375	19.131375	Too small

1 Solve the equation, $x^3 - x^2 = 200$, by trial and improvement to find the solution correct to 1 dp.

2 Write an equation to solve each problem. Use trial and improvement to find the exact answer.
   a The area of this rectangle is $77\,cm^2$.
   What are the length and width?

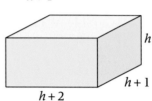

   b The volume of this cuboid is $990\,cm^3$.
   What are the dimensions?

3 **a** Solve $x^2 + 2x = 15$ using trial and improvement in order to find the *exact* solution of this equation.
   b Repeat **a** using an algebraic method and comment on which method you feel is most efficient.

4 Use trial and improvement to solve these equations, giving your answer to the stated degree of accuracy.
   a $x^2 - x = 69$ (to 1 decimal place)
   b $2x^3 + x = 197$ (to 1 decimal place)
   c $p(p + 1) = 100$ (to 2 decimal places)
   d $w^3 - 2w = 70$ (to 2 decimal places)

5 Write an equation to represent the given information. Use trial and improvement to find the value of the unknown in each case.
   a The product of three consecutive numbers is 85 140.
   b The area of a rectangle is $57\,cm^2$. Its length is 2 cm more than its width. Find the width to one decimal place.
   c The surface area of this cuboid is $500\,mm^2$. Find x to 1 decimal place.

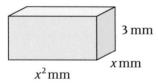

6 Find the number, to 2 dp, that satisfies this statement.

> The square of this number is 20 times its cube root.

Plotting curves

This spread will show you how to:

• Generate points and plot graphs of simple quadratic functions

Keywords
Curve
Minimum
Parabola
Quadratic

• The graph of a **quadratic** equation is a **parabola**, a U-shaped **curve**.

Example

Draw the graph of $y = x^2 + 1$.

You should plot at least 6 points to draw an accurate parabola.

First make a table of x and y values.

x	−4	−3	−2	−1	0	1	2	3	4
x^2	16	9	4	1	0	1	4	9	16
$y = x^2 + 1$	17	10	5	2	1	2	5	10	17

$(−4)^2 = 16$

Draw the x-axis from −4 to 4 and the y-axis from 0 to 18.
Plot the points, (−4, 17), (−3, 10) and so on, and join them in a smooth curve.

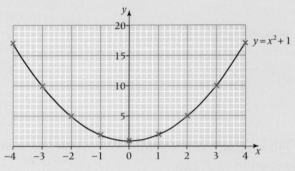

Example

Plot the curve $y = x^2 - x$ for $-2 \le x \le 2$ and find the coordinates of its minimum point.

Make a table of values.

x	−2	−1	0	1	2
x^2	4	1	0	1	4
$-x$	2	1	0	−1	−2
$y = x^2 + x$	6	2	0	0	2

Draw the x-axis from −2 to 2 and the y-axis from −1 to 7.

Plot the points (−2, 6), (−1, 2), ..., (2, 2).

Join the points in a smooth curve.
The **minimum** point is $\left(\frac{1}{2}, -\frac{1}{4}\right)$.

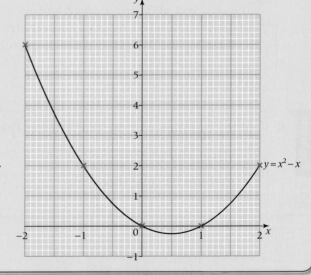

1 a Copy and complete this table to show if each graph will be a straight line or a parabola.

b Add an equation of your own in each column.

Straight line	Parabola

$y = 3x - 2$

$y = x^2 - 2$

$3x + 2y = 8$

$y = 10 + x^2$

$y = x^2 + 2x + 1$

$y = x$

2 a Draw axes labelled from -4 to 4 on the x-axis and -5 to $+15$ on the y-axis.

b Copy and complete this table for $y = x^2 - 2$.

x	−4	−3	−2	−1	0	1	2	3	4
x^2	16							9	
$y = x^2 - 2$	14							7	

c Plot the points that you have found in part **b** on your axes from part **a**. Join them to form a smooth parabola.

3 For each equation
   i Make a table with x-values from -4 to 4 and find the corresponding y-values.
   ii Draw an x-axis from -4 to 4 and a suitable y-axis.
   iii Plot the points and join them to form a parabola.
   iv Write the coordinate of the minimum point of each parabola.
   a $y = x^2 + 3$ **b** $y = 2x^2$
   c $y = 3x^2 - 1$ **d** $y = x^2 + x$

> In part **b**, square before you multiply by 2.

4 True or false? The point $(4, 10)$ lies on the graph $y = x^2 - 5$. Explain your answer.

5 a How do you think the graph $y = 10 - x^2$ will differ from those that you have plotted in questions **2** and **3**?

b Draw and complete a table of coordinates for $-3 \leqslant x \leqslant 3$.

c Plot your points. Was your prediction correct?

A03 Problem

6 a Plot the curve $y = x^2 + 5x + 6$ for $-4 \leqslant x \leqslant 4$.
   b Write the coordinates of the points where the graph intersects the x-axis.
   c Solve $x^2 + 5x + 6 = 0$ using factorisation.
   d Compare your answers to **b** and **c**. What do you notice? Why?

Further curve plotting

This spread will show you how to:

● Plot graphs of simple cubic and reciprocal functions

Keywords
Cubic
Reciprocal
S-shaped

● A **cubic** equation contains a term in x^3. It has a distinctive **S-shaped** graph.

Example

Draw the graph of $y = x^3 - 1$ and use it to estimate the value of y when $x = 2.5$.

First make a table of x and y values.

x	-3	-2	-1	0	1	2	3
x^3	-27	-8	-1	0	1	8	27
$y = x^3 - 1$	-28	-9	-2	-1	0	7	26

$(-3)^3 = -27$

Draw the x-axis from -3 to 3 and the y-axis from -30 to 30.
Plot the points, $(-3, -28)$, $(-2, -9)$ and so on, and join them in a smooth curve.

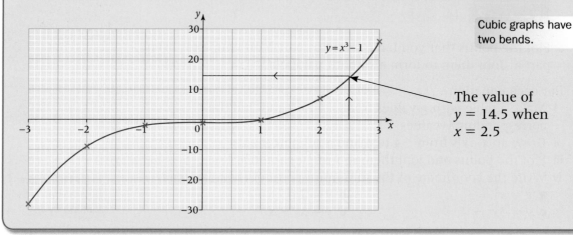

Cubic graphs have two bends.

The value of $y = 14.5$ when $x = 2.5$

A **reciprocal** equation contains a term in $\frac{1}{x}$.
It has a different-shaped curve.

Example

Draw the graph of $y = \frac{1}{x}$.
Use x-axis from -3 to $+3$.

x	-3	-2	-1	1	2	3
y	$-\frac{1}{3}$	$-\frac{1}{2}$	-1	1	$\frac{1}{2}$	$\frac{1}{3}$

Draw the x-axis from -3 to 3 and the y-axis from -1 to 1.
Plot the points.

Note that you cannot plot $x = 0$ as $\frac{1}{0}$ is not a defined value.

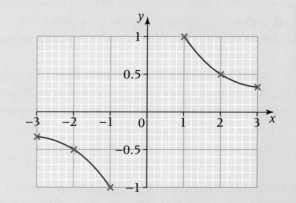

1 Match the graphs with their equations. For the equations that are left over, sketch the shape of their graphs.

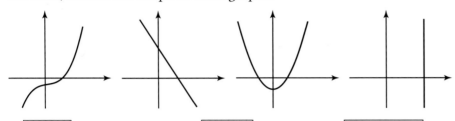

$x = 4$

$y = x^2 - x - 6$

$y = 5$

$y = x^3$

$y = 3 - 2x$

$y = \dfrac{1}{x}$

2 a Draw an x-axis from -3 to 3 and a y-axis from -30 to $+30$.
   b Copy and complete the table for $y = x^3 + 1$.

x	−3	−2	−1	0	1	2	3
y	−26						28

when $x = -3$, $y = (-3)^3 + 1 = -27 + 1$

> When using your calculator, remember brackets.

   c Plot the coordinates that you have found in part **b** on your axes from part **a**. Join them to form a smooth, S-shaped curve.
   d Use your graph to estimate the value of y when
      i $x = 1.5$ **ii** $x = 0.5^3 + 1$

3 For each equation, copy and complete the table of values. Plot these points on suitable axes and join them to form a smooth curve.

a $y = x^3 - 4$

x	−2	−1	0	1	2	3
x^3					8	
$x^3 - 4$					4	

b $y = \dfrac{2}{x}$

x	−2	−1	1	2	3
$\frac{1}{x}$				$\frac{1}{2}$	
y				1	

c $y = x^3 + x + 1$

x	−2	−1	0	1	2	3
x^3					8	
$x + 1$					3	
y					11	

d $y = \dfrac{3}{x} + 1$

x	−2	−1	1	2	3
$\frac{3}{x}$	$-\frac{3}{2}$				
y	$-\frac{1}{2}$				

4 a On a pair of axes for $-3 \leqslant x \leqslant 3$ and suitable y values, plot
      i $y = 8 - x^3$ **ii** $y = 2x^2 - 3x$
   b Use your graph to find the points where $8 - x^3 = 2x^2 - 3x$.

5 Repeat question **3b** plotting further points at $x = -\dfrac{1}{2}, -\dfrac{1}{3}, -\dfrac{1}{4}$
   and $\dfrac{1}{2}, \dfrac{1}{3}, \dfrac{1}{4}$.

This spread will show you how to:

- Find approximate solutions of equations from their graphs, including one linear and one quadratic and simple cubic equations

Keywords
Intersect
Solution

Quadratic and cubic equations can be solved by drawing a graph.

How can I solve $x^2 + 5x + 6 = 2$?

Draw a graph of $y = x^2 + 5x + 6$

x	−5	−4	−3	−2	−1	0
x^2	25	16	9	4	1	0
5x	−25	−20	−15	−10	−5	0
y	6	2	0	0	2	6

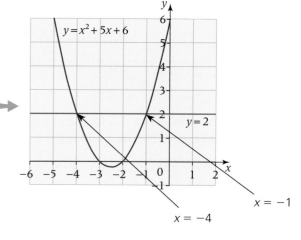

Since $y = x^2 + 5x + 6$,
then $x^2 + 5x + 6 = 2$ when $y = 2$.
Draw the line $y = 2$ on the graph.

The line $y = 2$ crosses the curve $y = x^2 + 5x + 6$ when $x = -1$ and $x = -4$;
$x^2 + 5x + 6 = 2$ when $x = -1$ and $x = -4$.

The **solution**(s) to $x^2 + 5x + 6 = 2$ are where $y = 2$ and $y = x^2 + 5x + 6$.
These are the points where the line and curve **intersect**.

Example

On the graph of $y = x^3 + x^2 - 6x$, where would you find the solutions to $x^3 + x^2 - 6x = 0$? Find them.

The solution of $x^3 + x^2 - 6x = 0$, is where the curve, $y = x^3 + x^2 - 6x$, and the line, $y = 0$, intersect.

The graph
$y = x^3 + x^2 - 6x$
intersects the
x-axis at $(-3, 0)$,
$(0, 0)$ and $(2, 0)$,
so the solutions
of $x^3 + x^2 - 6x = 0$
are $x = -3$, $x = 2$
and $x = 0$.

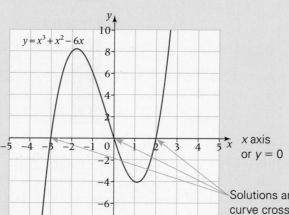

$y = 0$ is the same line as the x-axis.

Notice the characteristic 'S' shape of the cubic equation.

Solutions are where curve crosses the x-axis.

You can check by substituting −3, 2 and 0 for x in the equation.

1   Some graphs are drawn on these axes.

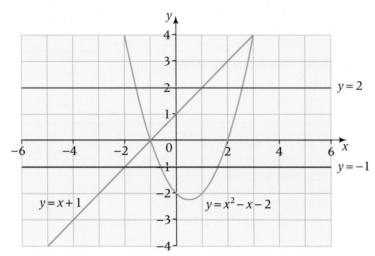

Use the graphs to find the approximate solutions of
a $x^2 - x - 2 = 2$ **b** $x^2 - x - 2 = -1$ **c** $x^2 - x - 2 = x + 1$

2   The graph shows $y = x^2 - 2x - 3$.

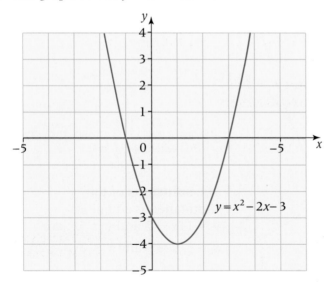

Copy the graph and, by adding lines, use it to find the approximate solutions of
a $x^2 - 2x - 3 = 1$ **b** $x^2 - 2x - 3 = -3$ **c** $x^2 - 2x - 3 = -4$
d $x^2 - 2x - 3 = x - 2$ **e** $x^2 - 2x - 3 = 1 - x$ **f** $x^2 - 2x - 3 = 0$

A03 Problem

3   Draw appropriate graphs to find the approximate solutions of
a $x^2 - 2 = 5$ **b** $x^3 + x = 2x - 1$
c $2x^2 - x = 0$ **d** $x^3 - x^2 = 2$

Solving quadratic and linear simultaneous equations

This spread will show you how to:

- Find the approximate solutions of simultaneous equations from graphs, when one is linear and one quadratic

You can solve two simultaneous equations, one of which is **linear** and the other **quadratic**, by drawing a graph.
Look to the points of intersection.

Example

Solve the simultaneous equations $y = 11x - 2$ and $y = 5x^2$.

Make a table of values for each equation and plot the graphs.

$y = 11x - 2$

x	−1	0	1
11x	−11	0	11
y	−13	−2	9

$y = 5x^2$

x	−3	−2	−1	0	1	2	3
5x²	45	20	5	0	5	20	45

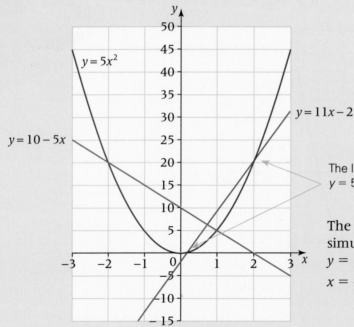

The line $y = 11x - 2$ crosses the curve $y = 5x^2$ where $x = \frac{1}{2}$ and $x = 2$.

The **solutions** of the simultaneous equations $y = 11x - 2$ and $y = 5x^2$ are $x = \frac{1}{5}$, $y = \frac{1}{5}$ and $x = 2$, $y = 20$.

Example

Using $y = 5x^2$ solve $5x^2 - 5 = 5 - 5x$.

Rearrange $5x^2 - 5 = 5 - 5x$
$$5x^2 = 10 - 5x$$
Draw $y = 10 - 5x$.
Points of intersection are $x = -2$ and $x = 1$.

1 Use the graphs to find approximate solutions of
  a $x^2 + 2x - 3 = 0$ **b** $x^2 + 2x - 3 = x + 1$
  c $x^2 + 2x - 5 = 0$ **d** $x^2 + x = 0$

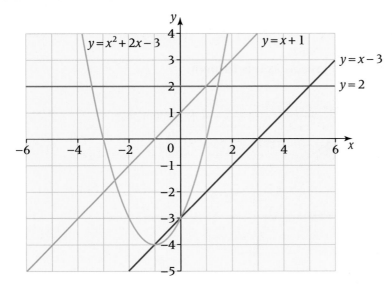

2 Given that the graph of $y = x^2 + 4x - 2$ is already drawn, which one line would you need to draw in order to solve

You do not have to draw them.

  a $x^2 + 4x - 2 = 3$ **b** $x^2 + 4x - 2 = 0$
  c $x^2 + 4x - 2 = 2x + 1$ **d** $x^2 + 4x = 6$
  e $x^2 + 5x = x + 4$ **f** $x^2 + 2 = 6x$

A03 Problem

3 The graph of $y = x^3 + x^2 - 1$ has been drawn below. Copy the graph and add on graphs of your choice in order to find the approximate solutions of
  a $x^3 + x^2 - 1 = 2$ **b** $x^3 + x^2 = x + 2$
  c $x^3 + x^2 - 1 = -2$ **d** $x^3 + x^2 - x = 0$
  e $x^3 + x^2 - 2x + 4 = 0$ **f** $x^3 - 1 = 1 - 2x^2$

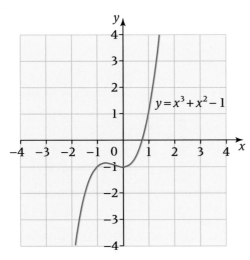

This spread will show you how to:

• Plot graphs of simple quadratic functions

• You can **model** some real-life situations with **quadratic** graphs.

This **parabola** shows the height of a javelin as it is thrown.

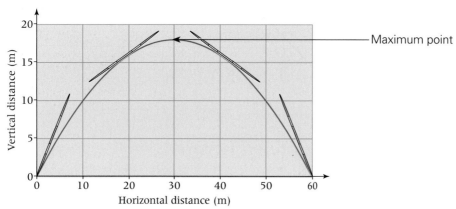

Maximum point

The **maximum** point on the graph shows that the javelin reaches a height of 18 m when it is 30 m away from the throwing line.

Example

One part of a roller coaster is modelled using the equation $y = \dfrac{x^2 - 20x}{10} + 10$ where x is the horizontal distance and y is the vertical distance from the start of the section.

a Complete this table of values and use it to plot the graph that shows the path of the roller coaster.

x	0	5	10	15	20
x^2	0	25			
$-20x$	0	-100			
$\dfrac{x^2 - 20x}{10}$	0	-7.5			
y	10	2.5			

b How long is this part of the roller coaster ride?

a

x	0	5	10	15	20
x^2	0	25	100	225	400
$-20x$	0	-100	-200	-300	-400
$\dfrac{x^2 - 20x}{10}$	0	-7.5	-10	-7.5	0
y	10	2.5	0	2.5	10

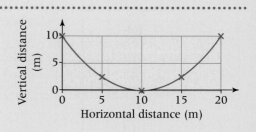

b 20 m

1 The graph $y = 2x^2 + x - 4$ is shown. What are the coordinates of the minimum point of the graph?

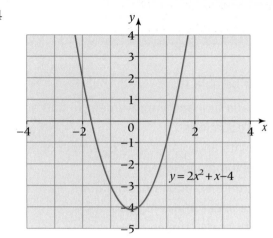

2 a Copy and complete the table of values for the graph $y = x^2 + x + 1$.

x	−3	−2	−1	0	1	2	3
x^2	9						
y	7			3			

b Plot the points for x and y and join them to form a smooth parabola.

c What is the approximate minimum value of $x^2 + x + 1$ and for what value of x does it occur?

3 A ball is thrown into the air.
The formula, $y = 20x - 4x^2$, shows its height, y metres, above the ground x seconds after it is thrown.

a Copy and complete the table of values to show the height of the ball during its first five seconds.

Time(x)	0	1	2	3	4	5
20x						
$4x^2$						
Height (y)						

b Use the table to plot a graph to show the ball's height against time.

c Use your graph to find
  i the maximum height reached by the ball and the time at which it reaches this height
  ii two times when the ball is 12 metres above the ground
  iii the interval of time when the ball is above 15 metres.

Summary

Check out

You should now be able to:

- Use systematic trial and improvement to find approximate solutions to equations
- Generate points and plot graphs of simple quadratic, simple cubic and reciprocal functions
- Find approximate solutions of equations from their graphs, including one linear and one quadratic
- Draw and interpret distance-time graphs
- Draw and interpret non-linear graphs modelling real-life situations
- Construct straight line and quadratic graphs from real-life problems

Worked exam question

a Complete the table of values for $y = 2x^2 - 4x$

x	−2	−1	0	1	2	3
y	16		0			6

(2)

b On the grid, draw the graph of $y = 2x^2 - 4x$ for values of x from −2 to 3

(2)

c **i** On the same axes, draw the straight line $y = 2.5$

ii Write down the values of x for which $2x^2 - 4x = 2.5$ (2)

(Edexcel Limited 2007)

a

When $x = 2$ $y = 2 \times 2^2 - 4 \times 2 = 0$
When $x = 1$ $y = 2 \times 1^2 - 4 \times 1 = -2$
When $x = -1$ $y = 2 \times (-1)^2 - 4 \times -1 = 6$

x	−2	−1	0	1	2	3
y	16	6	0	−2	0	6

b

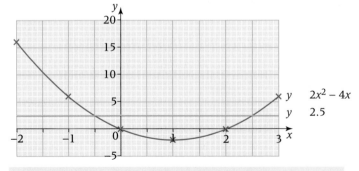

Draw a smooth curve using the points from part **a**.

y $2x^2 - 4x$
y 2.5

c

i See above   **ii** −0.5 and $x = 2.5$

The question asks for values of x (i.e. more than one)

Exam questions

1 The equation $x^3 + 5x = 20$
has a solution between 2 and 3
Use a trial and improvement method to find this solution.
Give your answer correct to one decimal place.
You must show ALL your working.

(4)

2 Here are six temperature graphs.

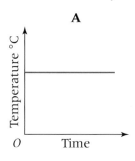

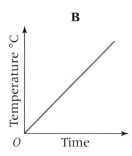

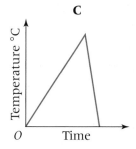

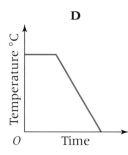

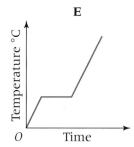

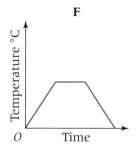

Each sentence in the table describes one of the graphs.
Write the letter of the correct graph next to each sentence.
The first one has been done for you.

The temperature starts at 0°C and keeps rising.	**B**
The temperature stays the same for a time then falls.	
The temperature rises and then falls quickly.	
The temperature is always the same.	
The temperature rises, stays the same for a time and then falls.	
The temperature rises, stays the same for a time and then rises again.	

(3)

(Edexcel Limited 2008)

Functional Maths 8: Business

One out of every two small businesses goes bust within its first two years of trading. Mathematics can be applied to reduce the risk of failure for a business as well as to maximise its profits.

A manager needs to know how much cash is coming into and going out of the business.

Accountants must set a suitable budget that includes realistic performance targets, and limits expenditure to what the business can afford.

Example

Annie sells hand made cards at a monthly craft fair. She has two ranges of cards; standard and deluxe.

The production costs and selling prices per card are:

	Materials used	Time to make	Wages paid	Selling price	Profit
Standard	£0.30	15 minutes	£1.00	£2.55	£1.25
Deluxe	£0.20	30 minutes	£2.00	£3.60	£1.40

This is Annie's cash flow budget for her first three craft fairs (some of the information is missing):

	January (£)	February (£)	March (£)
Standard card sales	45.90	40.80	
Deluxe card sales	43.20		32.40
TOTAL INCOME	89.10	91.20	93.60
Materials used	7.80	7.60	9.00
Wages			
Craft fair fees	10.00	10.00	10.00
Advertising	5.00	5.00	5.00
TOTAL EXPENDITURE	64.80		
NET CASH SURPLUS/DEFICIT	24.30		
CASH BALANCE BROUGHT FORWARD	-	24.30	
CASH BALANCE TO CARRY FORWARD	24.30		

How many of each type of card did Annie sell in each of the three months?

Calculate her spend on materials and wages for each month.

The net surplus (profit) or net deficit (loss) is calculated using the formula Balance = Income – Expenditure
Copy the table and complete the missing values.

On separate copies of the table template, show how the cash flow could change if

* the craft fair fees were increased to £15
* the cost of the materials used to make each type of card increased by 40%
* Annie sold the cards at a discount price of 20% off each type of card.

Investigate how other changes to costs/income might affect Annie's cash flow.

Managers can use mathematical models to make decisions about their business.
These techniques are widely used in production planning to obtain the maximum profit or
to incur the minimum cost in a given situation.

Real life problem → model

Solution to problem ← mathematical model ← model solution

Annie wants to know how much of each type of card she should produce in order to maximise her profits in a particular month. She has £9.00 cash available for materials and a maximum of £50 to spend on wages for that month.

Use s to represent the number of standard cards, and d to represent the number of deluxe cards.

For the material costs, you have $0.3s + 0.2d \leq 9$

For the wage costs, you have $s + 2d \leq 50$

The aim is to maximise the profit, £P, where

$P = 1.25s + 1.40d - 15$

The '– 15' is for the fixed costs.

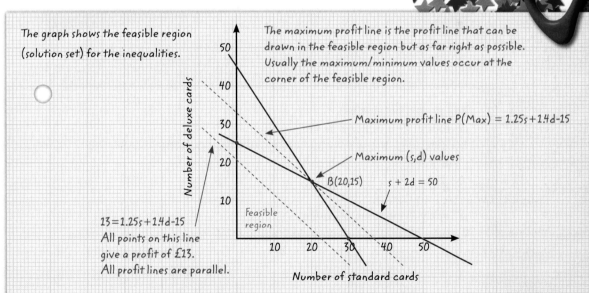

The graph shows the feasible region (solution set) for the inequalities.

The maximum profit line is the profit line that can be drawn in the feasible region but as far right as possible. Usually the maximum/minimum values occur at the corner of the feasible region.

Maximum profit line P(Max) = 1.25s + 1.4d − 15

Maximum (s,d) values

B(20,15)   s + 2d = 50

Feasible region

Number of deluxe cards

Number of standard cards

$13 = 1.25s + 1.4d - 15$
All points on this line give a profit of £13.
All profit lines are parallel.

Annie should make 20 standard and 15 deluxe cards, which would give her a profit of £31.00.

Given these constraints, investigate how changes to the production costs and selling prices might affect the maximum possible profit.

For another month, Annie has only £8.00 cash available for materials and a maximum of £40 to spend on wages. Using the method shown, calculate the maximum possible profit and the number of standard and deluxe cards Annie should make to achieve this value.

How do you think you would need to adapt the method to find the minimum amount Annie would need to spend on materials to guarantee a specified minimum profit and wage for a given month?

Trigonometry

Pilots have to learn how to navigate without the use of GPS. They use the stars as landmarks (celestial navigation) and calculate the angles two separate stars make with the horizon. They look up the position of the stars on their charts and then they can use trigonometry to calculate their position.

What's the point?

Once a right-angled triangle can be seen in a particular situation or problem, a mathematician only needs two pieces of information from which any other lengths or angles can then be calculated.

You should be able to

■ **rearrange simple formulae**

1 Rearrange these equations to make x the subject.

   a $y = \frac{x}{6}$ **b** $y = \frac{x}{5}$ **c** $y = \frac{x}{10}$

   d $y = \frac{2}{x}$ **e** $y = \frac{5}{x}$ **f** $y = \frac{8}{x}$

■ **apply Pythagoras' theorem**

2 Use Pythagoras' theorem to find the missing side in these triangles.

a
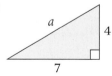

b
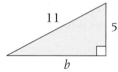

c
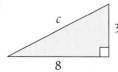

d
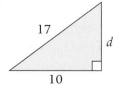

What I need to know	What I will learn	What this leads to
G5 Apply Pythagoras' theorem	■ Use trigonometry and Pythagoras' theorem in right angled triangles	**A-level** Maths, Physics
G6 Use bearings and scale drawings		Engineering, Surveying

Rich task

A company needs to design logos in the shape of regular polygons. Each polygon will have the same side length of 10 cm.

10 cm

10 cm

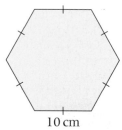
10 cm

Calculate the area of an equilateral triangle, a regular pentagon, *etc*.

Tangent ratio

This spread will show you how to:

- Understand, recall and use trigonometry in right-angled triangles
- Understand and use similarity and ratio to find missing angles and lengths in triangles

Keywords
Adjacent
Opposite
Right-angled
 triangle
Tangent

These two **right-angled triangles** with angle 30° are similar.

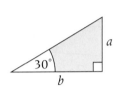 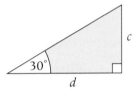

In similar triangles, corresponding pairs of angles are equal.

$\frac{c}{a} = \frac{d}{b}$ which you can rearrange to $\frac{c}{d} = \frac{a}{b}$

The ratio holds for all right-angled triangles with an angle 30°.

This result is true for all similar right-angled triangles.

Corresponding pairs of sides are in the same ratio.

$\frac{e}{f} = \frac{g}{h}$

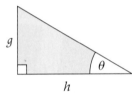 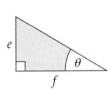

θ, the Greek letter, theta, is often used for angles.

For a right-angled triangle with angle θ, the ratio $\dfrac{\textbf{opposite}\ \text{side}}{\textbf{adjacent}\ \text{side}}$ is constant.

This ratio is called the **tangent** ratio.

Opposite side is opposite the angle θ. Adjacent side is adjacent (next) to angle θ.

- Tan $\theta = \dfrac{\text{opposite side}}{\text{adjacent side}}$
- You can use the tangent ratio in any right-angled triangle.

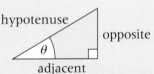

Example

Find the missing sides.

a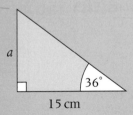

b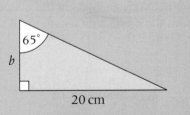

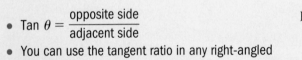

a $\tan 36° = \dfrac{a}{15}$

$15 \times \tan 36° = a$

$a = 10.9\,\text{cm} \ (1\,\text{dp})$

b Third angle $= 180° - (90° + 65°)$
$= 25°$

$\tan 25° = \dfrac{b}{20}$

$b = 20 \tan 25°$
$= 9.3\,\text{cm} \ (1\,\text{dp})$

Use the tan button on your calculator.

tan 3 6 =

Make sure it is in degree mode.

1 Find the missing side in each triangle.
Give your answers to 3 significant figures.

In some triangles you will need to find the
third angle to use in your calculations.

a
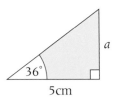

b
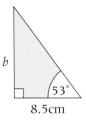

c
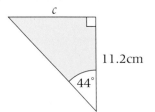

d
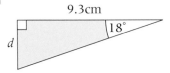

e
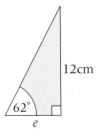

f
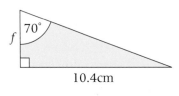

g

h
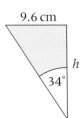

i
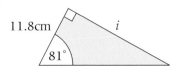

j
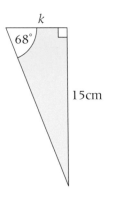

k
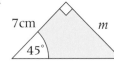

l
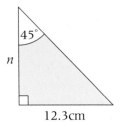

2 In question **1**, what sort of triangles are **k** and **l**?
How could you find the missing sides *m* and *n* without using the
tangent ratio?

This spread will show you how to:

- Understand, recall and use trigonometry in right-angled triangles
- Use the trigonometric functions of a scientific calculator

Keywords
Cosine
Hypotenuse
Right-angled
  triangle
Sine
Tangent

The **tangent** ratio is:

$$\text{Tan } \theta = \frac{\text{opposite side}}{\text{adjacent side}}$$

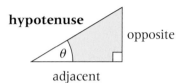

The hypotenuse is the longest side, opposite the right angle.

There are two other ratios you can use in **right-angled triangles**.

- **Sine** ratio

$$\sin \theta = \frac{\text{opposite side}}{\text{hypotenuse}}$$

- **Cosine** ratio

$$\cos \theta = \frac{\text{adjacent side}}{\text{hypotenuse}}$$

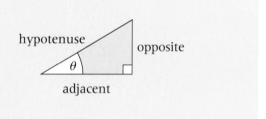

Label the sides you want to find and the side you know.
Remember that opposite and adjacent refer to the angle.

Use the $\boxed{\sin}$ key on your calculator.

Example

Find the missing sides.

a

a $\sin 32° = \dfrac{a}{14}$

$14 \times \sin 32° = a$

$a = 7.42\,\text{cm}$ (3 sf)

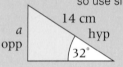

opp and hyp so use sine.

Use the $\boxed{\cos}$ key on your calculator.

b

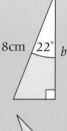

b $\cos 22° = \dfrac{b}{8}$

$8 \times \cos 22° = b$

$b = 7.42\,\text{cm}$ (3 sf)

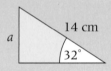

adj and hyp so use cosine.

c

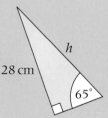

c $\sin 65° = \dfrac{28}{h}$

$h = \dfrac{28}{\sin 65}$

$h = 30.9\,\text{cm}$ (3 sf)

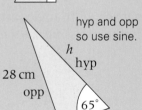

hyp and opp so use sine.

To find the hypotenuse, you will always need to divide by either sin or cos.

1 Find the missing side in each triangle.
Give your answers to 3 significant figures.

In some triangles you will need to find the
third angle to use in your calculations.

a

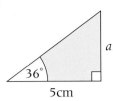

b

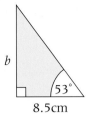

c

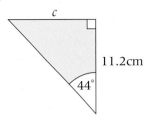

d

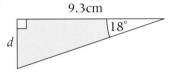

e

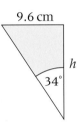

f

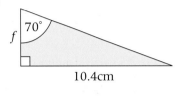

g

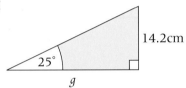

h

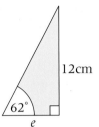

i

j

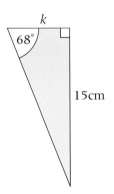

k

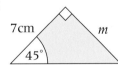

l

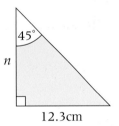

2 In question **1**, what sort of triangles are **k** and **l**?
How could you find the missing sides *m* and *n* without using the
tangent ratio?

This spread will show you how to:

- Understand, recall and use trigonometry in right-angled triangles
- Use the trigonometric functions of a scientific calculator

Keywords
Cosine
Hypotenuse
Right-angled
  triangle
Sine
Tangent

The **tangent** ratio is:

$$\text{Tan } \theta = \frac{\text{opposite side}}{\text{adjacent side}}$$

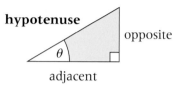

The hypotenuse is the longest side, opposite the right angle.

There are two other ratios you can use in **right-angled triangles**.

- **Sine** ratio

$$\sin \theta = \frac{\text{opposite side}}{\text{hypotenuse}}$$

- **Cosine** ratio

$$\cos \theta = \frac{\text{adjacent side}}{\text{hypotenuse}}$$

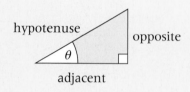

Label the sides you want to find and the side you know.
Remember that opposite and adjacent refer to the angle.

Example

Find the missing sides.

a

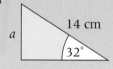

b

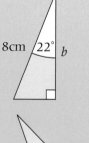

c

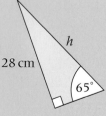

a $\sin 32° = \dfrac{a}{14}$

$14 \times \sin 32° = a$

$a = 7.42 \text{ cm (3 sf)}$

opp and hyp so use sine.

b $\cos 22° = \dfrac{b}{8}$

$8 \times \cos 22° = b$

$b = 7.42 \text{ cm (3 sf)}$

adj and hyp so use cosine.

c $\sin 65° = \dfrac{28}{h}$

$h = \dfrac{28}{\sin 65}$

$h = 30.9 \text{ cm (3 sf)}$

hyp and opp so use sine.

Use the [sin] key on your calculator.

Use the [cos] key on your calculator.

To find the hypotenuse, you will always need to divide by either sin or cos.

1 Find the missing side in each of these right-angled triangles.
Give your answer to 3 significant figures.

a
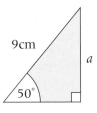
9cm *a* 50°

b
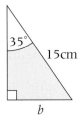
35° 15cm *b*

c
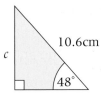
10.6cm *c* 48°

d
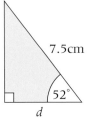
7.5cm 52° *d*

e
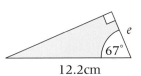
e 67° 12.2cm

f
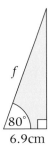
f 80° 6.9cm

g
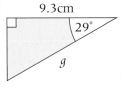
9.3cm 29° *g*

h

42° *h* 6.4cm

i
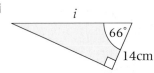
i 66° 14cm

j
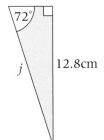
72° *j* 12.8cm

k
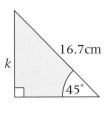
16.7cm *k* 45°

l
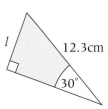
l 12.3cm 30°

m
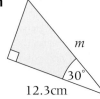
m 30° 12.3cm

n
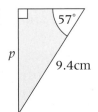
57° *p* 9.4cm

o
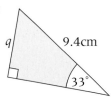
q 9.4cm 33°

Finding angles in right-angled triangles

This spread will show you how to:

- Understand, recall and use trigonometry in right-angled triangles
- Use the trigonometric functions of a scientific calculator

Keywords
Cosine
Right-angled
   triangle
Sine
Tangent

You can use the **sine**, **cosine** and **tangent** ratios in a **right-angled triangle**.

$$\sin \theta = \frac{\text{opp}}{\text{hyp}} \qquad \cos \theta = \frac{\text{adj}}{\text{hyp}} \qquad \tan \theta = \frac{\text{opp}}{\text{adj}}$$

- You can use the inverse functions $\sin^{-1}$, $\cos^{-1}$ and $\tan^{-1}$ to find the angle if you know two sides.

Always start your calculation by labelling opposite side and adjacent side in relation to the angle.

Find the $\boxed{\sin^{-1}}$, $\boxed{\cos^{-1}}$ and $\boxed{\tan^{-1}}$ keys on your calculator. They may be 2nd functions, or you may need to use the $\boxed{\text{INV}}$ key.

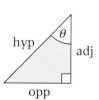

Example

Find the missing angles.
Give your answers to the nearest degree.

a

10.5 cm

x

b

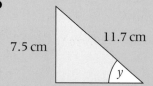

7.5 cm   11.7 cm

y

c

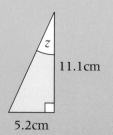

z

11.1 cm

5.2 cm

a You have adjacent and hypotenuse, so use cosine.

$$\cos x = \frac{6}{10.5}$$

$$x = \cos^{-1}\left(\frac{6}{10.5}\right)$$

$$x = 55° \text{ (to the nearest degree)}$$

On your calculator, use the brackets keys for the fraction.

b You have opposite and hypotenuse, so use sine.

$$\sin y = \frac{7.5}{11.7}$$

$$y = \sin^{-1}\left(\frac{7.5}{11.7}\right)$$

$$y = 40° \text{ (to the nearest degree)}$$

c You have opposite and adjacent, so use tan.

$$\tan z = \frac{5.2}{11.1}$$

$$z = \tan^{-1}\left(\frac{5.2}{11.1}\right)$$

$$z = 25° \text{ (to the nearest degree)}$$

Tan can be a fraction >1. sin and cos are always fractions <1.

1 Find the missing angle in each triangle.

Give your answers to 3 significant figures.

a
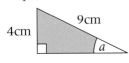
9cm
4cm
a

b
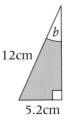
b
12cm
5.2cm

c
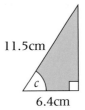
11.5cm
c
6.4cm

d

17.2cm
d
6.6cm

e
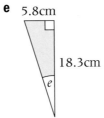
5.8cm
18.3cm
e

f
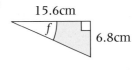
15.6cm
f
6.8cm

g
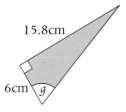
15.8cm
6cm *g*

h
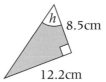
h 8.5cm
12.2cm

i
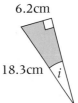
6.2cm
18.3cm *i*

j

19cm
j
23.4cm

k
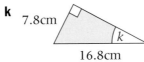
7.8cm
k
16.8cm

l
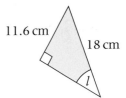
11.6 cm
18 cm
l

m
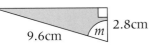
2.8cm
9.6cm *m*

n

16cm
10.5cm
n

o
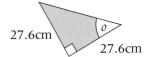
27.6cm
o
27.6cm

p
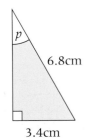
p
6.8cm
3.4cm

q
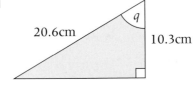
q
20.6cm
10.3cm

Pythagoras' theorem and trigonometry

This spread will show you how to:

- Understand, recall and use Pythagoras' theorem and trigonometry in 2-D problems

Keywords
Cosine
Pythagoras' theorem
Sine
Tangent

Keywords
Cosine
Pythagoras' theorem
Sine
Tangent

You use **Pythagoras' theorem** in a right-angled triangle when you know two sides and want to find the third.

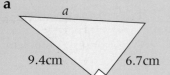

Example

Use Pythagoras' theorem to find the missing sides in these triangles.

a
a

9.4cm   6.7cm

b
b

8cm

24.3cm

Pythagoras' theorem is $a^2 + b^2 = c^2$

a $a^2 = 9.4^2 + 6.7^2 = 133.25$
   $a = \sqrt{133.25} = 11.5\,cm\ (1\,dp)$

b $b^2 + 8^2 = 24.3^2$
   $b^2 = 24.3^2 - 8^2 = 526.49$
   $b = \sqrt{526.49} = 22.9\,cm\ (1\,dp)$

You can use **sine**, **cosine** and **tangent** ratios in a right-angled triangle:
- When you know a side and an angle and want to find another side
- When you know two sides and want to find an angle.

$\mathbf{S}$in $\theta = \dfrac{\mathbf{O}pp}{\mathbf{H}yp}$

- $\sin \theta = \dfrac{opp}{hyp}$ $\cos \theta = \dfrac{adj}{hyp}$ $\tan \theta = \dfrac{opp}{adj}$

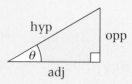

$\mathbf{C}$os $\theta = \dfrac{\mathbf{A}dj}{\mathbf{H}yp}$

$\mathbf{T}$an $\theta = \dfrac{\mathbf{O}pp}{\mathbf{A}dj}$

Example

a Calculate the length FG.
b Calculate the size of angle GEH.

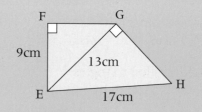

9cm   13cm

17cm

a Using Pythagoras in triangle EFG:
$9^2 + FG^2 = 13^2$
$FG^2 = 169 - 81 = 88$
$FG = \sqrt{88} = 9.4\,cm$ (to 1 dp)

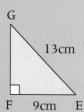

13cm

F   9cm   E

b In triangle GEH:
$\cos \theta = \dfrac{13}{17}$
$\theta = \cos^{-1}\left(\dfrac{13}{17}\right)$
$\theta = 40°$ (to the nearest degree)

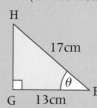

17cm

G   13cm   E

You have adjacent and hypotenuse, so use cosine.

1 Find the missing lengths.

a

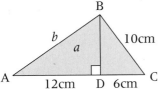

Use BCD, then ABD

b

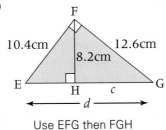

Use EFG then FGH

c

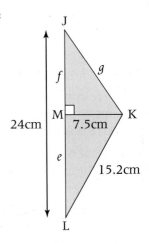

2 Find the missing angles.

a

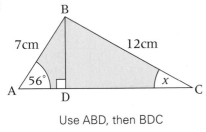

Use ABD, then BDC

b

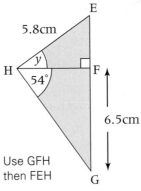

Use GFH then FEH

c

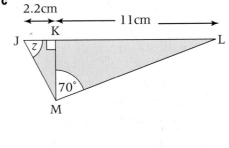

3 ABC and ACD are right-angled triangles.
   a Find AC.
   b Find angle CAD.

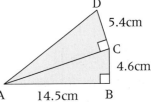

4 JKL and JLM are right-angled triangles.
   a Find JL.
   b Find angle JML.

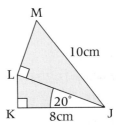

5 PQR and PRS are right-angled triangles.
   a Find PR.
   b Find RQ.

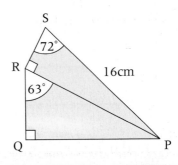

Trigonometry in problem solving

This spread will show you how to:

- Understand, recall and use Pythagoras' theorem and trigonometry in 2-D problems
- Use bearings, maps and scale drawings in problem-solving

 In more complex problems involving lengths and angles, it helps to sketch the situation.

Example

PQRS is a parallelogram. PQ = 9.4 cm. QR = 7.8 cm. Angle PQR = 47°. Find the area of the parallelogram.

....

Draw a diagram.

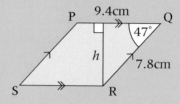

First find the vertical height, h, between PQ and RS.

h is at right-angles to PQ and RS.

$\sin 47° = \dfrac{h}{7.8}$

$h = 7.8 \times \sin 47° = 5.704\ldots$

Area of PQRS = $h \times b$ = 5.704… $\times$ 9.4 = 53.6 cm² (to 3 sf)

Do not round intermediate values in the calculation.

Example

A boat sails from a harbour on a **bearing** of 072° to a buoy 12 km away. Then it changes direction and sails 20 km to a lighthouse due east of the harbour.

a On what bearing does the boat sail from the buoy to the lighthouse?
b How far is it from the lighthouse back to the harbour?

....

Draw a diagram.

H is the harbour.
B is the buoy.
L is the lighthouse.

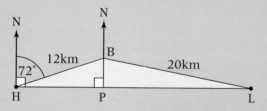

Draw in BP to divide HBL into two right-angled triangles.

a $\angle BHP = 90° - 72° = 18°$

$\sin 18° = \dfrac{BP}{12}$

$BP = 12 \times \sin 18° = 3.708\ldots$

$\cos PBL = \dfrac{3.708\ldots}{20}$

$\angle PBL = \cos^{-1}\left(\dfrac{3.708\ldots}{20}\right)$

$= 79.314\ldots°$

Bearing of L from B
$= 180° - 79.314\ldots°$
$= 101°$ (nearest degree)

b $BP = 3.708\ldots$

$PL^2 = 20^2 - 3.708\ldots^2$ Pythagoras' theorem

$PL = \sqrt{386.249\ldots}$

$= 19.653\ldots$

$PH^2 = 12^2 - 3.708\ldots^2$

$PH = \sqrt{140.291\ldots}$

$= 11.412\ldots$

$HP = PL + PH$

$= 11.412\ldots + 19.653\ldots$

$= 31.1$ km (to 3 sf)

1 Find the area of a parallelogram with side lengths 5 cm and 11 cm and smaller angle 64°.

2 A rhombus has side lengths 9 cm and smaller angle 52°.
Find the area of the rhombus.

3 Find the area of a rhombus with sides 7 cm and smaller angle 40°.

4 A chord AB with length 10 cm is drawn inside a circle with centre O and radius 7 cm.
Find the angle AOB.

5 A chord PQ is drawn inside a circle with centre O and radius 8 cm such that angle POQ = 80°.
Find the length of the chord PQ.

6 A chord ST is drawn inside a circle with centre O and radius 9 cm such that angle SOT = 110°.
Find the length of the chord ST.

7 An isosceles triangle has side lengths 9 cm, 9 cm and 6 cm.
Find the angle between the two equal sides.

A02 Functional Maths

8 Jenny walks 4 km on a bearing 052°. She changes direction and walks a further 5 km to finish due east of her starting point.
Find how far Jenny is from her starting point.

9 Liz leaves home and cycles 16 km on a bearing 215° to a lake.
She changes direction and cycles 12 km to a wood which is due south of her home.
a On what bearing does she cycle from the lake to the wood?
b How far does she have to cycle home?

10 A flag pole TP, with T at the top, is held upright by two ropes, TX and TY, fixed on horizontal ground at X and Y.
Angle PXT = 23°. Angle PYT = 36°. TX = 10 m.
Find the length of TY.

A03 Problem

11 Ali and Pete are estimating the height of a phone mast.
Ali stands 15 m from the mast and measures the angle of elevation to the top as 60°.
Pete stands 25 m from the mast and measures the angle of elevation to the top as 46°.
Can they both be correct? Discuss.

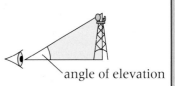

angle of elevation

Summary

Check out

You should now be able to:

- Understand, recall and use Pythagoras' theorem in 2-D problems
- Understand, recall and use trigonometric relationships in right-angled triangles to solve 2-D problems
- Use the trigonometric functions of a scientific calculator
- Use bearings, maps and scale drawings in problem solving

Worked exam question

A lighthouse, L, is 3.2 km due West of a port, P.
A ship, S, is 1.9 km due North of the lighthouse, L.

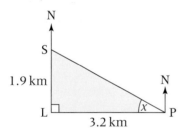

Diagram NOT accurately drawn.

a Calculate the size of the angle marked x.
Give your answer correct to 3 significant figures. (3)

b Find the bearing of the port, P, from the ship, S.
Give your answer correct to 3 significant figures. (1)

(Edexcel Limited 2005)

a

$$\tan x = \frac{1.9}{3.2}$$
$$x = \tan^{-1}\left(\frac{1.9}{3.2}\right)$$
$$x = 30.7°$$

Write out an equation with the trig ratio.

The angle must be rounded to 3 significant figures.

b

$$90° + 30.7° = 120.7°$$
$$= 121°$$

The angle must be rounded to 3 significant figures.

Exam questions

1

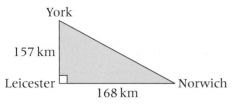

York

157 km

Leicester

168 km

Norwich

Diagram NOT accurately drawn.

The diagram shows three cities.
Norwich is 168 km due East of Leicester.
York is 157 km due North of Leicester.

Calculate the distance between Norwich and York.
Give your answer correct to the nearest kilometre. (3)

(Edexcel Limited 2006)

2 Here is a right-angled triangle.

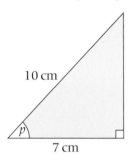

10 cm

p

7 cm

Diagram NOT accurately drawn.

a Calculate the size of the angle marked *p*.
Give your answer correct to 1 decimal place. (3)

Here is another right-angled triangle.

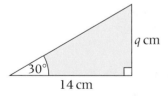

30°

14 cm

q cm

Diagram NOT accurately drawn.

b Calculate the value of *q*.
Give your answer correct to 1 decimal place. (3)

GCSE formulae

In your Edexcel GCSE examination you will be given a formula sheet like the one on this page.
You should use it as an aid to memory. It will be useful to become familiar with the information on this sheet.

Volume of a prism = area of cross section $\times$ length

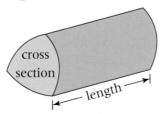

Volume of sphere = $\frac{4}{3}\pi r^3$
Surface area of sphere = $4\pi r^2$

Volume of cone = $\frac{1}{3}\pi r^2 h$
Curved surface area of cone = πrl

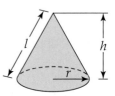

Area of trapezium = $\frac{1}{2}(a + b)h$

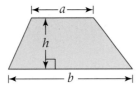

In any triangle ABC

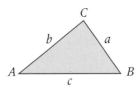

Sine rule $\dfrac{a}{\sin A} = \dfrac{b}{\sin B} = \dfrac{c}{\sin C}$

Cosine rule $a^2 = b^2 + c^2 - 2bc \cos A$

Area of triangle = $\frac{1}{2} ab \sin C$

The Quadratic Equation

The solutions of $ax^2 + bx + c = 0$ where $a \neq 0$, are given by

$$x = \frac{-b \pm \sqrt{(b^2 - 4ac)}}{2a}$$

N1 Check in

1. **a** Four thousand **b** Four hundred
   c Four tenths **d** Four thousandths

2. **a**

$$-2.5$$
$$-3 \quad\quad 0 \quad\quad\quad +4$$

-5 -4 -3 -2 -1 0 1 2 3 4 5

   b -3, -2.4, -1.8, 0, +1.5, +5

N1.1

1. **a** 0.1, 0.3, 1, 1.3, 2, 3.1
   b 6.07, 7.06, 27.6, 77.2, 607
2. **a** 682.8, 862.6, 6000.8, 6008, 8000.6
   b 47.9, 49.7, 74.9, 79.4, 94.7, 97.4
3. **a** 167 **b** 248 **c** 7.16
   d 10.95 **e** 2430 **f** 2813
4. **a** 21.4 **b** 6.73 **c** 410.6
   d 20.07 **e** 0.6025 **f** 8.6
5. **a** 4.52 **b** 5.5 **c** 16.8 **d** 16.8
6. **a** 7.03, 7.08, 7.3, 7.38, 7.8, 7.83
   b 2.18, 2.4, 4.18, 4.2, 8.24, 8.4
7. **a** 18.7, 18.16, 17.6, 17.16, 16.7, 16.18
   b 13.2, 13.145, 2.5, 2.38, 1.1, 1.06
8. **ai** 3050 **aii** 3000 **bi** 1760 **bii** 1800
   ci 290 **cii** 300 **di** 50 **dii** 100
   ei 40 **eii** 0 **fi** 740 **fii** 700
9. **a** 3000 **b** 1000 **c** 0
   d 25 000 **e** 16 000 **f** 168 000
10. **ai** 39.1 **aii** 39.11 **bi** 7.1 **bii** 7.07
   ci 5.9 **cii** 5.92 **di** 512.7 **dii** 512.72
   ei 4.3 **eii** 4.26 **fi** 12.0 **fii** 12.01
   gi 0.8 **gii** 0.83 **hi** 26.9 **hii** 26.88
11. **ai** 0.1 **aii** 0.07 **aiii** 0.070
   bi 15.9 **bii** 15.92 **biii** 15.918
   ci 128.0 **cii** 128.00 **ciii** 127.998
   di 887.2 **dii** 887.17 **diii** 887.172
   ei 55.1 **eii** 55.14 **eiii** 55.145
   fi 0.0 **fii** 0.01 **fiii** 0.007
12. **a** 1306 **b** 2.085 **c** 1085 **d** 2.487
   e 0.0008 **f** 6.19 **g** 0.04513 **h** 0.0045

N1.2

1. **a** -2 **b** -1 **c** -2 **d** -2 **e** -5 **f** -5
2. **a** +1 **b** +3 **c** +4 **d** +9 **e** +12 **f** +5
3. **a** -5 **b** -2 **c** -4 **d** +2 **e** +4 **f** +7
4. **a** -2 **b** + **c** +, - **d** -, -
5. **a** +22 **b** -12 **c** -2 **d** -14
   e +12 **f** +12 **g** +19 **h** -15
   i +11 **j** -61 **k** +344 **l** +49
6. **a** -10.8 **b** +6.3 **c** +13.5 **d** -0.7
   e -38.2 **f** +112.7
7. **a** £5.49 **b** £172.38
8. **a** -13 °C **b** 15 °C **c** 41 °C

N1.3

1. **a** -15 **b** -18 **c** -21 **d** -56 **e** -36 **f** -12
2. **a** +16 **b** +16 **c** +15 **d** +42 **e** +56 **f** +81
3. **a** -25 **b** -32 **c** -72 **d** -20 **e** +30 **f** +49
   g +16 **h** -20 **i** -18 **j** +26 **k** -42 **l** -48
4. **a** -3 **b** -4 **c** -7 **d** -2 **e** +19 **f** +5
5. **a** -2 **b** -5 **c** +5 **d** +4 **e** -22 **f** -1
   g +40 **h** +4 **i** +5 **j** -17 **k** -3 **l** +27
6. **a** + **b** -6 **c** +45 **d** -12
7. **a** +450 **b** -150 **c** +63 **d** -25
   e -0.73 **f** +0.092

8. **a** -49 **b** +63 **c** +3.77 **d** -619.7
   e +140.9 **f** +0.09
9. **a** -36 **b** +0.1 **c** -0.98 **d** -0.0087
   e +0.0073 **f** -0.00006
10. **a** -55 **b** +0.48 **c** +5.266 **d** -156
   e +0.0082 **f** -0.50005
11. **a** +0.18 **b** -27 **c** -7 **d** -380

N1.4

1. **a** 30 **b** 30 **c** 50 **d** 210
   e 780 **f** 23 780
2. **a** 6 **b** 4 **c** 22 **d** 39
   e 18 **f** 454
3. **a** 200 **b** 200 **c** 100 **d** 700
   e 1400 **f** 134 600
4. **a** 2000 **b** 13 000 **c** 8000 **d** 11 000
   e 78 000 **f** 156 000
5. **a** 0.3 **b** 0.7 **c** 0.3 **d** 0.2
   e 4.6 **f** 105.4
6. **a** 0.32 **b** 0.46 **c** 15.30 **d** 104.68
   e 16.45 **f** 0.00
7. **a** 480 **b** 1200 **c** 490 **d** 14 000
   e 530 **f** 15 000
8. **a** 0.36 **b** 0.42 **c** 0.057 **d** 0.0047
   e 1.4 **f** 0.0000042
9. **a** 200 **b** 2000 **c** 5 **d** 10
   e 0.0005 **f** 100 000
10. **a** 0.62 **b** 0.57 **c** 0.56 **d** 380
   e 550 **f** 7 300 000
11. **a** $400 \div 20$ **b** 40×40 **c** $1000 \div 90$
   d $4000 + 10\,000$ **e** $100 + (2000 \div 50)$
12. **a** 20 **b** 1600 **c** 11 **d** 14 000 **e** 140
13. **a** 16.9047619, estimate is slightly high
   b 1677, estimate is close
   c 11.44565217, estimate is close
   d 16 040, estimate is slightly low
   e 153.3846154, estimate is close

N1.5

1. **a** 1 **b** 1.2 **c** 1.5 **d** 8
   e 1.4 **f** 7.5 **g** 16 **h** 15
2. **a** 40 **b** 10 **c** 20 **d** 80
   e 0.2 **f** 10 **g** 10 **h** 5000
3. **a** 0.8 **b** 0.8 **c** 21 **d** 0.21
   e 4.8 **f** 40 **g** 0.9 **h** 4
4. as Q3
5. **a** $4 \div 2$ **b** $6 \div 5$ **c** $12 \div 5$ **d** $2 \div 1000$
6. **a** 4×2 **b** 6×4 **c** 16×100 **d** 15×20
7. **a** 32 **b** 30 **c** 35 **d** 1
   e 0.32 **f** 200 **g** 1 **h** 720
   i 0.375 **j** 28 **k** 1 **l** 7
8. **a** 80 **b** 1000 **c** 520 **d** 500
9. **a** 78.72 to 4 sf **b** 1063 to 4 sf
   c 509.2 **d** 548.4 to 4 sf
10. **a** 9.3 **b** 700 **c** 12.2 **d** 10.6
   e 500 **f** 130 **g** 570 **h** 0.65
11. as Q10

N1.6

1. **a** 38.76 **b** 2.61 **c** 8.201 **d** 22.52
2. **a** 2.12 **b** 4.19 **c** 2.04 **d** 5.7
3. **a** 155.09 **b** 0.45 **c** 24.32 **d** 14.72
4. **a** 3.97 **b** 0.095 **c** 12.44 **d** 58.34
5. **a** 2.81 **b** 0.757 **c** 88.01 **d** 1126.72
6. See Q1 to Q5

413

7   **a**   63.6   **b**   5.37   **c**   13.52   **d**   0.003 36
8   **a**   1.7   **b**   0.43   **c**   48   **d**   269
9   **a**   16.72   **b**   10.26   **c**   793.8   **d**   7.918
10   **a**   8.82   **b**   203   **c**   886   **d**   21.9
11   See Q7 to Q10

N1.7

1   **a**   59   **b**   37   **c**   2   **d**   26
2   **a**   1   **b**   10   **c**   43   **d**   110
3   **a**   +   **b**   −   **c**   ×   **d**   ×
4   **a** $(11 - 1) \times 5 = 50$ **b** $(12 + 3) \div 3 = 5$
      c $12 - (4 - 1) = 9$ **d** $8 \div (4 + 4) + 1 = 2$
5   **a**   2   **b**   5   **c**   110   **d**   14
6   **a**   196   **b**   4   **c**   18   **d**   289
7   **a**   121   **b**   39   **c**   18   **d**   5.29
8   **a**   10   **b**   18
9   **a** ≈ 14.72 **b** ≈ 102.38 **c** ≈ 286.63 **d** ≈ 9.90

N1 Summary

1   200 (Accept 190 to 210)
2   20 000 (Accept 20 000 to 22 800)
3   **a**   9 Accept (8.4 to 9)
     b   0.05 (Accept 0.05 to 0.08)
4   **a**   Pam (A reason must be given)
      $20^2 = 400$
      $16^2 = 256$
      $400 - 256 = 144$
      $\sqrt{144} = 12$
     b   2.75
5   **a**   −6
     b   4
     c   2.5

G1 Check in

1   **ai**   16 cm   **aii**   160 mm   **aiii**   12 cm²
    bi   18 cm   **bii**   180 mm   **biii**   20.25 cm²
    ci   19.1 cm   **cii**   191 mm   **ciii**   14 cm²
2   54 cm²

G1.1

1   **a**   28 cm²   **b**   22.26 cm²   **c**   26.1 cm²
    d   7440 mm²   **e**   10 290 mm²
2   **a**   7.5 cm²   **b**   11.76 cm²   **c**   126 mm²
    d   6 cm²   **e**   14 cm²   **f**   7.2 cm²
    g   10.8 cm²
3   **ai**   40 cm   **aii**   72 cm²
    bi   33 cm   **bii**   32 cm²
    ci   36 cm   **cii**   44 cm²
4   479 cm²

G1.2

1   **a**   15 cm²   **b**   33.48 cm²   **c**   45.6 cm²
    d   31.5 cm²   **e**   13.34 cm²
2   **a**   12 cm²   **b**   20 cm²   **c**   20.5 cm²
    d   600 mm²   **e**   1250 mm²
3   205.5 cm²

G1.3

1   **ai**   **aii**

bi   **bii**

ci   **cii**

di   **dii**

ei   **eii**

fi   **fii**

2   **a**   **b**

     c   **d**

     e   **f**

G1.4

1 a 142 cm² b 98 cm² c 114 cm²
  d 65.6 cm² e 96 cm² f 102 mm²
2 a 330 cm² b 468 cm²
3 a 8 m² b 18.02 m²

G1.5

1 a 105 cm³ b 60 cm³ c 72 cm³
  d 36 cm³ e 64 cm³ f 28 mm³
2 a 270 cm³ b 600 mm³
3 1080 cm³

G1.6

1 a 2.5 litres/s b 1.6 litres/s
2 1200 litres
3 a 30 ml b 24 ml c 12.6 ml
4 ai 96 km aii 12 km aiii 16 km
  bi 3 hours bii 1 hour 30 minutes
  biii 10 minutes

5

Distance	Time taken	Average speed
120 km	1.5 hours	80 km/h
250 miles	4 hours	62.5 mph
4 km	15 minutes	16 km/h
120 m	24 seconds	5 m/s
300 m	15 seconds	20 m/s
0.4 km	160 seonds	2.5 m/s or 12 km/h
3 km	125 seconds	24 m/s
20 km	20 minutes	60 km/h

6 a 57 040 kg b 34 000 kg c 61 824 kg
  d 125 cm³ e 2500 cm³ f 0.002 m³
7 a 8000 kg/m³ b 4500 kg/m³ c 2700 kg/m³

G1 Summary

1

2 432 cm²
3 600 kmph

A1 Check in

1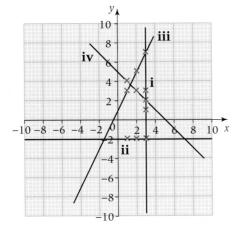

2 a $6\frac{1}{2}$ b $5\frac{1}{4}$ c $4\frac{5}{7}$ d $-9\frac{1}{11}$
  e $\frac{7}{3}$ f $\frac{15}{2}$ g $\frac{23}{4}$ h $-\frac{35}{8}$
3 a $n = 2$ b $m = 1$ c $p = \frac{1}{2}$

A1.1

1 $y = 2x + 3$, $y = 7 - 3x$, $y = 5x$, $y = 7$, $2x + 7y = 8$, $x = -2$

2 ai

x	0	1	2
y	2	5	8

aii

x	0	2	3
y	−2	0	2

aiii

x	0	5	1
y	2	0	1.6

b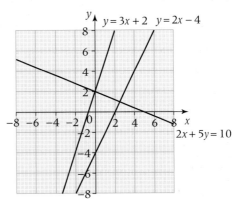

3 a

x	1	2	3	4	5
y	12	19	26	33	40

b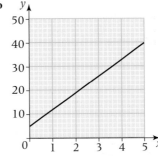

c £26
d $y = 7x + 5$

4 a No, $2 \times 3 + 1 \neq 8$
  b For example (3, 7)

5 a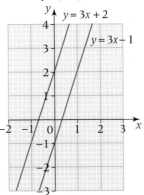

b They are parallel so will never intersect.
c $y = 3x + k$, where k is any number other than −1 or 2.

6 Need to do 40 chores to receive the same amount under both options. If you do less than 40 chores, better to take £5 option; if more, take £3 option.

A1.2

1 a $y = 7$ is horizontal, $x = 9$ and $x = -0.5$ are vertical, $y = 2x - 1$ is diagonal, $y = x^2 + x$ is none of these

2 a $y = 3$ **b** $y = \frac{3}{4}$ **c** $y = -3$

   d $x = -2$ **e** $x = \frac{1}{4}$ **f** $x = 2.5$

3 a Vertical line cutting x-axis at $x = 5$
   b Horizontal line cutting y-axis at $y = 2$
   c Vertical line cutting x-axis at $x = 1.6$
   d Horizontal line cutting y-axis at $y = -3$
   e Horizontal line cutting y-axis at $y = 1$
   f Vertical line cutting x-axis at $x = 1\frac{1}{4}$

4 a $(5, 2)$ **b** $(4, -3)$ **c** $(-2, 9)$ **d** $(-2, -4)$

5 a For example $x = 4$, $x = 7$, $y = 1$, $y = 6$
   b For example $x = 4$, $x = 7$, $y = 3$, $y = 6$
   c For example $x = 4$, $y = 3$, $4y + 3x = 36$

6 a $(1, 2)$
   b For example, below $y = 5$, above $y = 3$, right of $x = 2$, on the line $y = x + 1$

A1.3

1 a $-2, 1$ **b** $3, 2$ **c** $1, -3$ **d** $\frac{1}{2}, -1$

2 a Any line sloping up from left to right.
   b Any line through $(0, 3)$.
   c Any line that goes up 1 unit for every 4 across.
   d Any line that slopes down and goes through $(0, -2)$.
   e The line $y = 3x + 5$.
   f The line $y = \frac{2}{3}x + 1$.

3 a

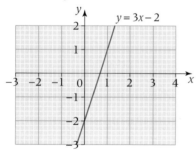

   b $3, -2$
   c Number before x gives gradient, number after x term gives intercept.
   d $y = 2 - \frac{1}{2}x$

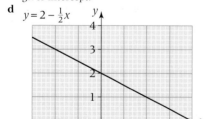

     Gradient $\frac{1}{2}$, intercept 2

4 a $\frac{2}{3}$
   b Divide difference in y-coordinates by difference in x-coordinates.
   c 2

A1.4

1 a $y = 2 - 4x$ **b** $y = 3x + 1$
   c $y = 4x - 2$ **d** $y = x$

2

Equation	Gradient	Direction	Intercept
$y = 4x + 3$	4	Positive	3
$y = 3x + 4$	3	Positive	4
$y = 9x - 2$	9	Positive	-2
$y = 4x - 5$	4	Positive	-5
$2y = 8x + 6$	4	Positive	3

3 ai Gradient = 1   Intercept = 2
   aii $y = x + 2$
   bi Gradient = 2   Intercept = 3
   bii $y = 2x + 3$
   ci Gradient = 3   Intercept = 0
   cii $y = 3x$
   di Gradient = -2   Intercept = 3
   dii $y = -2x + 3$

4 a $y = 3x + k$, where k is any number
   b $y = 3 - 2x$ **c** $y = -3x + 1$

5 a $y = 3x + 4$ **b** $y = 3x + 4$

A1.5

1 $y = 3x + 5$, $y = 5x - 2$, $y = 7 - 2x$, $2y = x + 18$, $4y = -x - 12$, $y = 4$, $y = x$

2 a $y = 4x$, $y = 12 - 2x$ **b** $y = 4x$, $y = 5x - 1$

3 a $y = 7x + 5$ **b** $y = \frac{1}{2}x + 3$ **c** $y = 4x - 4$
   d $y = 3x - 5$ **e** $y = 5 - 2x$ **f** $y = \frac{1}{4}x - 2$
   g $y = 4x + 1$ **h** $y = x + 2$ **i** $y = 8x - 6$

4 a 3   **b** $y = 3x + 5$ **c** -2, $y = 16 - 2x$

5 a $(0, -2.5)$ **b** $\left(\frac{5}{9}, 0\right)$ **c** $(2, 6.5)$

6 $3x + 2y = 12$

A1.6

1 ai 32 km   **aii** 37.5 miles
   b Charlie   **c** $y = 1.6x$

2 a

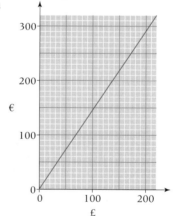

bi Approximately €138   **bii** Approximately £59

3 a

Number of people	1	2	3	4	5	6	7	8	9	10	11	12	13	14	15
Cost (£)	18	21	24	27	30	33	36	39	42	45	48	51	54	57	60

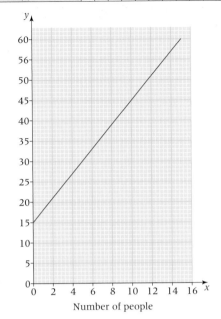

Number of people

b

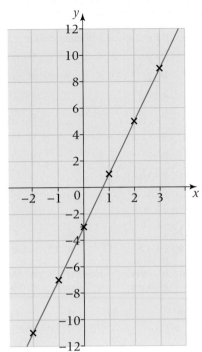

  b 7
    c This would mean that a fraction of a person stayed in the tent.
    d $c = 15 + 3p$, where c = cost per night and p = number of people
    e £96
4 Power Up! is cheaper for less than 10 units; they cost the same for 10 units; Sparks Are Us! is cheaper for more than 10 units.

A1.7

1 A: $y = 4x + 3$, B: $y = 2x + 1$, C: $y = \frac{1}{2}x + 3$,
    D: $y = \frac{1}{3}x + 1$

2 a $y = 6x + 50$; m is 6, meaning that for every year you age, you grow by 6 cm; c is 50, meaning that when you are born you are 50 cm tall, this is reasonable.
  b $y = \frac{1}{2}x + 20$; m is $\frac{1}{2}$, meaning that for every two years a person needs one more driving lesson; it is not sensible to interpret c as no one takes a driving lesson at birth.

3 a Pocket money is not given to very young children or adults, so only makes sense between $x = 4$ and $x = 21$
  b Does make sense to interpret the equation for any value of x, as temperature can take any value.

A1 Summary

1 a

x	-2	-1	0	1	2	3
y	-11	-7	-3	1	5	9

2 a 4
  b (0, 3)

Case study 1: Weather
  16°C; 2.5°C; −8.9°C

England:
Range = 38.5 − −26.1 = 64.6°C

Wales:
Range = 35.2 − −23.3 = 58.5°C

Scotland:
Range = 32.9 − −27.2 = 60.1°C

N. Ireland:
Range = 30.8 − −17.5 = 48.3°C

N2 Check in

1 a 6 squares shaded   **b** 8 squares shaded
  c 9 squares shaded   **d** 10 squares shaded
  e 7 squares shaded
2 a $\frac{1}{2}$ **b** $\frac{3}{4}$ **c** $\frac{4}{5}$ **d** $\frac{19}{20}$ **e** $\frac{3}{4}$
3 a 0.75   **b** 0.4   **c** 0.7   **d** 0.45   **e** 0.17
4 a $\frac{1}{2}$ **b** $\frac{1}{4}$ **c** $\frac{3}{10}$ **d** $\frac{4}{5}$ **e** $\frac{9}{20}$
5 a $20 \times 30 = 600$ **b** $360 \div 60 = 6$
  c $1200 - 800 = 400$ **d** $7000 + 6000 = 13000$

N2.1

1 a 3 squares, 1 and 2 shaded different colours
  b 5 squares, 1 and 3 shaded different colours
2 a $\frac{3}{5}$ **b** 1   **c** $\frac{5}{7}$ **d** 1
3 a 5 squares, 1 and 2 shaded different colours
  b 4 squares, 1 and 3 shaded different colours
  c 7 squares, 2 and 3 shaded different colours
  d 8 squares, 3 and 5 shaded different colours
4 a $\frac{1}{3}$ **b** $\frac{3}{5}$ **c** $\frac{2}{3}$ **d** $\frac{3}{5}$
5 a 6 squares, 2 and 3 shaded different colours
  b 10 squares, 6 and 3 shaded different colours
6 a $\frac{3}{10}$ **b** $\frac{5}{6}$ **c** $\frac{11}{20}$ **d** $\frac{3}{8}$

7 a 10 squares, 2 and 1 shaded different colours
b 6 squares, 4 and 1 shaded different colours
c 20 squares, 8 and 3 shaded different colours
d 8 squares, 1 and 2 shaded different colours

8 a $\frac{1}{8}$ **b** $\frac{1}{2}$ **c** $\frac{3}{4}$ **d** $\frac{5}{16}$

9 a $\frac{5}{21}$ **b** $\frac{2}{15}$ **c** $\frac{1}{18}$ **d** $\frac{7}{20}$

10 a $1\frac{1}{10}$ **b** $1\frac{7}{12}$

11 a $1\frac{1}{6}$ **b** $1\frac{3}{10}$ **c** $1\frac{2}{15}$ **d** $1\frac{1}{14}$

12 a $1\frac{1}{4}$ **b** $1\frac{4}{5}$ **c** $1\frac{5}{8}$ **d** $4\frac{1}{4}$

13 a $\frac{7}{4}$ **b** $\frac{23}{16}$ **c** $\frac{14}{9}$ **d** $\frac{18}{7}$

14 a $3\frac{14}{15}$ **b** $4\frac{7}{12}$ **c** $7\frac{13}{14}$ **d** $7\frac{55}{63}$
e $1\frac{7}{20}$ **f** $\frac{3}{4}$ **g** $1\frac{19}{20}$ **h** $4\frac{13}{14}$

N2.2

1 a $3\frac{1}{2}$ **b** $1\frac{3}{5}$ **c** $3\frac{1}{3}$ **d** $\frac{2}{3}$ **e** $2\frac{2}{5}$

2 a 10 **b** 21 **c** 12 **d** 13 **e** 22

3 a 24 m **b** 36 m

4 a 15 litres **b** 180 cl **c** 48 cl **d** 250 ml

5 a 30 m **b** 8 km **c** 16 mm **d** 90 m

6 a $1\frac{2}{3}$ **b** $5\frac{1}{2}$ **c** $4\frac{1}{2}$ **d** $\frac{2}{3}$ **e** $9\frac{1}{3}$

7 a $2\frac{2}{3}$ miles **b** $33\frac{1}{3}$ miles
c $12\frac{1}{2}$ miles **d** $3\frac{3}{4}$ miles

8 a 6.67 m **b** 5.33 mm
c 8.73 cm **d** 22.3 km

9 40.83 m³ to 2 dp

10 a 5 hours 15 minutes **b** 13 hours 45 minutes
c 16 hours 48 minutes **d** 24 hours 30 minutes

11 a £4.67 **b** £16.67 **c** £12.75 **d** £29.33.

N2.3

1 a $\frac{1}{4}$ **b** $\frac{1}{6}$ **c** $\frac{1}{10}$ **d** $\frac{1}{12}$

2 a 5 **b** 9 **c** 2 **d** 3

3 a $8\times\frac{1}{5}$ **b** $6\times\frac{1}{4}$ **c** $9\times\frac{1}{5}$ **d** $17\times\frac{1}{3}$

4 a $\frac{1}{4}$ **b** 2 **c** 3

5 a

b
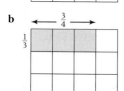

c
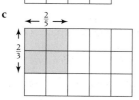

d

6 a $\frac{3}{20}$ **b** $\frac{4}{27}$ **c** $\frac{1}{14}$ **d** $\frac{1}{4}$
e $\frac{20}{63}$ **f** $\frac{1}{12}$ **g** $\frac{12}{65}$ **h** $\frac{1}{15}$

7 a $\frac{3}{4}$ **b** $\frac{3}{4}$ **c** $\frac{4}{5}$ **d** $\frac{4}{5}$
e $\frac{3}{7}$ **f** $1\frac{4}{5}$ **g** $3\frac{3}{7}$ **h** $2\frac{3}{4}$

8 a $\frac{5}{32}$ **b** $\frac{1}{8}$ **c** $\frac{2}{21}$ **d** $\frac{1}{48}$
e 20 **f** 9 **g** $12\frac{1}{2}$ **h** $25\frac{2}{3}$

9 a $\frac{1}{3}$ **b** $\frac{1}{4}$ **c** $\frac{1}{6}$ **d** $\frac{1}{4}$
e $\frac{8}{9}$ **f** $\frac{45}{56}$ **g** $\frac{2}{9}$ **h** $\frac{8}{21}$

10 a $1\frac{1}{8}$ **b** $1\frac{1}{10}$ **c** 3 **d** 3
e $4\frac{19}{24}$ **f** $1\frac{3}{14}$ **g** $1\frac{21}{22}$ **h** $4\frac{4}{27}$

N2.4

1 a 0.5 **b** 0.75 **c** 0.4 **d** 0.1 **e** 0.2 **f** 0.25

2 a 50% **b** 75% **c** 40% **d** 10% **e** 20% **f** 25%

3 a 62.5% **b** 80% **c** 87.5% **d** 60% **e** 37.5% **f** 12.5%

4 See Q3

5 a 0.0625 **b** 0.28 **c** 0.056 **d** 0.075 **e** 0.4375 **f** 0.03125

6 a 6.25% **b** 28% **c** 5.6% **d** 7.5% **e** 43.75% **f** 3.125%

7 a $0.\dot{3}$ **b** $0.1\dot{6}$ **c** $0.\dot{6}$ **d** 0.1428 57 **e** $0.\dot{1}$ **f** $0.8\dot{3}$

8 a $33.\dot{3}\%$ **b** $16.\dot{6}\%$ **c** $66.\dot{6}\%$ **d** 14.285 714% **e** $11.\dot{1}\%$ **f** $83.\dot{3}\%$

9 See Q7 and Q8

10 a $0.42857\dot{1}$ **b** 0.1875 **c** 0.2125 **d** $0.\dot{5}$ **e** $0.1\dot{6}$ **f** $0.7142 8\dot{5}$

11 a Terminating (only prime factor of denominator is 5)
b Terminating (only prime factors of denominator are 2 and 5)
c Recurring (denominator has a prime factor of 11)
d Recurring (denominator has prime factors of 3 and 7)
e Terminating (only prime factor of denominator is 5)
f Terminating (only prime factor of denominator is 2)

12 The decimal does recur. The restricted number of digits on the calculator display, and the rounding of the final digit, obscure the recurring pattern.

N2.5

1 a 0.43 **b** 0.86 **c** 0.94 **d** 0.455 **e** 0.0375 **f** 1.05

2 a $\frac{1}{2}$ **b** $\frac{1}{4}$ **c** $\frac{1}{5}$ **d** $\frac{1}{8}$ **e** $\frac{3}{4}$ **f** $\frac{9}{10}$

3 a $\frac{51}{100}$ **b** $\frac{43}{100}$ **c** $\frac{413}{1000}$ **d** $\frac{719}{1000}$ **e** $\frac{91}{100}$ **f** $\frac{871}{1000}$

4 a $\frac{49}{100}$ **b** $\frac{53}{100}$ **c** $\frac{73}{100}$ **d** $\frac{81}{100}$ **e** $\frac{37}{100}$ **f** $\frac{19}{100}$

5 **a** $\frac{8}{25}$ **b** $\frac{11}{20}$ **c** $\frac{11}{25}$

  d $\frac{31}{200}$ **e** $\frac{16}{25}$ **f** $\frac{53}{200}$

6 **a** $\frac{11}{20}$ **b** $\frac{31}{50}$ **c** $\frac{21}{25}$

  d $\frac{13}{20}$ **e** $\frac{18}{25}$ **f** $\frac{37}{200}$

7 $\frac{8}{11}$, 11 not multiple of 2 or 5

8 **a** $0.\dot{1}$ **b** $0.\dot{5}$ **c** $0.7\dot{5}$

  d $0.3\dot{4}\dot{6}$ **e** $0.76\dot{5}$

9 **a** $\frac{2}{9}$ **b** $\frac{2}{3}$ **c** $\frac{25}{99}$

  d $\frac{3}{11}$ **e** $\frac{545}{999}$ **f** $\frac{605}{999}$

10 **a** $\frac{47}{90}$ **b** $\frac{1}{18}$ **c** $\frac{249}{550}$

  d $\frac{752}{9000}$ **e** $\frac{2}{3}$ **f** $\frac{8197}{99\,900}$

11 $\frac{13\,717\,421}{1\,111\,111\,111}$

N2 Summary

1 **a** $\frac{1}{4}$

  b $1\frac{1}{20}$

  c $\frac{1}{3}$ is not 0.3, but 0.333 recurring

  OR

  0.3 is $\frac{3}{10}$ not $\frac{1}{3}$

2 $3\frac{1}{3}$

3 **a** $\frac{14}{15}$

  b $3\frac{3}{4}$

4 $5\frac{4}{15}$

5 **a** $\frac{2}{3}$

  b 3 is not a factor of a power of 10

D1 Check in

1 **a** Primary: Nicola, Secondary: Maddy
  b Primary data is data collected by yourself. Secondary data has already been collected by someone else.
2 **ai** There are two groups containing 4 hours.
  aii There are no groups containing 2 hours.
  b Use inequalities for the groups, e.g. $0 \leqslant t < 2$, $2 \leqslant t < 4$, 4 or more.
3 **a** $22\frac{1}{2}$ **b** 21 **c** 6 **d** 21

D1.1

1 **a** Assumes you visit cinema. Answers may differ at different times of year so could say on average. No answer choices given.
  b On average how many times do you go to the cinema in one month? Once or less often, 2 or 3 times a month, 4 times or more often.
  c On average how much do you spend when you go to the cinema? Less than £5, £5 to £10, more than £10.
2 **ai** Leading.
  aii What is your favourite sport? Tennis, swimming, football, rugby, cricket, other.
  b On average, how many times a week do you play sport? Never, 1 or 2 times, 3 or 4 times, more than 4 times.
3 **a** May not like either/not enough choices.
  b What is your favourite flavour of crisps? Plain, cheese and onion, salt and vinegar, smokey bacon, ketchup, other.
  ci 'lots' is vague; options do not cover all possible answers; needs a time frame.
  cii How many times have you visited the tuck shop in the last month? Never, once or twice, three to five times, more than five times.
4 **ai** Does not cover all possible answers.
  aii How far would you travel to see your favourite band? Less than 1 mile, 1 mile to 5 miles, between 5 and 10 miles, 10 miles or more.
  b How much would you pay for a ticket to see your favourite band? Less than £5, £5 to £10, £10.01 to £15, more than £15.

D1.2

1 Obviously visit cinema so not typical of population.
2 People in an athletics club will probably play sport more often than those who aren't.
3 **a** It is not representative of the people who use the school tuck shop.
  b It is not representative of the whole school.
  c Pick names out of a hat, or take every 20th person on a list of all the people in the school.
4 **a** His friends are not representative of the whole population, they might particularly like (or dislike) travelling to see bands.
  b People listening to MP3 players may be more interested in music than is typical.
5 All girls/all friends so may have same taste in music/small sample.
6 Cars passing at similar time/small sample.
7 **a** People at bus stops are more likely to take the bus to work.
  b Biased against people who are ex-directory, don't have a landline or aren't in when phoned.

D1.3

1 Two-way table with number of visits per week and amount spent.
2 Two-way table with number of times per week play sport and gender.
3 Two-way table with crisp flavour and Year groups.
4 Two-way table with distances and money.
5 Two-way table with favourite bands and numbers of CDs.
6 Two-way table with colour and make of car.
7 Two-way table with mode of travel and time taken.
8 **a** 500
  bi 20.8% **bii** 24% **biii** 39.2%

D1.4

1

ai 7	**aii** 6	**aiii** 5.82	**aiv** 6	**av** 2
bi 75	**bii** 63	**biii** 60.1	**biv** 63	**bv** 27
ci 8	**cii** 96	**ciii** 95.6	**civ** 96	**cv** 2
di 71	**dii** 22, 37	**diii** 40.4	**div** 37	**dv** 38
ei 26	**eii** 88, 89	**eiii** 84.2	**eiv** 87	**ev** 7
fi 72	**fii** 27	**fiii** 46.9	**fiv** 34	**fv** 37
gi 8	**gii** 105	**giii** 105.2	**giv** 105	**gv** 3

2 Range is unduly affected by one extreme value, which IQR ignores.
3 Mode is the lowest value.
4 **a** 1, 6, 8, 2, 8, 5, 6, 9, 3, 5, 7, 4, 4, 5, 5
  bi 8 **bii** 5 **biii** 5.2 **biv** 5 **bv** 3
  c Range and IQR stay same.
  d Answers are as for Q1 less 100.
  e Adding 100 does not affect spread of values but does affect averages.
5 **ai** 200, 200 **aii** 100, 100 **aiii** 100, 100
  aiv 100, 100 **av** 2, 2
  b All measures the same although sets of numbers are different.
6 **i** 77 **ii** 22, 37 **iii** 44.5 **iv** 37 **v** 46

D1.5
1 80.4 minutes
2 7.325 hours
3 43 lessons
4 78.6%
5 7.33
6 8.29

D1.6
1	**ai**	5	**aii**	6	**aiii**	5.8	**aiv**	4	**av**	2
	bi	5	**bii**	5	**biii**	4.96	**biv**	4	**bv**	2
	ci	4	**cii**	6	**ciii**	6.59	**civ**	6	**cv**	4
	di	6 and 7	**dii**	6	**diii**	5.77	**div**	6	**dv**	3

2 42.5
3 2.46

D1 Summary
1 **a** The mode is the most common number of rooms.
The mode is 7 rooms, as 7 rooms has the highest frequency.
b 6.325 rooms
c 7 rooms
d Beccy's median.
Beccy's sample size is greater than Ali's.

G2 Check in
1 $a = 56°$, $b = 112°$, $c = 235°$, $d = 160°$
2 $e = 72°$, $f = 63°$, $g = 118°$, $h = 75°$

G2.1
1 **ai**

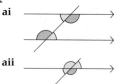

aii

bi

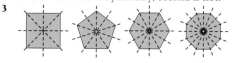

bii

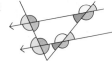

ci

cii

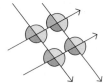

2 **a** $a = 17°$, angles on a straight line; $b = 17°$, alternate angles; $c = 163°$, corresponding angles
b $d = 125°$, alternate angles; $e = 105°$, vertically opposite angles

c $f = 134°$, vertically opposite angles; $g = 134°$, corresponding angles; $h = 24°$, angles in a triangle
3 **a** 180°
b x (bottom left) and z (bottom right)
c angles in the triangle are x, y, z (part **b**) and add up to 180° (part **a**)
4 **a** $a = 25°$, $b = 155°$, $c = 25°$, $d = 155°$
b Parallelogram
5 $a = 80°$, $b = 110°$, $c = 70°$, $d = 30°$, $e = 150°$

G2.2
1

Shape	△	□	⬠	⬡	⬡	⬡
Number of sides	3	4	5	6	8	10
Number of triangles the shape splits into	1	2	3	4	6	8
Sum of the interior angles in the shape	180°	360°	540°	720°	1080°	1440°
Size of one interior angle	60°	90°	108°	120°	135°	144°
Size of one exterior angle	120°	90°	72°	60°	45°	36°

2 **a** $x = 80°$ **b** $x = 77°$, $2x = 154°$
c $x = 55°$, $3x = 165°$
3 **a** $x = 125°$, $y = 50°$ **b** $x = 110°$, $y = 65°$
4 **ai** 120° **aii** 108° **aiii** 118°
b The sum of the two opposite angles
5 **a** 12   **b** 8

G2.3
1 **a** Yes **b** No **c** Yes
2 **a**

2 **b** First is a line of symmetry, second is not.
3

4 Cuboid: planes through middle of sides and through diagonally opposite edges.
Triangular prism: planes vertically through middle and from edges through opposite side.
5 One horizontal plane of symmetry halfway up the cylinder and infinitely many vertical planes of symmetry through the diameters of the circle on top of the cylinder.

G2.4
1 22.5 cm²
2 **a** Square with sides 7.1 cm **b** 50 cm²
3 **a** All 42 cm²
b Halve the product of the diagonals.
4 60 cm²
5 Both shapes have been split into congruent triangles; all sides equal in both cases.
6 **a** True; all squares have 2 pairs of sides of equal length and 4 equal angles.
b False; not all kites have diagonals which bisect each other.
c False; not all rhombuses have 4 equal angles.

G2.5

1 **a** $a = 63°$, angle at centre is double the angle at the circumference
   b $b = 46°$, angle at centre is double the angle at the circumference
   c $c = 118°$, angle at centre is double the angle at the circumference
   d $d = 32°$, angles in same arc
   e $e = 78°$, angles in same arc
   f $f = 103°$, angle at centre is double the angle at the circumference
   g $g = 122°$, angles in same arc
   h $h = 21°$, angles in same arc
   i $i = 140°$, angle at centre is double the angle at the circumference
   j $j = k = l = 38°$, angles in same arc
   k $m = 47°$, $n = 62°$, angles in same arcs
   l $p = 75°$, $q = 25°$, angles in same arcs
   m $r = 90°$, angle in same arc
   n $s = 49°$, $t = 22°$, angles in same arcs
   o $u = 54°$, $v = 45°$, angles in same arcs
   p $w = 180°$, angle at centre is double the angle at the circumference
   q $x = 45°$, angles in isosceles triangle
   r $y = 45°$, angle at centre is double the angle at the circumference

G2.6

1 **a** $a = 90°$, angle in a semicircle
   b $b = 61°$, angles in a triangle
   c $c = 43°$, angles in a triangle
   d $d = 135°$, opposite angles in cyclic quadrilateral
   e $e = 76°$, opposite angles in cyclic quadrilateral
   f $f = 15°$, angles in a triangle
   g $g = 143°$, $h = 100°$, opposite angles in cyclic quadrilateral
   h $i = 124°$, $j = 54°$, opposite angles in cyclic quadrilateral
   i $k = 56°$, $l = 124°$, angles in a triangle, angles on a straight line
   j $m = 125°$, $n = 250°$, angles at centre are double the angles at the circumference
   k $p = q = 106°$, opposite angles in cyclic quadrilateral, angles on same arc
   l $r = 50°$, angle in a semicircle
   m $s = 35°$, angle in a semicircle
   n $t = 62°$, $u = 118°$, angles on a straight line, opposite angles in a cyclic quadrilateral
   o $v = 74°$, angles on a straight line, opposite angles in a cyclic quadrilateral
   p $w = x = 90°$, opposite angles in a cyclic quadrilateral
   q $y = 45°$, angles in isosceles triangle
   r $z = 100°$

G2.7

1 **a** $a = 90°$, angle between tangent and radius
   b $b = 33°$, angle between tangent and radius, $c = 57°$, angles in a triangle in a semicircle
   c $d = 42°$, angle between tangent and radius, $e = 48°$, angles in a triangle in a semicircle
   d $f = 26°$, angle between tangent and radius, $g = 26°$, angles in a triangle in a semicircle
   e $h = 51°$, angle between tangent and radius, angles in a triangle in a semicircle
   f $i = 16°$, $j = 4$ cm; 2 tangents drawn from a point to a circle
   g $k = 44°$, angles in a quadrilateral
   h $l = 40°$, $m = 70°$, angles in a quadrilateral and angle at centre is double the angle at the circumference

i $n = 16°$, $p = 82°$, angles in a quadrilateral and angle at centre is double the angle at the circumference
j $q = 30°$, $r = 105°$, angles in a quadrilateral and angle at centre is double the angle at the circumference
k $s = 100°$, $t = 50°$, angles in a quadrilateral and angle at centre is double the angle at the circumference
l $u = 116°$, $v = 58°$, angles in a quadrilateral and angle at centre is double the angle at the circumference

G2 Summary

1 **ai** $x = 110°$ **aii** Corresponding angles are equal
2 $150°$
3 **ai** $90°$ **aii** Angle in a semicircle is $90°$
    bi $65°$ **bii** Angle at the centre is twice the angle at the circumference

Case study 2: Sandwich shop

1

Day	Number of Customers	
	Week 1	Week 2
Monday	50	54
Tuesday	68	60
Wednesday	47	53
Thursday	58	57
Friday	52	56
Saturday	76	70
TOTAL	351	350

a Daily average (mean) = 58.4
b Weekly average (mean) = 350.5
c Without Saturdays;
   weekly average (mean) = 277.5
d With coach trip on second Wednesday;
   weekly average (mean) = 362.5
   From frequency polygon:
   busiest time 1pm to 2pm;
   quietest time 9am to 10am
Customer numbers vary from hour to hour.
There is a peak around lunchtime.
Customer numbers build at the start of the day and tail off towards the end of the day, with a lull in the early afternoon.
This seems to be quite a suitable pattern for an average day.

2 Sales for second week:
   Ham 91, Cheese 63, Humous 35, Tuna 52.5, Chicken 108.5

Product	Stock (packs)	Portions per pack	Portions left	Stock needed	Amount to order
Bread	6	20	120	351	12
Ham	2.5	10	25	91	7
Cheese	3	10	30	63	4
Humous	2	8	16	35	3
Tuna	1.5	14	21	53	3
Chicken	1	10	10	109	10

Estimate of stock left on Wednesday morning:

Bread 3, Ham -6 (will have none left), Cheese 9, Humous 4, Tuna 3, Chicken -27
(will have none left)
The shop will not be able to cater for the coach trip. There would be enough tuna and cheese, but not enough of the other ingredients (including the bread). The manager could calculate a percentage
surplus to order so that unexpected customers can be catered for.
This value would need to be calculated so that the resulting waste was minimised.

N3 Check in

1 Gill gets £400, Paul gets £100
2 33.3% decrease
3 a Batch B, it has a greater proportion of black paint
  b 9 litres black, 21 litres white

N3.1

1 a 17:13 b 11:19 c 14:15
2 11A, 7:23; 11B, 13:17; 11C, 16:13
3 a 1:1 b 2:5 c 2:3 d 3:4
  e 1:2 f 1:1 g 1:3 h 4:3
4 a 3:2 b 4:1 c 1:2 d 1:4
  e 2:3 f 3:2 g 2:3 h 3:2
5 a 24, 16 b 27, 18 c 54, 36
  d 60, 40 e 150, 100 f 1200, 800
  g 4500, 3000 h 5550, 3700
6 a 16 hours, 8 hours b 120°, 60°
  c 30 minutes, 15 minutes d €240, €120
  e 164 cm, 82 cm f 80 kg, 40 kg
  g 54 km, 27 km h 36 miles, 18 miles
7 a £180, £180 b £120, £240
  c £240, £120 d £135, £225
8 a 5:3 b Karla: £13.75, Wayne: £8.25
9 a £200 b Annie: £120, Ben: £180

N3.2

1 a 1:2 b 1:2:2 c 3:1:2
  d 5:10:4 e 2:5:3 f 3:7:2
2 a £80, £160 b £96, £144
  c £150, £90 d £40, £80, £120
  e £150, £30, £60 f £140, £40, £60
3 a 50 m, 200 m b 100 m, 150 m
  c 71.43 m, 178.57 m d 214.29 m, 35.71 m
  e 115.38 m, 134.62 m f 160.71 m, 89.29 m
4 Peter: £2700, Bob: £1500, Yasmin: £3300
5 Ann: £377.78, Charles: £283.33, Edward: £188.89
6

Quantity	Ratio	Share 1	Share 2	Share 3
200 km	2:3:5	40 km	60 km	100 km
38 kg	1:2:3	6.33 kg	12.67 kg	19 kg
450 cm	2:3:8	69.2 cm	104 cm	277 cm
£720	4:5:10	£151.58	£189.47	£378.95
95 litres	1:6:7	6.79 litres	40.71 litres	47.5 litres

7 Mango: 444 ml, pineapple: 333 ml, passion fruit: 222 ml
8 a £119.05, £116.67, £214.29
  b £125, £166.67, £208.33
9 Robert: £12660, Kathleen: £16880

N3.3

1 a 75 g b 4.5 g c 100 g d 125 g
2 a 90 ml b 228 ml c 1.2 litres d 168 cm³
3 a 60 g b 36 mm c 380 g d 31.2 km
4 a $\frac{1}{10}$ b $\frac{3}{40}$ c $\frac{1}{5}$ d $\frac{3}{16}$
5 a 10% b 7.5% c 20% d 18.75%
6 ai $\frac{3}{20}$ aii 15%
  bi $\frac{1}{25}$ bii 4%

ci $\frac{2}{5}$ cii 40%
di $\frac{8}{125}$ dii 6.4%
7 a 73.536 mm b £315 c £9.88 d 840 kg
8 a $\frac{3}{25}$ b 12%
9 a 200 ml b 3 litres c 1.8 litres
10 a £2100 b $\frac{8}{15}$, 53.3%

N3.4

1 a 2:3 b $\frac{2}{5}$
2 a 1:2 b $\frac{1}{3}$
3 a 1:3:2 bi $\frac{1}{6}$ bii $\frac{1}{2}$ biii $\frac{1}{3}$
4 a $\frac{1}{2}$ b $\frac{1}{3}$ c $\frac{2}{5}$ d $\frac{3}{8}$ e $\frac{4}{7}$ f $\frac{5}{7}$
5 a $\frac{1}{2}$ b $\frac{2}{3}$ c $\frac{3}{5}$ d $\frac{5}{8}$ e $\frac{3}{7}$ f $\frac{2}{7}$
6 2:3
7 a £600 b 3:7
8 £770
9 $\frac{1}{2}$
10 £5333.33, $\frac{8}{21}$

N3.5

1 10%
2 a 4:7 b 36.4%
3 Nickeline: 80%, US nickel coinage: 75%, Medal bronze: 93%
4 60%
5 44.4%
6 a 80% b 3:1
7 a £1140, £2660 b 30%
8 a $324.47, $175.53 b 35.1%
9 a 333 g, 104 g, 63 g (to 0 dp)
  b 66.7%, 20.8%, 12.5%

N3.6

1 9 mph
2 £2.10 per metre
3 32 mph
4 89.25 mph
5

Speed (km/h)	Distance (km)	Time
105	525	5 hours
48	106	2 hours 12.5 minutes
$37\frac{1}{3}$	84	2 hours 15 minutes
86	215	2 hours 30 minutes
37.1	65	1 hour 45 minutes

6 a 5 g/cm³ b 87.88 g
7 a 7.59 g/cm³ to 3 sf b 105
8 a 9.46 kg to 3 sf b 6.15 litres to 3 sf
9 1358 m

N3 Summary

1 a 90 g
  b 400 ml
2 a 3:2
  b 45 apples
3 a 46 g
  b 1.2 g per cm³
4 640 seconds or $10\frac{2}{3}$ minutes

A2 Check in

1 a 45 b 52 c −26 d 196
  e 7 f 13 g −50 h 30
2 a 9 + 6 = 15 b 8 + 7 = 15
  c 5 × 3 = 15 d 27 − 12 = 15

3 **a** 3 **b** 4 **c** 10 **d** 6
e 4 **f** 25 **g** 33 **h** 7

A2.1

1 a $5w$ **b** $\frac{6}{k}$ **c** y^2 **d** $6ab$ **e** $8k^3$
2 a 20 **b** 4 **c** 36 **d** 22 **e** 108
3 a $11a + 6b$ **b** $2t + 26$ **c** $x - 12y$
d $p^2 + 14p$ **e** $20xy$ **f** $7ab$
4 Abdul
5 a $28mn$ **b** $12m^2$ **c** $10p$
d 2 **e** $24abc$ **f** $6k^3$
g $4b$ **h** $9c$

6

$4p + 7q$	$4p + 7q$	$\mathbf{4p + 7q}$
$6mn$	$\mathbf{10mn}$	$6mn$
$2d$	$\mathbf{2}$	$2d$
$2n - 8$	$\mathbf{2n}$	$2n - 8$

7 ai $8p + 16$ **aii** $32p$ **b** $3x, 2y$

8

$9b - 2a$	$12a^2$	$4b$
$2ab$	$5p^3 + 7p^2 + 10p$	$13abc$
$5m - 4$	$60m^3$	$\frac{2}{a^2}$

A2.2

1 a $4n + 20$ **b** $6b - 42$
c $a^2 + 3a$ **d** $ab - ac$
e $8x + 12y - 16z$ **f** $2h^2 + 18h$
2 a $-3k - 27$ **b** $-2h + 10$
c $-w + 4$ **d** $-t + p$
e $-k^2 - 7k$ **f** $-18m + 9k - 36$
g $-x^2 + x + 8$ **h** $-2x^2 - 6$
i $-3 + 3x$
3 a $10c + 62$ **b** $23x + 67$
c $2x^2 + 10x$ **d** $17t^2 + 32t$
e $7x - 45$ **f** $2x - 11$
g $2m - 26$ **h** $-11g + 33$
i $p + 14$ **j** $2q - 7$
4 $-5x^2 + 44x$
5 a $3(2x - 1)$ **b** $6x - 3$
c $6x - 3 = 15$, which gives $6x - 18 = 0$
6 a For example, $8(3x + 2)$
b For example, $2(2x + 3) + 5(4x + 2)$
7 $\frac{1}{2}(2y + 2)\, 2y = 2y^2 + 2y$

A2.3

1 a $x^2 + 5x + 6$ **b** $p^2 + 11p + 30$
c $w^2 + 5w + 4$ **d** $c^2 + 10c + 25$
e $x^2 + 2x - 8$ **f** $y^2 + 5y - 14$
g $t^2 + 4t - 12$ **h** $x^2 - 7x + 10$
i $y^2 - 14y + 40$ **j** $w^2 - 3w + 2$
k $p^2 - 10p + 25$ **l** $q^2 - 24q + 144$
2 a $6x^2 + 17x + 7$ **b** $10p^2 + 19p + 6$
c $6y^2 + 11y + 4$ **d** $4y^2 + 24y + 36$
e $10t^2 + 12t - 16$ **f** $15w^2 + 42w - 9$
g $6x^2 - 6y^2$ **h** $9m^2 - 24m + 16$
i $6p^2 - pq - 40q^2$ **j** $4m^2 - 12mn + 9n^2$
3 a $(x - 3)(x + 6) = x^2 + 3x - 18$
b $(2m - 3)^2 = 4m^2 - 12m + 9$
4 a $a^2 + 2ab + b^2$ **b** 16
c For example,
$2.5^2 + 2 \times 2.5 \times 3.5 + 3.5^2 = 6^2$
5 a $(3x - 1)(2x + 3) = 75$, which gives
$6x^2 + 7x - 78 = 0$

b $(3x - 4)(5x + 2) = 2 \times 75$, which gives
$15x^2 = 14x^2 + 158$

A2.4

1 a $2(x + 2)$ **b** $3(y - 2)$
c $12(p + 3q)$ **d** $5(5w - 1)$
e $x(6y + w)$ **f** $b(a - 2c)$
g $q(pr + rt - sw)$ **h** $x(5y - 1)$
i $2x(y + 3)$ **j** $2a(2b - 3a)$
k $5p(5p - 2)$ **l** $7x(1 + 2y)$
m $2a(c + 2a - 4)$ **n** $5m(3n - 1 + 2m^2)$
o $6p(p^3 - 2)$
2 Correct factorisations are: Clare $5x(1 + 2y)$,
Ben $3p(2q + 1)$, Vicky $7p(3 + 2q)$
3 a $23(x + y)$ **b** $(a - b)(a - b + 5)$
c $(q + r)(6 - (q + r)^2)$ **d** $7(pt - w)$
4 a $(a + b)(x + y)$ **b** $(c + b)(d + m)$
c $(a + b)(a + 2)$ **d** $(c - m)(d + e)$
5 a $4(x - 1)$ **b** $20(b + 2)$
6 a 6 **b** 16.5 **c** 58.6 **d** 33.2
7 $4(3x + 2) - 2(2x - 1) = 8x + 10 = 2(4x + 5)$

A2.5

1 a $(x + 2)(x + 4)$ **b** $(x + 3)(x + 7)$
c $(x + 4)(x + 7)$ **d** $(x + 3)(x + 8)$
e $(x - 2)(x - 6)$ **f** $(x - 3)(x - 6)$
g $(x - 9)(x - 4)$ **h** $(x + 4)(x - 3)$
i $(x - 7)(x + 5)$ **j** $(x + 9)(x - 3)$
k $(x - 16)(x + 2)$ **l** $(x + 20)(x - 2)$
2 a $x^2 + 9x - 22$ **b** $(x + 11)(x - 2)$
3 a $(x + 12)(x - 6)$ **b** $(x - 12)(x + 2)$
c $(x - 15)(x - 5)$ **d** $(x + 16)(x - 4)$
e $(x - 8)(x + 8)$ **f** $(x - 4)(x - 25)$
4 a $(x + 2)(x + 19)$ **b** $x(5x + 5 + y)$
c $(x + 11)^2$ **d** $(x + 9)(x - 2)$
e $(p + 3)(p + 11)$ **f** $x(2x + 3y)$
5 $(x + 5)(x + 4) = 12$
$x^2 + 9x + 20 = 12$
$x^2 + 9x + 8 = 0$
$(x + 1)(x + 8) = 0$
6 a $2(x + 4)(x + 7)$ **b** $x(x - 8)(x + 3)$
c $x(x - 4)(x + 4)$
7 $(2.3 + 1.7)^2 = 4^2 = 16$

A2.6

1

Identities	Equations	Formulae
$3x(x + 1) = 3x^2 + 3x$	$3x + 1 = 10$	$C = 2\pi r$
$y + y = y^2$	$2x + 5 = 3 - 7x$	$A = \frac{1}{2}(a + b)h$
$20 - x = -(x - 20)$	$2x^2 = 50$	$a^2 + b^2 = c^2$

2 a £17 **b** 9
c You can't have half a person.
3 a 53.6 °F **b** 0 **c** 4
d 108, 5 or −5 **e** −14 **f** 1
4 a 33 cm²
b Any a, b such that both are positive and $a + b = 10$

A2.7

1 a $P = 4s$ **b** $P = 26 + 2y$
2 a $P = 0.64d + 12$
b $T = 10 + 35p$
c $B = 10n + 30m + 560$
3 a, b, c are lengths of the base of the triangle, height
of the triangle and length of the prism. Correct as
multiplies area of cross-section by length.

4 a $V = \pi r^2 h$
b $2\pi r^2$ represents the combined area of the top and bottom circles. The $2\pi rh$ term represents the curved section; if opened out flat it would be a rectangle measuring $2\pi r$ in length (the circumference of the circle) and h in height.

A2 Summary
1 a $2a + 6b$
  b $x(x - 6)$
  c $3x - 2x^3$
  d $4x(3y + x)$
2 a m^5
  b n^4
3 a a^3
  b $15x - 10$
  c $3y^2 + 12y$
  d $5x - 2$
  e $x^2 + x - 12$
4 a $k = 15$
  b $y = -7$
5 a $C = 90 + 0.5m$
  b 300 miles

D2 Check in
1

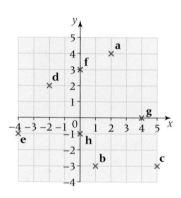

2 ai €2.8 **aii** €6.3 **bi** £2.85 **bii** £5
3 a 40   **b** 60   **c** 15   **d** 50
   e 30   **f** 150   **g** 120   **h** 630

D2.1
1 a

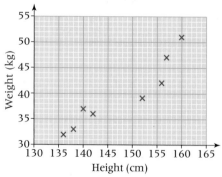

b Positive correlation
c As weight increases, so does height.

2 a

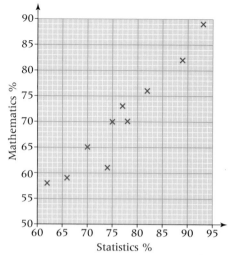

b Positive correlation
c As percentage achieved in statistics increases, so does percentage achieved in maths.

3 a

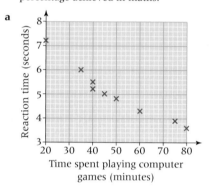

b Negative correlation
c As time spent playing computer games increases, reaction time decreases.

4 a

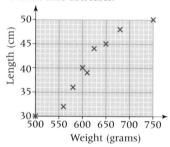

b Positive correlation
c As weight of fish increases, so does length.

D2.2
1 a

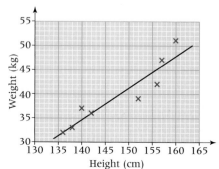

bi 38 kg **bii** 156 cm

2 a

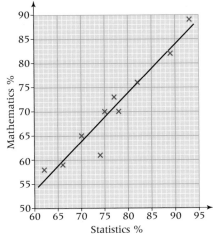

bi 74% **bii** 86%
c Score of 46% outside range of data collected

3 a

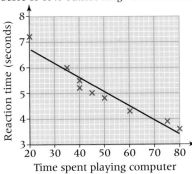

bi 3.9 seconds **bii** 26 minutes
c Outside range of data collected

4 a

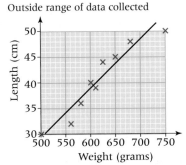

bi 48 cm **bii** 635 g

D2.3

1 a

```
4 | 5 8
5 | 1 1 2 4 9 9
6 | 1 2 3 6 9
7 | 0 1 4 5 7 8
8 | 1 2 9
9 | 3
```
Key: 4|5 means 45%

bi 48%   **bii** 66%
biii 54%, 77%   **biv** 23%
c Most students scored between 50% and 80%.

2 a

```
2 | 6 8 9
3 | 0 3 7 8 9
4 | 0 1 3 3 4 7 9
5 | 2 5 8 9
6 | 0 1 3 6 7 9
7 | 1 3 5 7
8 | 0 4
```
Key: 3|4 means 34 minutes

bi 58 minutes
bii 52 minutes
biii 39 minutes, 67 minutes
biv 28 minutes
c Most people spend 30 to 70 minutes playing computer games each day.

3 a

```
0 | 3 5 6 7 8 9
1 | 0 1 2 2 5 6 7 8 9
2 | 0 1 2 4 6 7 7 9 9
3 | 1 2 3 4 5 7
4 | 0 1
```
Key: 2|7 means 27 minutes

bi 38 minutes
bii 20.5 minutes
biii 11.5 minutes, 30 minutes
biv 18.5 minutes
c Most people take between 10 and 30 minutes to complete the crossword.

4 a

```
3 | 2 6 7 9
4 | 2 5 7 8 9
5 | 1 2 4 6 6 6 7 8 9
6 | 0 1 3 6
7 | 0
```
Key: 3|6 means 36 kg

bi 38 kg   **bii** 54 kg
biii 45 kg, 59 kg   **biv** 14 kg
c Most boys weigh between 40 kg and 60 kg.

5 a

```
14 | 6 7 9
15 | 0 0 2 2 3 4 5 5 7 8 9
16 | 0 2 3 5 7 8
17 | 1 2 2
```
Key: 14|2 means 142 cm

bi 26 cm   **bii** 157 cm
biii 152 cm, 165 cm   **biv** 13 cm
c Most girls are between 150 cm and 160 cm tall.

6 a

```
8  | 8 9
9  | 1 2 4 8
10 | 1 3 4 5 6 7 8 8 9
11 | 0 2 4 6 6 7 7 8 9
12 | 1 5 6 7 9
13 | 1 3
```
Key: 11|4 means IQ of 114

bi 45   **bii** 110
biii 103, 119   **biv** 16
c Most students have an IQ between 100 and 120.

7 a

```
0 | 3 6 8 9 9
1 | 0 1 2 5 6 7 8 9
2 | 1 1 2 3 3 5 7
3 | 0 2 3
```
Key: 3|2 means 32 minutes

bi 30 minutes
bii 18 minutes
biii 10 minutes, 23 minutes
biv 13 minutes
c Most people took 10 to 20 minutes

D2.4

1 a

```
        A                           B
                    8  7 │ 3
    9  7  5  4  4  3  2  1 │ 4 │ 1  4  8
       8  6  6  5  3  2  1 │ 5 │ 3  3  4  6  7  7
                      2  1 │ 6 │ 1  3  5  6  9
                          │ 7 │ 0  2  2  5  9
```
Key: 2|5|3 means 52% in A, 53% in B

bi A: 49%, B: 61%
bii A: 10%, B: 17%
biii A: 24%, B: 38%
c Results for test A are generally lower but less varied than for test B.

2 a

```
      Men                          Women
                     │14│ 8  9
               9  8 │15│ 0  1  2  3  4  5  5  6  7  8  8
    9  9  8  7  7  5  0 │16│ 1  2  2  5  6  7  9
    8  8  7  6  5  4  2  1 │17│ 2  4  8
          8  4  3  3  2  0 │18│
```
Key: 2|16|3 means 162 cm for men, 163 cm for women

bi Men: 174 cm, women: 158 cm
bii Men: 13 cm, women: 13 cm
biii Men: 30 cm, women: 29 cm
c Men are taller than women and heights are evenly spread in both groups.

3 a

```
          X                          Y
                   9 │ 8 │ 7
                 9  8  6 │ 9 │ 3  7  9
            9  8  6  5  5  4  3 │10│ 2  3  4  6  7  7  9
    8  8  7  7  6  5  5  2  0  0 │11│ 3  4  4  4  6  8  8  8  9  9
          9  6  4  3  3  2  1 │12│ 1  1  4  6  6  8  9
                4  1  0 │13│ 0  1  2
```
Key: 7|9|6 means for 97 in X, 96 in Y

bi X: 116, Y: 116
bii X: 18, Y: 18
biii X: 45, Y: 45
c Average IQ and spread are same in X and Y.

4 a

```
        Boys                        Girls
    8  8  8  6  2  2 │ 3 │
          7  5  4  2 │ 4 │ 0  3  4  6  8
    9  8  7  6  3  2 │ 5 │ 2  2  5  6  9
          8  6  4  2  1 │ 6 │ 2  3  3  5  6  7
                   2  1 │ 7 │ 2  3  4  6  7
                       │ 8 │ 0  2
```
Key: 5|4|6 means 4.5 minutes for boys,
4.6 seconds for girls

bi Boys: 5.3 seconds, girls: 6.3 seconds
bii Boys: 2.4 seconds, girls: 2.1 seconds
biii Boys: 4 seconds, girls: 4.2 seconds
c Boys have a faster reaction time than girls, variation in times is similar.

5 a

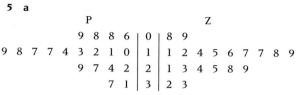

```
          P                          Z
        9  8  8  6 │ 0 │ 8  9
    9  8  7  7  4  3  2  1  0 │ 1 │ 1  2  4  5  6  7  7  8  9
          9  7  4  2 │ 2 │ 1  3  4  5  8  9
                7  1 │ 3 │ 2  3
```
Key: 4|3|6 means 34 minutes for P, and 36 minutes for Z

bi P: 17 minutes, Z: 18 minutes
bii P: 14 minutes, Z: 11 minutes
biii P: 31 minutes, Z: 25 minutes
c The average time taken is similar but times for P are more varied than those for Z.

D2.5

1
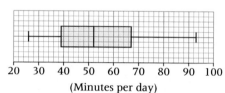
(%)

2
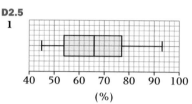
(Minutes per day)

3
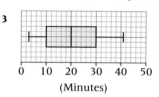
(Minutes)

4
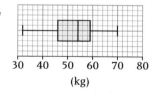
(kg)

5
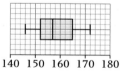
(cm)

6
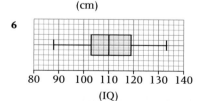
(IQ)

7
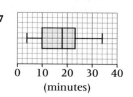
(minutes)

D2.6

1 a

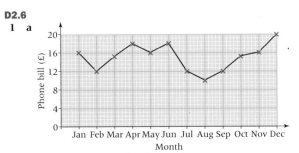

b Typical phone bills are about £15.
These fall during the Summer months and peak in December.

2 a

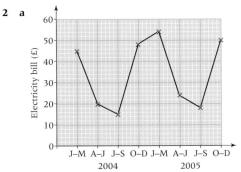

b Electricity bills are highest in the Winter months and lowest in the Summer months. This annual pattern repeats itself; there is a slight trend for bills to rise from year-to-year.

3 a

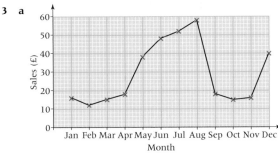

b Icecream sales grow steadily during Spring and Summer but drop sharply in the Autumn. Sales are low during Autumn and Winter except for a peak in December.

4 a

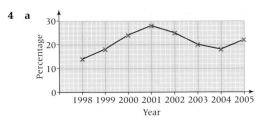

b The percentage of students using the library grow steadily from about 15% in 1998 to 28% in 2001. It has since fallen back to around 20%.

5 a

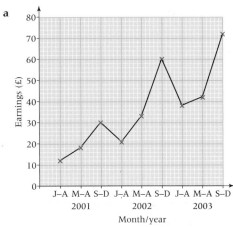

b Christable's earnings have grown steadily during the three years. Her earnings peak in the Winter months.

6 a

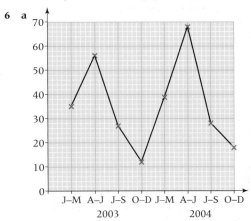

b Steve's expenses grow to a peak in Spring and fall back in Winter. The level of expenses appears to be growing from year-to-year.

D2 Summary

1 1 30 mm
2 a Plot points at (65, 100) and (80, 110)
  b Positive correlation
  c 108 cm   (Accept 106 to 110)

Case study 3: Recycling

Chart text, from DEFRA:
There has been a change in the composition of recycled waste over time. In 1997/8 paper and card was the largest component, making up 37% of the total, followed by compost (20 per cent) and glass (18 per cent). In 2007/08 compost was the largest component (36.1 per cent of the total) with the next largest being paper and card (18.1 per cent) followed by co-mingled (17.7 per cent).

Co-mingled collections - the collection of a number of recyclable materials in the same box or bin, for example paper, cans, plastics - have become more widespread in recent years.'

Three largest components
1997/98 paper and card, compost and glass
2007/08 compost, paper and card and co-mingled waste.

25.3 million tonnes of household waste was collected in England in 2007/08. Total amount of waste collected for re-use, recycling and composting 8.84 million tonnes. 34.9% of this waste was collected for re-use, recycling or composting.

The amount of household waste not re-used, recycled or composted was 16.4 million tonnes, a decrease of 7.0 % from 2006/07. The amount of household waste not re-used, recycled or composted in 2006/07 was 17.7 million tonnes making the total amount of household waste collected in 2006/07 was 25.76 million tonnes.

This decrease equates to 324 kg per person of residual household waste and shows progress towards the 2010 target, in the Waste Strategy 2007, of reducing this amount to 15.8 million tonnes.

New can weighs 55.8 g (to 3 s.f)
Weight has decreased by approximately 50%

Volume of container = 1296 cm³
Volume of six tomatoes = 679 cm³

48% of the available volume is empty.

N4 Check in
1 a 30%   b 65%   c 72.5%
   d 105%   e 6%
2 a 0.15   b 0.065   c 0.125
   d 0.975   e 1.08
3 a 0.75   b 0.4   c 0.7
   d 0.45   e 0.17
4 a $\frac{1}{2}$ b $\frac{1}{4}$ c $\frac{3}{10}$
   d $\frac{4}{5}$ e $\frac{9}{20}$
5 a £35   b 20 m   c 54°
   d 120 cm   e £54   f 90 mm

N4.1
1 a 10.5   b 126   c 240   d 300
   e 132   f 99
2 a £225   b 990 m   c 2520 kg   d €144
3 a 135   b 91   c 744   d 18
   e 387   f 192
4 a 1104 mm   b   1952 kg
   c €1443   d £493
5 a 0.5   b 0.6   c 0.25   d 0.51
   e 0.64   f 0.22   g 0.15   h 0.7
   i 0.07   j 0.085   k 0.0015   l 0.0001
6 a 5.7   b 200   c 15.93   d 99.84
   e 16.81   f 20
7 a £3.84   b £53.55   c £13.95   d £170.10
   e £28.16   f £15.90
8 £425
9 £73.70
10 649
11 £49 090.60

N4.2
1 a 108   b 72   c 49.5   d 618
   e 600   f 700
2 a 25.2   b 36   c 51   d 37.5
   e 228   f 64.35
3 a 342   b 264   c 1008   d 325
   e 1372.5   f 1109.2
4 a 672   b 532   c 272   d 301.5
   e 136.5   f 124.8
5 a 1.2   b 1.3   c 1.45   d 1.85
   e 1.065

6 a 0.6   b 0.4   c 0.65   d 0.28
   e 0.815
7 a 56.71   b 40.32   c 719.2   d 246
8 a £33 600   b £19 317.15
   c £27 348   d £56 021
9 a £351   b £559   c £725   d £714.10
10 They could have more than doubled in price, which gives a price increase of more than 100%.
11 A price fall by more than 100% leads to a negative cost, which is impossible.

N4.3
1 a £5   b £24
   c £315   d £227.50
2 a £262.50   b £416.16
   c £1348.32   d £1061.21
3 a 1.06   b 1.1236
4 a £530   b £561.80
5 a 1.157 625   b 1.370 086 663
6 a £5788.13   b £1096.07
7 Option a
8 a 0.85   b Multiply by 0.85 again

N4.4
1 a 80%   b 28%   c 12%   d 50%   e 50%
2 a 60%   b 48%   c 60%
3 a 51.35%   b 28.57%   c 3.83%
4 a Science: 88.3%, Maths: 86.7% (both to 3 sf)
   b Jason dropped a greater proportion of the total marks in maths than he did in science.
5 £5
6 a 50   b 40   c 80   d 110
7 a £6   b £80
8 a Francesca   b £78.03
9 She needs to find 20% of the original price, which is not the same as 20% of the sale price.

N4.5
1 a £125   b £560   c 46 m   d 54 cm
2 a £45   b 20 cm   c 360 cm   d 102 mm
3 a 1.2   b 0.85   c 1.06   d 0.95
   e 0.94   f 0.83
4 a £59.52   b 51.3 kg   c 10.088 seconds
   d 198 m   e €260.40   f $300.90
5 a £50   b £30.94
6 £560
7 a £321.63   b £1749.60

N4 Summary
1 a 19.9%
   b 4.4 million km²
2 £3244.80
3 £22
4 24.72 litres

A3 Check in
1 a 10, 12   b 70, 64   c 16, 22
   d −2, −5   e 48, 96   f $\frac{1}{6}, \frac{1}{7}$
2 a $4(n-1), 2n+7, 15-n, 2n^2 = \frac{9}{n}+15$
3 Square numbers. They form a pattern of squares when drawn.

A3.1
1 a 29, 34   b 65, 58
   c 16, 21   d 999 999, 9 999 999
   e 13, 21   f 3.375, 1.6875
2 a 7, 13   b 8, 16
   c 93, 91, 89   d 8, 125

3 a 3, 6, 9, 12, 15
b 2, 4, 8, 16, 32
c 2, 3, 5, 7, 11
d 121, 144, 169, 196, 225

4 a 6, 10, 15
b 1, 3, 6, 10, 15, 21, 28, 36, 45, 55
c They form a pattern of squares when drawn.
d They form a pattern of cubes when drawn.

5 a 110 **b** $\frac{10}{11}$ **c** 100 **d** 100 000

6 a $6667^2 = 44\ 448\ 889$, $66\ 667^2 = 4\ 444\ 488\ 889$
b 4 444 444 444 488 888 888 889
c 666 666 667

A3.2

1 a 10, 18, 26, 34, 42 **b** 1, 6, 11, 16, 21
c 7, 14, 21, 28, 35 **d** 8, 6, 4, 2, 0
e −2, 1, 6, 13, 22 **f** 2, 8, 18, 32, 50
2 a 8, 7, 6, 5, 4 **b** 6, 12, 20, 30, 42
c $1, \frac{1}{2}, \frac{1}{3}, \frac{1}{4}, \frac{1}{5}$ **d** −1, −8, −27, −64, −125
e −4, −9, 0, 35, 108 **f** 1, 16, 81, 256, 625
3 a n^2, n^3 **b** $18 + 2n$, $n(n+5)$
c n^3, $n(n+5)$
4 a 10, 14, 18, 22, 26, 30, 34, 38; $T(n) = 4n + 6$ (for $n \le 7$)
b $T(n) = n^2 + 1$, until $T(n)$ is over 49
5 a For example, n^2, $6n − 5$
b For example, n^2, $6n − 5$
c No

A3.3

1 a $5n − 1$ **b** $2n − 1$ **c** $2n + 8$
d $0.5n + 0.5$ **e** $2n − 6$ **f** n
g $13n$ **h** $10n − 6$ **i** $12 − 2n$
j $105 − 5n$ **k** $50\frac{1}{4} − \frac{1}{4}n$ **l** $79 − 4n$

2 Many possibilities
3 a False **b** False **c** False
4 a $8n − 20$ **b** $3n + 2$ **c** $2n − 46$
5 a $\frac{2n+1}{3n+4}$ **b** $\frac{2n+8}{33-3n}$ **c** $\frac{n+6}{n^2}$ **d** $\frac{n^3}{13-2n}$

6 a $\frac{1}{n}$ **b**

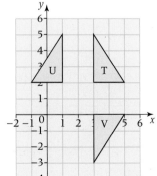

c It gets closer and closer to zero without ever actually reaching it.

A3.4

1 a For each square added, 3 sides added, and in first case there is an extra side
b Number of black tiles is matched by number of white tiles in middle of figure, with 4 extra whites at side
c Length of rectangle is always one more than width
2 a $W = 2B + 6$
bi $B = 2W + 2$ **bii** $L = 3W + 1$
ci $P = n(n − 1)$ **cii** $H = \frac{n(n-1)}{2}$
3 a $C = 4$, $E = 4(n − 2)$, $M = (n − 2)^2$
b $C = 4$, $E = 2(m − 2) + 2(n − 2)$, $M = (m − 2)(n − 2)$

A3.5

1 a $21 − 7 = 14$, which is even
b 5 squared is 25, which is odd
c $3 \times 2 = 6$, which is even, and 2 is prime
d $3 \times 4 = 12$, which is even
2 a For example, $2 + 3 + 4 + 5 + 6 = 20$ which is 4×5
b $n + (n + 1) + (n + 2) + (n + 3) + (n + 4) = 5n + 10 = 5(n + 2)$, which is clearly a multiple of 5 for any n
3 a Even numbers $2m$ and $2n$, then $2m + 2n = 2(m + n)$ which is a multiple of 2 and hence even for any m, n
b $(2n)^2 = 4n^2$, which is always in the four times table for any n
c $2n + (2m + 1) = 2(m + n) + 1$, which is odd for any m, n
d $k(k + 1) − k = k^2 + k − k = k^2$, which is a square number for any k
4 a Any number between 0 and 1 inclusive
b Any number between 0 and 1 inclusive
5 a $k(k + 2) = k^2 + 2k$, $(k + 1)^2 = k^2 + 2k + 1 = k(k + 2) + 1$ for any k
b $(n + 4)(n + 7) − (n + 3)(n + 8)$
$= n^2 + 11n + 28 − n^2 − 11n − 24 = 4$
c $m^2 − n^2 = m^2 − mn + mn − n^2 = (m − n)(m + n)$ for any m, n
6 If angle ACB is x then angle OAC is also x (isosceles triangle) so angle AOC is $180 − 2x$ (angles in a triangle). If angle ABC is y then angle OAB is also y (isosceles triangle) so angle AOB is $180 − 2y$ (angles in a triangle).
Hence, $180 − 2x + 180 − 2y = 180$ (angles on a straight line)
$180 = 2x + 2y$
$90 = x + y$
i.e. angle CAB is $90°$

A3 Summary

1 a $6n − 2$
2 $d = 4n + 6$
3 $n = 1$ $n^2 + 4 = 5$
$n = 3$ $n^2 + 4 = 13$
$n = 5$ $n^2 + 4 = 29$
$n = 7$ $n^2 + 4 = 53$
$n = 9$ $n^2 + 4 = 85$ which is not prime.
OR
$n = 11$ $n^2 + 4 = 125$ which is not prime.
4 Any example of $2 \times$ prime = even number
For example, $2 \times 3 = 6$

G3 Check in

1 A(3, 2), B(5, 5), C(4, −1), D(−2, −3), E(1, 6)
2 a $x = 3$ **b** $x = 5$ **c** $y = 2$
d $y = −1$ **e** $y = x$

G3.1

1

2

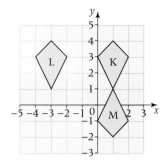

2

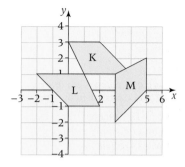

3

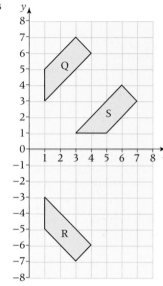

3

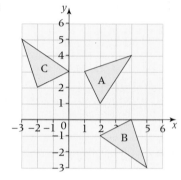

4 **a** and **b**

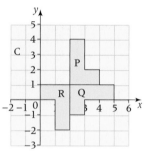

c It maps to R. Two 90° rotations are the same as one 180° rotation.

4 **a** and **b**

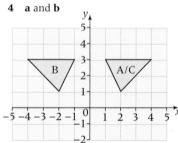

c A and C are the same.

G3.2

1

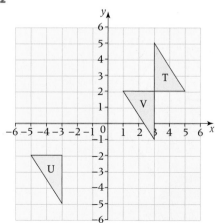

G3.3

1

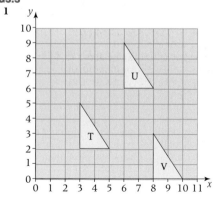

2

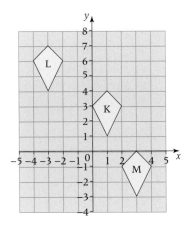

3 a

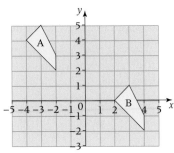

b B translates back to A. The translation has been reversed.

G3.4

1 a Reflection in y-axis
  b Rotation by 180° about (0, 0)
  c Reflection in x-axis
  d Rotation by 180° about (0, 0)
  e Reflection in y-axis
  f Reflection in y-axis

2 a Translation by $\begin{pmatrix} 16 \\ 2 \end{pmatrix}$ **b** Translation by $\begin{pmatrix} 5 \\ 3 \end{pmatrix}$
  c Translation by $\begin{pmatrix} 9 \\ 8 \end{pmatrix}$ **d** Translation by $\begin{pmatrix} -4 \\ -5 \end{pmatrix}$
  e Translation by $\begin{pmatrix} 7 \\ -6 \end{pmatrix}$ **f** Translation by $\begin{pmatrix} -7 \\ 6 \end{pmatrix}$

3 a Rotation by 90° clockwise about (0, 0)
  b Rotation by 180° about (0, 0)
  c Rotation by 90° anticlockwise about (0, 0)
  d Rotation by 90° clockwise about (0, 0)
  e Rotation by 90° clockwise about (0, 0)
  f Rotation by 180° about (0, 0)

G3.5

1

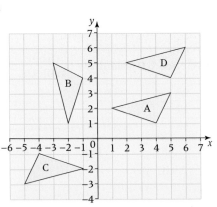

c Rotation 180° about (0, 0)
e Translation by $\begin{pmatrix} 1 \\ 3 \end{pmatrix}$

2

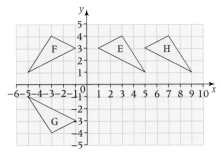

c Rotation 180° about (0, 0)
e Translation by $\begin{pmatrix} 4 \\ 1 \end{pmatrix}$

3

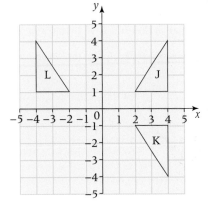

c Reflection in y-axis

4 a and b

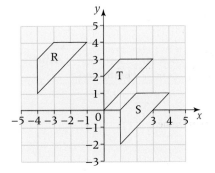

c Translation by $\begin{pmatrix} -4 \\ 1 \end{pmatrix}$

1 a, b

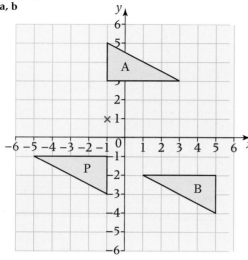

Triangle A has vertices at (−1, 5), (−1, 3) and (3, 3).
Triangle B has vertices at (1, −2), (5, −2) and (5, −4).

c

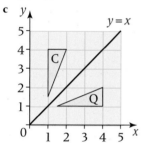

Triangle C has vertices at (1, 1.5), (2, 4) and (1, 4).

Case study 4: Sport
1 a 0.01s   **b** 0.1s

Rank	Lower bound (s)	Time (s)	Upper bound (s)	Athlete
1	9.575	9.58	9.584	Bolt
2	9.685	9.69	9.694	Gay
3	9.715	9.72	9.724	Powell
4	9.785	9.79	9.794	Greene
5	9.835	9.84	9.844	Bailey
	9.835		9.844	Surin
7	9.845	9.85	9.854	Burrell
	9.845		9.854	Gatlin
	9.845		9.854	Fasuba
10	9.855	9.86	9.864	Lewis
	9.855		9.864	Fredericks
	9.855		9.864	Boldon
	9.855		9.864	Obikwelu

D3 Check in

1 a $\frac{3}{5}$ **b** $\frac{3}{7}$ **c** $\frac{5}{8}$ **d** $\frac{7}{11}$
  e $\frac{5}{6}$ **f** $\frac{9}{20}$ **g** $\frac{23}{24}$ **h** $\frac{5}{8}$

2 a 40 **b** 40 **c** 21 **d** 56

3 a 0.7 **b** 0.75 **c** 0.375
  d 0.4 **e** 0.3 **f** 0.0625
  g 0.1 **e** 0.625

D3.1

1 0.53
2 0.65
3 a $\frac{5}{18}$ **b** 0 **c** $\frac{5}{9}$ **d** $\frac{1}{6}$ **e** $\frac{5}{6}$
4 a $\frac{7}{25}$ **b** $\frac{18}{25}$ **c** $\frac{9}{25}$ **d** $\frac{16}{25}$ **e** $\frac{2}{5}$ **f** $\frac{17}{25}$
5 0.27
6 a Gerry will not be chosen.
  b 0.4
7 a The chance of Rangers winning is double that of Rovers winning.
  b 0.2

D3.2

1 a $\frac{7}{20}$ **b** $\frac{13}{20}$ **c** $\frac{1}{20}$ **d** $\frac{19}{20}$ **e** $\frac{11}{20}$
  f $\frac{9}{20}$ **g** $\frac{8}{20} = \frac{2}{5}$ **h** $\frac{15}{20} = \frac{3}{4}$ **i** $\frac{16}{20} = \frac{4}{5}$
2 a $\frac{1}{9}$ **b** $\frac{8}{9}$ **c** $\frac{5}{9}$ **d** $\frac{1}{3}$ **e** $\frac{1}{9}$
  f $\frac{8}{9}$ **g** $\frac{2}{9}$ **h** $\frac{7}{9}$ **i** $\frac{4}{9}$
3 a 30
  bi $\frac{7}{30}$ **bii** $\frac{1}{2}$ **biii** $\frac{2}{5}$
  biv $\frac{7}{30}$ **bv** $\frac{3}{10}$ **bvi** $\frac{7}{10}$
4 a $\frac{1}{15}$ **b** $\frac{1}{3}$ **c** $\frac{23}{30}$ **d** $\frac{1}{2}$

D3.3

1 a

	France	Spain	UK	Total
Girls	3	4	11	18
Boys	3	8	3	14
Total	6	12	14	32

  bi $\frac{3}{16}$ **bii** $\frac{13}{16}$ **biii** $\frac{7}{16}$
  biv $\frac{2}{3}$ **bv** $\frac{9}{16}$ **bvi** $\frac{5}{6}$

2 a

	Orienteering	Paintballing	Quadbiking	Total
Girls	11	8	4	23
Boys	5	8	14	27
Total	16	16	18	50

  bi $\frac{8}{25}$ **bii** $\frac{17}{25}$ **biii** $\frac{16}{25}$ **biv** $\frac{17}{25}$
  ci 128 **cii** 184

3 a

	Science	Humanities	Other subjects	Total
Girls	29	18	32	79
Boys	27	3	11	41
Total	56	21	43	120

  bi $\frac{79}{120}$ **bii** $\frac{7}{40}$ **biii** $\frac{77}{120}$ **biv** $\frac{9}{40}$
  ci 280 **cii** 105

D3.4

1 140

2 120

3 $\frac{5}{8}$

4 a Frequency of 3 is double frequency of 1.

   b $\frac{17}{60}$ **c** 50

5 ai $\frac{11}{100}$ **aii** $\frac{8}{25}$ **aiii** $\frac{27}{100}$

   aiv $\frac{41}{100}$ **av** $\frac{87}{100}$

   bi 80   **bii** 120

6 a 104   **b** 114   **c** 25

D3.5

1 Expected number of tails is 160, which is not close to 114, so coin is probably not fair.

2 133 is close to expected number of black (140), so spinner is fair.

3 Yes, each outcome occurs with similar frequency.

4 No, 2 occurs more than twice as often as other outcomes.

5 a Relative frequency: $\frac{4}{10}, \frac{7}{20}, \frac{11}{30}, \frac{13}{40}, \frac{18}{50}, \frac{22}{60}, \frac{26}{70}, \frac{29}{80}, \frac{32}{90}, \frac{34}{100}$

   b $\frac{34}{100}$

   c Expected number of heads is 50, which is not close to 34, so coin could be biased.

D3 Summary

1 a $\frac{5}{12}$

   b $\frac{7}{12}$

2 a

	Walk	Car	Other	Total
Boy	15	25	14	54
Girl	22	8	16	46
Total	37	33	30	100

   b $\frac{37}{100}$ or 0.37

3 560 boxes

N5 Check in

1 a 120   **b** 138   **c** 90   **d** 265

2 a 4438   **b** 1977   **c** 857   **d** 14 224

3 a 147   **b** 1515   **c** 1040   **d** 350

4 a 11 700   **b** 78 408   **c** 205   **d** 67 564

N5.1

1 a 0.8   **b** 0.5   **c** 0.4   **d** 1.1

   e 0.4   **f** 1

2 a 5.8   **b** 6.5   **c** 7.4   **d** 4.1

   e 11.4   **f** 10

3 a 10   **b** 10.1   **c** 10.3   **d** 11.1

   e 16.4   **f** 7.5

4 a 7.77   **b** 5.25   **c** 3.6   **d** 3.9

   e 2.13   **f** 13.04

5 a 4.15   **b** 6.85   **c** 4.58   **d** 10.42

   e 12.14   **f** 2.04

6 a 0.7   **b** 0.6   **c** 0.3   **d** 3.4

   e 10.2   **f** 14.1

7 a 0.8   **b** 0.7   **c** 0.8   **d** 4.1

   e 10.7   **f** 14.4

8 a 0.74   **b** 1.05   **c** 1.51   **d** 7.14

   e 1.08   **f** 0.64

9 a 1.62   **b** 3.51   **c** 0.45   **d** 4.37

   e 14.52   **f** 11.66

10 a 1.72   **b** 3.61   **c** 0.65   **d** 4.67

   e 15.02   **f** 12.36

N5.2

1 a 12.4   **b** 12   **c** 13.1   **d** 22.5

   e 23.1   **f** 26.1

2 See Q1

3 a 13   **b** 1.871   **c** 201.321   **d** 45

   e 38.97   **f** 21.69

4 a 140.33   **b** 242.3   **c** 98.807   **d** 203.59

   e 161.002   **f** 102

5 See Q4

6 a 2.2   **b** 9.1   **c** 15   **d** 4.9

   e 8.4   **f** 2.9

7 See Q6

8 a 2.6   **b** 7.86   **c** 5.9   **d** 92.54

   e 0.97   **f** 24.27

9 a 13.896   **b** 19.45   **c** 359.79   **d** 7.683

   e 0.326   **f** 11.42

10 See Q9

11 a 4.1   **b** 40.2   **c** 11.288   **d** 5.968

   e 0.892   **f** 0.469

N5.3

1 a 4.8 **b** 4.8 **c** 0.48 **d** 48 **e** 48

2 a 9.1 **b** 9.1 **c** 0.91 **d** 91 **e** 91

3 a 42 **b** 16 **c** 7 **d** 104 **e** 2 **f** 28

4 a 4.2 **b** 1.6 **c** 0.7 **d** 10.4 **e** 2 **f** 0.28

5 a $9 \times 7 = 63, 63 \div 7 = 9, 63 \div 9 = 7$

   b $8 \times 6 = 48, 48 \div 6 = 8, 48 \div 8 = 6$

   c $7 \times 13 = 91, 91 \div 13 = 7, 91 \div 7 = 13$

   d $18 \times 15 = 270, 270 \div 15 = 18, 270 \div 18 = 15$

   e $3.5 \times 5 = 17.5, 17.5 \div 3.5 = 5, 17.5 \div 5 = 3.5$

   f $3.9 \times 2.4 = 9.36, 9.36 \div 2.4 = 3.9, 9.36 \div 3.9 = 2.4$

7 a 4.5 **b** 4500 **c** 45 **d** 0.45

   e 5 **f** 90 **g** 500

8 a 345.8 **b** 345.8 **c** 38 **d** 345.8

   e 9.1 **f** 0.0091

9 a 10 185 **b** 101.85 **c** 0.10185 **d** 0.35

   e 3.5 **f** 0.00291

10 a 29.61 **b** 0.2961 **c** 6300

N5.4

1 a 98   **b** 152   **c** 273

   d 323   **e** 308

2 See Q1

3 a 9.8   **b** 15.2   **c** 2.73

   d 0.0323   **e** 0.308

4 a 80   **b** 12   **c** 12

   d 21   **e** 22

5 See Q4

6 a 0.8   **b** 1.2   **c** 1.2

   d 0.21   **e** 0.22

7 a 24.91   **b** 4.284   **c** 105.84

   d 130.8985   **e** 42.9442

8 See Q7

9 a 3.87   **b** 0.775   **c** 0.916

   d 7.53   **e** 18.13

10 See Q9

11 a 3.45   **b** 4.15   **c** 7.74

   d 4.08   **e** 2.35

12 See Q11

13 a 5.26   **b** 28.88   **c** 1384.29

   d 175.56   **e** 28.65

14 See Q13

N5 Summary

1 a £217
2 £190.12
3 a 4560
  b 45.6
  c 2.4
4 £102.33
5 a 15.456
  b 0.15456
  c 3220

A4 Check in

1 a 9   b 4   c 30   d 10
2 a I think of a number and multiply it by 6.
  b I think of a number and multiply it by itself.
  c I think of a number, multiply it by 2 and then subtract 3.
  d I think of a number, subtract 4 from it and then multiply by 4.
  e I think of a number and divide it by 7.
  f I think of a number, square it and then multiply it by 2.
  g I think of a number, multiply it by 2 and then square.
  h I think of a number, multiply it by 3 and then subtract the result from 10.
3 a $3x + 27$ b $8x - 4$ c $6 - 12y$
  d $x^2 - 7x$ e $x^2 + 7x + 10$ f $y^2 + 4y - 21$
  g $w^2 + 8w + 16$ h $y^2 - 15y + 54$
4 a $(x+2)(x+3)$ b $(x-6)(x+4)$
  c $(x-3)^2$ d $(x-10)(x+10)$

A4.1

1

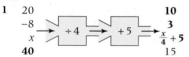

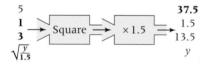

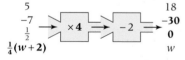

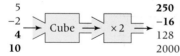

2 The starting numbers are different, so 10% of the final answer is different from 10% of the starting value.
3 a 5.5 b 5 c –3 d 40 e 100
4 Yes, –5, as 25 has two possible square roots
5 ai Add 3
  aii Multiply by 2, subtract 1
  aiii Add 5, divide by 2
  aiv Square, multiply by 3
  bi 14 bii 9 biii 43 biv 4 or –4
6 a –2 bi 7 bii 10 biii $-\frac{3}{16}$
7 a Many possibilities, e.g. ×2, ×2, ×2, +20, –7
  b Depends on **7a**

A4.2

1 a 5 b 6 c 10 d –2 e 5 or –5
  f 13 g $8\frac{1}{3}$ h $\sqrt[3]{16}$ i 100

2

1	3		2 4
0		3 4	0
4 5	5 1		
	6 3	2	5

3 sides = 1 and 14
4 51
5 a 5   b 1
6 a $2x - 10$ b $5x - 20 = 180$ c 40°, 70°, 70°

A4.3

1 a –1 b 2 c 3 d $\frac{13}{16}$
2 a $6a - 2 = 2a + 6$ b $3b - 14 = b$
  c $1 - 4c = 2 - 8c$ d $5d + 7d - 3 = 10 - d$
3 a $8x - 2 = 2x + 10$, $x = 2$ b $5x + 3 = 24 - 2x$, $x = 3$
  c $11 - 2x = 14 - 3x$, $x = 3$
4 a 40°, 60°, 80°
  b Square: 8 × 8, rectangle: 10 × 6
  c Mark: 160 cm, Miranda: 144 cm

A4.4

1 a $\frac{1}{3}$ b $\frac{1}{3}$ c $1\frac{1}{3}$ d $\frac{7}{8}$
  e –5 f $\frac{5}{11}$ g $-2\frac{1}{3}$ h $\frac{1}{3}$
2 a 5 b 10 c 28 d –1
3 a –8 b $\frac{3}{4}$ c $\frac{2}{11}$ d $-1\frac{1}{4}$
4 a –7 b 2 c –4 d 2
5 a $\frac{16}{x} = 10$, $x = 1.6$ b $\frac{12}{x+4} = 7$, $x = 2\frac{2}{7}$
  c $\frac{11}{x-3} = \frac{8}{x}$, $x = -8$
6 a 36 b 14 c 80
7 $x = 37\frac{2}{3}$
  Values for Set 1: $74\frac{1}{3}$, 115, $192\frac{1}{3}$, $263\frac{2}{3}$, 222, $-65\frac{1}{3}$
  Values for Set 2: 106, $196\frac{1}{3}$, $-24\frac{2}{3}$, 228, $162\frac{2}{3}$

A4.5

1 a $x = 6$ b $y = 14$ c $x = 6$ d $p = 8$
  e $t = 2$ f $b = 2$
2 a $x = 1.1$ b $y = -1\frac{5}{22}$ c $z = 1\frac{3}{4}$ d $a = 5\frac{1}{8}$
  e $a = 4.3$ f $x = 5$ g $x = -13$ h $y = 1\frac{5}{7}$
3 a $x = 2$ b $y = 3$ c $z = \frac{3}{8}$ d $a = -\frac{3}{7}$
  e $b = -1$ f $c = -2.5$
4 a $40 - 3x = 22$, $x = 6$
  b $2(x+6) = 6(x-5)$, $x = 10.5$
  c $(x+7)^2 = (x-5)^2$, $x = -1$
5 a $(x-4)(x+11) = (x+3)^2$, $x = 53$
  b $26 - 4x = 15 - 2x$, $x = 5.5$

A4.6

1 a $2x^2 = 50$, $x^2 = 100$, $x^2 = -49$, $x^2 = 169$
  b $2x^2 = 50$, $x = 5$ or -5
    $x^2 = 100$, $x = 10$ or -10
    $x^2 = 169$, $x = 13$ or -13
2 a $(x+2)(x+6)$ b $(x+3)(x+8)$
  c $(x+9)(x+4)$ d $(x+11)(x+5)$
  e $(x-3)(x-4)$ f $(x-4)(x-6)$
  g $(x+7)(x-4)$ h $(x+3)(x-5)$
3 a $x = -3$ or -4 b $x = -2$ or -6

c $x = -5$ **d** $x = -7$ or 2
e $x = 5$ or -1 **f** $x = 2$ or 3
g $x = 3$ or 7 **h** $x = 8$ or -5
4 a $x = 0$ or 8 **b** $x = 0$ or -4
c $x = 0$ or 6 **d** $y = 0$ or -5
e $x = 0$ or 9 **f** $x = 0$ or 12
g $x = 0$ or -4 **h** $x = 0$ or 6
5 No two integers add up to 7 and multiply to give 11.
6 a She should make the right-hand side equal to zero first.
b $(x+4)(x-2) - 7 = 0$, $x^2 + 2x - 15 = 0$,
$(x+5)(x-3) = 0$, $x = -5$ or 3
7 $x = 1$ or 5

A4 Summary

1 a $x = 7.5$ **b** $y = 2.4$
2 a $y = 3$ **b** $t = 4.6$
3 $x = 3.5$
4 a The opposite lengths of the rectangle are equal.
b $x = 5.5$ **c** 57 cm
5 a $(x+2)(x+4)$ **b** $x = -2$, $x = -4$

Case study 5: Art

1 a 30
b 4.44 m
c length = 40 hands; height = 29.6 hands
d 30 cm squares
e area scale factor 900
2 ai $B = (2, 60)$ **aii** $A = (1, 210)$ **aiii** $C = (3, 300)$
b $D = (1.5, 0)$
3 a Circle radius = 4
b $a = 4$
c 9 units
d $k = 0.025$
5 a The central part of the crop circle has constant radius and forms a circle.
b A spiral shape starts with radius equal to that of the inner circle and extends outwords as the radius increases.
The circle and spiral are joined by a criss-cross pattern formed by radial and tangent lines to the inner circle.

G4 Check in

1

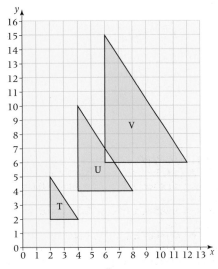

2 a 1 : 3 **b** 1 : 3 **c** 2 : 1 **d** 5 : 2
3 a $x = 14$ **b** $x = 15$ **c** $x = 10$ **d** $x = 18$

G4.1

1

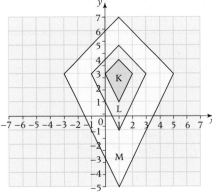

2

3

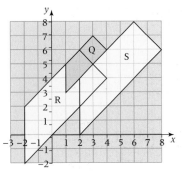

d Twice as large

4

d Three times as large

G4.2

1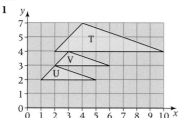

2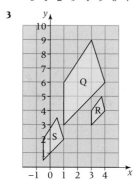

3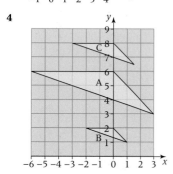

4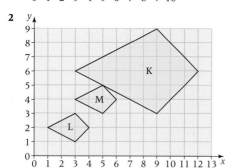

G4.3

1 a Enlargement with centre (0, 0) and scale factor 3
   b Enlargement with centre (0, 0) and scale factor $\frac{1}{3}$
2 ai Enlargement with centre (−3, 0) and scale factor 2
   aii Enlargement with centre (−3, 0) and scale factor $\frac{1}{2}$
   b 16 units
3 a Enlargement with centre (0, 0) and scale factor 2
   b Enlargement with centre (0, 0) and scale factor $\frac{1}{2}$
4 ai Enlargement with centre (0, 0) and scale factor $\frac{1}{2}$
   aii Enlargement with centre (0, 0) and scale factor 2
   aiii Enlargement with centre (0, 0) and scale factor 4
   b 4 times

G4.4

1 $a = 6$ cm, $b = 4.5$ cm
2 $c = 6$ cm, $d = 2.5$ cm

3 $e = 5$ cm, $f = 2.4$ cm
4 15 cm
5 a 3 : 2   **b** $5\frac{1}{3}$ cm   **c** 22 cm, $14\frac{2}{3}$ cm
   d 3 : 2, same as ratio of lengths of sides

G4.5

1 a 4.5 cm   **b** 9 cm
2 a 2.67 mm   **b** 4.5 mm
3 a 6 cm   **b** 5 cm   **c** 15 cm
4 a MN = 6.3 cm, JN = 6 cm   **b** 15.5 cm
5 a 19.35 cm   **b** 32.1 cm

G4 Summary

1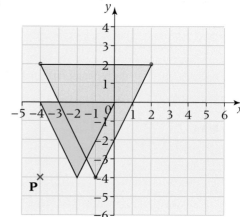

Vertices at (−1, −4), (−4, 2) and (2, 2)
2 a 20 cm   **b** 6 cm

N6 Check in

1 ai 1, 2, 3, 6
   aii 1, 2, 3, 4, 6, 12
   aiii 1, 2, 4, 7, 14, 28
   aiv 1, 2, 3, 4, 6, 9, 12, 18, 36
   b 2, 3, 5, 7, 11, 13, 17, 19, 23, 29, 31, 37, 41, 43, 47
2 a 3^2 **b** 4^5 **c** 6^3 **d** 5^4
3 a $3 \times 3 \times 3$ **b** 6×6
   c $4 \times 4 \times 4 \times 4 \times 4$ **d** $8 \times 8 \times 8$
   e $7 \times 7 \times 7 \times 7$ **f** $5 \times 5 \times 5 \times 5 \times 5$
4 a 16   **b** 27   **c** 16
   d 125   **e** 128   **f** 7

N6.1

1 a 12   **b** 8   **c** 4   **d** 6
2 a $77 = 7 \times 11$ **b** $51 = 3 \times 17$
   c $65 = 5 \times 13$ **d** $91 = 7 \times 13$
   e $119 = 7 \times 17$ **f** $221 = 13 \times 17$
3

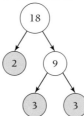

4 a $2^2 \times 3^2$ **b** $2^3 \times 3 \times 5$
   c 2×17 **d** 5^2
   e $2^4 \times 3$ **f** $2 \times 3^2 \times 5$
   g 3^3 **h** $2^2 \times 3 \times 5$
5 a $2^2 \times 263$ **b** $2^9 \times 5$
   c $2 \times 3^2 \times 5 \times 7$ **d** $3 \times 5^2 \times 11$
   e $5 \times 11 \times 13$ **f** $7 \times 11 \times 13$

g 3×73
h 17^2
i $2^3 \times 5 \times 71$
j $5 \times 7^2 \times 11$
k $7 \times 13 \times 19$
l $2 \times 3^2 \times 11 \times 17$
m $2^2 \times 11 \times 13 \times 17$
n $2 \times 5 \times 7 \times 13^2$
o $2^2 \times 23 \times 31$
p $3^3 \times 13 \times 29$

6 a $2 \times 3 \times 35$, $2 \times 5 \times 21$, $2 \times 7 \times 15$, $3 \times 7 \times 10$, $3 \times 5 \times 14$, $5 \times 6 \times 7$
 b $1 \times 1 \times 121$, $2 \times 3 \times 35$, $2 \times 5 \times 21$, $2 \times 7 \times 15$, $3 \times 7 \times 10$, $3 \times 5 \times 14$, $5 \times 6 \times 7$

N6.2

1 a $6^2 = 36$, so 6 joins to itself.
 b 1, 2, 3, 4, 6, 8, 12, 16, 24, 48

 c 12
2 a 1 b 1 c 3 d 2 e 8 f 10
3 a Multiples of 12 = 12, 24, 36, 48, ...
 Multiples of 9 = 9, 18, 27, 36, 45, ...
 b 36
4 a 20 b 36 c 30 d 60 e 70 f 40
5 8
6 a $5,600$ b $4,432$ c $12,720$
 d $14,168$ e $5,105$ f $10,220$

N6.3

1 a 4 b 10 c 24 d 12
 e 20 f 35 g 60 h 60
2 a 6 squares, 3 shaded
 b 6 squares, 4 shaded
 c 15 squares, 9 shaded
 d 20 squares, 15 shaded
3 a $\frac{20}{60}$ b $\frac{15}{60}$ c $\frac{40}{60}$ d $\frac{24}{60}$
4 a $\frac{18}{24}$ b $\frac{8}{24}$ c $\frac{9}{24}$ d $\frac{10}{24}$
5 a $\frac{20}{30}$ b $\frac{18}{42}$ c $\frac{35}{45}$ d $\frac{25}{40}$
6 a $\frac{2}{10}$ and $\frac{3}{10}$, $\frac{3}{10}$ b $\frac{8}{12}$ and $\frac{9}{12}$, $\frac{3}{4}$
 c $\frac{6}{15}$ and $\frac{5}{15}$, $\frac{2}{5}$ d $\frac{21}{30}$ and $\frac{20}{30}$, $\frac{7}{10}$
7 a Two sets of 10 squares; 2 and 3 shaded
 b Two sets of 12 squares; 8 and 9 shaded
 c Two sets of 15 squares; 6 and 5 shaded
 d Two sets of 30 squares; 21 and 20 shaded
8 a 30 b 12 c 24 d 28
9 a $\frac{6}{12}, \frac{8}{12}, \frac{9}{12}$ b $\frac{4}{20}, \frac{15}{20}, \frac{7}{20}$
 c $\frac{3}{24}, \frac{14}{24}, \frac{16}{24}$ d $\frac{56}{84}, \frac{64}{84}, \frac{24}{84}$
10 a $\frac{2}{15}, \frac{1}{5}, \frac{2}{3}$ b $\frac{1}{4}, \frac{7}{20}, \frac{2}{5}$
 c $\frac{5}{14}, \frac{3}{8}, \frac{3}{7}$ d $\frac{2}{7}, \frac{2}{3}, \frac{5}{6}$
11 a $\frac{1}{2}, \frac{2}{5}, \frac{3}{10}, \frac{1}{4}$ b $\frac{4}{5}, \frac{1}{3}, \frac{3}{20}, \frac{1}{10}$
 c $\frac{3}{4}, \frac{17}{40}, \frac{2}{5}, \frac{3}{8}$ d $\frac{5}{6}, \frac{5}{8}, \frac{7}{12}, \frac{11}{24}$

N6.4

1 a 3^2 b 2^3 c 3^3 d 5^4
 e 7^3 f 10^3 g 6^4 h 5^3
2 a $3 \times 3 \times 3 \times 3$
 b 5×5
 c $7 \times 7 \times 7 \times 7$
 d $10 \times 10 \times 10 \times 10 \times 10$
 e $4 \times 4 \times 4 \times 4 \times 4 \times 4 \times 4 \times 4 \times 4$
 f $6 \times 6 \times 6$
 g $2 \times 2 \times 2 \times 2 \times 2$
 h $9 \times 9 \times 9$

3 a 16 b 64 c 32 d 100
 e 1000 f 27 g 8 h 9

4

Index form	Product	Value
10^6	$10 \times 10 \times 10 \times 10 \times 10 \times 10$	1 000 000
10^5	$10 \times 10 \times 10 \times 10 \times 10$	100 000
10^4	$10 \times 10 \times 10 \times 10$	10 000
10^3	$10 \times 10 \times 10$	1 000
10^2	10×10	1 00
10^1	10	1 0

5

Index form	Product	Value
2^{10}	$2 \times 2 \times 2 \times 2 \times 2 \times 2 \times 2 \times 2 \times 2 \times 2$	1024
2^9	$2 \times 2 \times 2 \times 2 \times 2 \times 2 \times 2 \times 2 \times 2$	512
2^8	$2 \times 2 \times 2 \times 2 \times 2 \times 2 \times 2 \times 2$	256
2^7	$2 \times 2 \times 2 \times 2 \times 2 \times 2 \times 2$	128
2^6	$2 \times 2 \times 2 \times 2 \times 2 \times 2$	64
2^5	$2 \times 2 \times 2 \times 2 \times 2$	32
2^4	$2 \times 2 \times 2 \times 2$	16
2^3	$2 \times 2 \times 2$	8
2^2	2×2	4
2^1	2	2

6 a 9^2 b 5^3 c 2^7 d 10^5
 e 3^4 f 7^3
7 a 3^6 b 2^9 c 4^8 d 5^5
 e 8^8 f 3^8 g $2^7 \times 3^4$ h $5^4 \times 7^7$
8 a 2^4 b 11^2 c 3^3 d 5^3
 e 13^2 f 5^4 g 3^5 h 2^8
9 a $2^2 \times 13$ b $2^2 \times 3^2$ c 2×5^2 d $2^3 \times 3$
 e 2×3^2 f $2^4 \times 3$ g $2^2 \times 3 \times 5$
 j $2^4 \times 3^2$
10 a $16,448$ b $19,266$

N6.5

1 a 7^3 b 3^3 c 5^3 d 6^4
 e 5^2 f 8^5 g 9^6 h 8^8
2 a 6^5 b 4^9 c 2^{13} d 11^7
 e 1^{30} f 7^{12} g 3^{12} h 9^{10}
3 a 7^2 b 8^4 c 3 d 9^3
 e 4^6 f 1 g 12^2 h 1
4 a 8^5 b 5^5 c 2^6 d 9^2
 e 8^8 f 7^7 g 4^{10} h 11
5 a 3 b 5^{10} c 4^2 d 7^2
 e 8^{11} f 9^{10}
6 a 4^2 b 6^2 c 9^2 d 8
 e 5^3 f 6^5 g 8^2 h 10^0
7 a $3^3 \times 4^4$ b $7^2 \times 8^3$ c $5^6 \times 6^4$ d $3^6 \times 4^9$
 e $2^3 \times 5^2$ f $7^4 \times 9^7$ g $5^3 \times 8^5$ h $3^8 \times 8^7$
 i $2^2 \times 9^5$
8 a $5^2 \times 8^3$ b $6^2 \times 7^2$ c $5^2 \times 6^2$ d $5^3 \times 7^6$
 e $3^3 \times 8^2$ f $4^5 \times 5^2$ g $6^4 \times 7^2$ h $4^4 \times 7^5$
9 a x^7 b x^3 c x^{15} d x^{12}
 e $x^{12} y^2$ f $x^{15} y^5$ g $x^8 y^3$ h $x^2 y^2 z$

N6.6

1 a 5 b 6 c 1 d 1
 e 1 f 41 g 1 h 0
2 a 10 b 4 c 7 d 2
 e 8 f 3 g 11 h 8
 i 12 j 2 k 3 l 10
3 a 6 b 36 c 9 d 81
4 a 3^{-1} b 5^{-1} c 7^{-1} d 11^{-1}
 e 2^{-1} f 5^{-1} g 10^{-1} h 3^{-1}
5 a $\frac{1}{9}$ b 1 c 3 d 9
 e 81 f 729 g 59 049 h $\frac{1}{729}$
6 a 7^{-2} b 9^{-2} c 2^{-2} d 2^{-3}

e 2^{-5} f 3^{-4} g 5^{-3} h 6^{-4}

7 a $\frac{1}{8^2}$ b $\frac{1}{7^3}$ c $\frac{1}{5^2}$ d $\frac{1}{9^4}$
  e $\frac{1}{3^2}$ f $\frac{1}{9^3}$ g $\frac{1}{4^5}$ h $\frac{1}{6^6}$

8 a $\frac{1}{16}$ b $\frac{1}{4}$ c 1 d 2
  e 4 f 16 g 64 h $\frac{1}{2}$

9 a $3^{-\frac{1}{2}}$ b $5^{-\frac{1}{2}}$ c $7^{-\frac{1}{2}}$ d $11^{-\frac{1}{2}}$

10 a 2^4 b 2^6 c 3^6 d 4
   e 5^8 f 4^{-6} g 7^{12} h 5^{-4}

11 a 2^{-2} b 3^{-1} c 4^{-6} d $3^{-1} \times 5^{-1}$
   e $5^4 \times 7^{-4}$

N6.7

1 a 2×10^2 b 8×10^2 c 9×10^3
  d 6.5×10^2 e 6.5×10^3 f 9.52×10^2
  g 2.358×10 h 2.5585×10^2 i 3×10^{-1}
  j 4.7×10^{-3} k 7.8×10^{-5} l 4.485×10^{-1}

2 a 500 b 3000 c 100 000
  d 250 e 4900 f 3 800 000
  g 750 000 000 000
  h 8 100 000 000 000 000 000

3 a 1×10^{-2} km b 2×10^{-3} g
  c 5×10^{-6} m d 1.1×10^{-2} litre

4 a 6×10^2 b 4.5×10^4
  c 6.5×10^0 d 5×10^6
  e 2.8×10^{-1} f 4×10^{-2}
  g 1.35×10^{-3} h 1.2×10^{-7}

5 a 4×10^5 b 9×10^7
  c 2.5×10^8 d 2.4×10^{13}
  e 5×10^{-1} f 9.2×10^{-8}
  g 2×10^{-2} h 4.2×10^{-8}

6 a 2×10^2 b 2×10^4
  c 5×10 d 7.5×10^2

7 a 5.2×10^{-1} b 4.6×10^{-2}
  c 2.09×10^{-2} d 1.3×10^{-2}

8

Planet	Mean distance from Sun (m)	Light travel time
Mercury	5.79×10^{10}	3 minutes 13 seconds
Earth	1.50×10^{11}	8 minutes 20 seconds
Mars	2.28×10^{11}	12 minutes 40 seconds
Jupiter	7.78×10^{11}	43 minutes 13 seconds
Pluto	5.90×10^{12}	5 hours 27 minutes 47 seconds

N6 Summary

1 $m = 3, n = 5$
2 ai $2 \times 2 \times 3 \times 5$ or $2^2 \times 3 \times 5$
  aii $2 \times 2 \times 2 \times 2 \times 2 \times 3$ or $2^5 \times 3$
  b 12 c 480
3 a $2 \times 3 \times 3 \times 7$ or $2 \times 3^2 \times 7$
  b 42
4 96
5 a 12 b 48
6 a $\frac{1}{6}, \frac{3}{8}, \frac{1}{2}, \frac{2}{3}, \frac{3}{4}$
  b $\frac{3}{5}, 65\%, \frac{2}{3}, 0.72, \frac{3}{4}$
7 a 64 b 3 c 12
8 a 1 b 2 c $\frac{1}{4}$
9 a 4.56×10^5 b 3.4×10^{-4}
  c 1.6×10^8
10 a 2.8×10^4 b 0.000542
11 1.4×10^{10}

D4 Check in

1 a 62 b 70 c 60 d 45
  e 102 f 108

2 a £50 b £70
3 2 days

D4.1

1 ai $10 < t \le 15$ aii $15 < t \le 20$ aiii 16.1
  bi $10 < t \le 20$ bii $20 < t \le 30$ biii 23.5
  ci $5 < t \le 10$ cii $10 < t \le 15$ ciii 14.7
  di $15 < t \le 25$ dii $25 < t \le 35$ diii 30
2 a $20 < b \le 30$ b £24.17 to 3 sf
3 a $165 \le h < 170$ b 164.3 cm c $165 \le h < 170$

D4.2

1 a

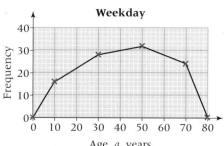

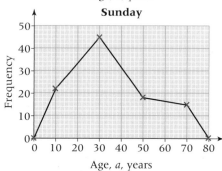

b Weekday: $40 < a \le 60$, Sunday: $20 < a \le 40$
c Generally, people visiting on a weekday are older.

2 a

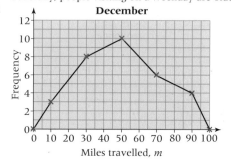

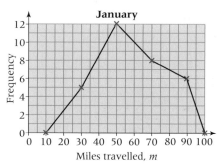

b December: $40 < m \le 60$, January: $40 < m \le 60$
c Less variation in miles travelled in January, less short journeys. The most common journey length does not change.

3 a

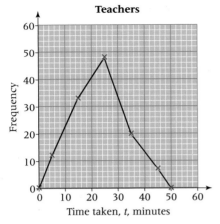

Teachers

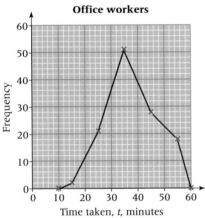

Office workers

b Teachers: $20 < t \leq 30$, office workers: $30 < t \leq 40$

c On average, office workers take longer travelling home.

D4.3

1 a

Height, h, cm	Cumulative frequency
$h < 145$	0
$h < 150$	7
$h < 155$	32
$h < 160$	78
$h < 165$	95
$h < 170$	100

b

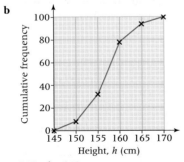

c $155 \leq h < 160$

2 a

Age, A, years	Cumulative frequency
$A < 20$	0
$A < 30$	18
$A < 40$	55
$A < 50$	106
$A < 60$	134
$A < 70$	150

b

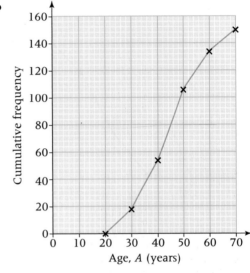

c $40 \leq A < 50$

3 a

Time, t, minutes	Cumulative frequency
$t < 10$	4
$t < 20$	15
$t < 30$	44
$t < 40$	81
$t < 50$	108
$t < 60$	120

b

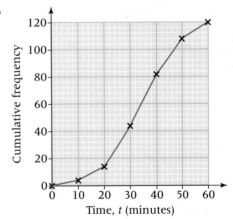

c $30 \leq t < 40$

4 a

Weight, w, grams	Cumulative frequency
$w < 1500$	0
$w < 2000$	9
$w < 2500$	31
$w < 3000$	68
$w < 3500$	88
$w < 4000$	100

b

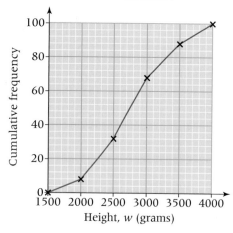

c $2500 \leqslant w < 3000$

5 a

Height, h, cm	Cumulative frequency
$h < 40$	0
$h < 60$	2
$h < 80$	19
$h < 100$	47
$h < 120$	86
$h < 140$	110
$h < 160$	120

b

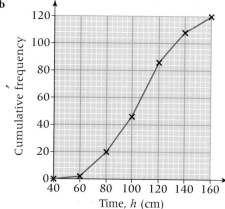

c $100 \leqslant h < 120$

6 a

Amount, £p	Cumulative frequency
$p < 10$	16
$p < 20$	30
$p < 30$	53
$p < 50$	70
$p < 70$	85
$p < 100$	100

b

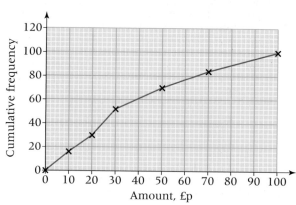

c $20 \leqslant p < 30$

D4.4

	a	b	ci	cii
1	157 cm	6 cm	18	12
2	44 cm	17 cm	36	30
3	34 minutes	18 minutes	30	26
4	2750 g	750 g	18	10
5	107 cm	36 cm	98	86
6	£28	£39	30	10

D4.5

1 The boys' results are higher than the girls', on average. The middle half of the girls' results is less varied than that of the boys. The range is the same for the girls and the boys.

2 Farmer Jenkins' sunflowers are shorter than Farmer Giles', on average.
The middle half of Farmer Jenkins' sunflowers vary in height more than those of Farmer Giles.
The range of heights of Farmer Jenkins' sunflowers is greater than that of Farmer Giles'.

3 Boys have higher mobile phone bills, on average. The middle half of the mobile phone bills varies the same for girls and boys. The range of mobile phone bills is the same for girls and boys.

D4.6

1 Min 145 cm, LQ 153.5 cm, median 157 cm, UQ 159.5 cm, max 170 cm

2 Min 20 cm, LQ 35 cm, median 44 cm, UQ 52 cm, max 70 cm

3 Min 0 minutes, LQ 25 minutes, median 34 minutes, UQ 43 minutes, max 60 minutes

4 Min 1500 g, LQ 2350, median 2750 g, UQ 3100 g, max 4000 g

5 Min 40 cm, LQ 88 cm, median 107 cm, UQ 124 cm, max 160 cm

6 Min £0, LQ £17, median £28, UQ £56, max £100

7 a Min 11 years, LQ 13.1 years, median
14.3 years, UQ 15.4 years, max 18 years
b Min 0 minutes, LQ 23 minutes, median
31 minutes, UQ 42 minutes, max 70 minutes

D4.7

1 On average, waiting times are higher at the dentist. The range of waiting times is greater at the doctor. The middle half of waiting times varies more at the doctor than at the dentist. The waiting times at the doctor are symmetrical, but those at the dentist are negatively skewed.

2 The average reaction time is the same for boys and girls. The range of reaction times is greater for girls. The middle half of reaction times varies less for girls than for boys. Reaction times for girls are symmetrical, but for boys they are negatively skewed.

3 On average, results are the same in the English and French tests. The ranges of results are the same. The middle half of results varies more in the English test. The English test results are negatively skewed, but the French test results are symmetrical.

4 On average, 17-year-old girls make longer phone calls than 13-year-olds. The range of the length of calls made is greater for 17-year-olds.
The middle half of the calls made varies more for 13-year-olds than for 17-year-olds.   The lengths of calls made by 13-year-olds is negatively skewed, but those for 17-year-olds are symmetrical.

D4 Summary

1 Points plotted at (5, 4), (15, 13), (25, 17), (35, 19) and (45, 7) joined in order by straight lines.

2 a 73 kg (Accept 72 to 74)
b 15 kg (Accept 14 to 17)

Case study 6: Holiday

1 £32.63 (assuming 13 weeks)
2 £4 per hour, £52 (assuming 13 weeks)
3 CDs £2.50 each, DVDs £3.50 each
£88.50
4 £1 : 1.31 Eur
90.78 euros for £70 and 1% commission
£1 : 10.37 FRF
£1 : 3.07 DEM
$°C = \frac{2}{5}°F - \frac{16}{5}$

A5 Check in

1 $A = lw$, area of a rectangle; $A = \pi r^2$, area of a circle;
$a^2 + b^2 = c^2$, length of sides in a right-angled triangle;
$V = lwh$, volume of a cuboid;
$A = \frac{1}{2}bh$, area of a triangle

2 a $x = 5\frac{2}{3}$ **b** $x = 2\frac{1}{2}$ **c** $x = \pm 5$ **d** $x = 16$
e $x = 2$ **f** $x = 2$

3 a   b

x	1	2	3
y	2	8	11

x	−1	0	6
y	14	12	0

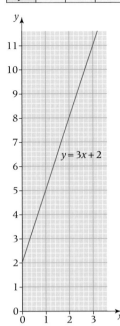

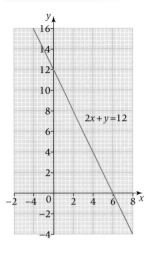

A5.1

1 a $x = \frac{C - b}{a}$ **b** $x = M + b + c$
c $x = Kt + qt$ **d** $x = \frac{W - t}{y}$
e $x = Hp - z$ **f** $x = q + \frac{D}{p}$ or $\frac{D + pq}{p}$
g $x = AB + ct$ **h** $x = \frac{Y - c}{m}$

2 Sebastian expanded the brackets to begin with but James didn't.

3 a $Y = \sqrt{c}$ **b** $y = 4(k + 2)$
c $y = \frac{M + t}{xz}$ **d** $y = 4x^2 \frac{T}{-K}$
e $y = 3R\sqrt{}\, p - 2$ **f** $y = R$
g $y = \frac{3R}{az}$ **h** $y = p^3$

4 He should have divided by m before taking cube root.

5 $\frac{8(D + k)}{ab} = c$
$8(D + k) = abc$
$D + k = \frac{1}{8}abc$
$d = \frac{1}{8}abc - k$

6 a 86 °F   **b** $C = \frac{5(F + 40)}{9}$ **c** $-35\frac{5}{9}$ °C

A5.2

1 a $p = \frac{m + q}{x}$ **b** $p = R\sqrt{}\, w + r$
c $p = (m - h)^3$ **d** $p = t(h + g)$
e $p = 2(q - r)$ **f** $p = \sqrt{\frac{k}{b}}$
g $p = \frac{z}{aw}$ **h** $p = (2x + y)^2$

2 a $x = k - w$ **b** $x = \frac{t - p}{a}$
c $x = \frac{b - y}{t}$ **d** $x = \frac{a - m}{n}$ or $\frac{an - m}{n}$
e $x = \frac{k}{w}$ **f** $x = \frac{t}{m}$

g $x = \frac{h}{g-p}$ **h** $x = \sqrt{\frac{p}{k}}$

3 a 36 mph **b** $t = \frac{d}{s}$

  c 1 hour 26 minutes to nearest minute

4 $k = \frac{t}{p-q}$

5 D $x = \frac{2(p-y)}{ab}$

  A $abx = 2(p-y)$

  E $\frac{1}{2}abx = (p-y)$

  B $y + \frac{1}{2}abx = p$

  C $y = p - \frac{1}{2}abx$

6 a $g = \frac{4\pi^2 p}{T^2}$ **b** 9.8

7 a $h = \frac{v}{\pi r^2}$ **b** $r = R^{-\frac{v}{\pi h}}$

8 $b = \frac{2A}{h} - a$

A5.3

1 **a, b, e**

2 a $x = 4, y = 3$ **b** $x = 2, y = -3$
  c $x = -1, y = 4$ **d** $x = 3, y = 0.5$
  e $m = 5, n = 2$ **f** $x = -2, y = 1$

3 a $x = 7, y = -1$ **b** $x = 3, y = 0.5$
  c $a = 3, b = -1$ **d** $x = 2, y = 5$
  e $x = 3.25, y = -\frac{2}{7}$ **f** $p = 2, q = 7$

4 a $x = 5, y = -2$ **b** $x = 2, y = 3$
  c $a = -2, b = 3$ **d** $v = 4, w = 2$
  e $p = 2, q = 2$ **f** $x = 5, y = 2$

5 a $a = 9\frac{2}{3}, b = 1\frac{1}{3}$ **b** $v = 7, w = 4$

6 a 17p **b** 6.4 cm

A5.4

1 a $x = 12, y = -16$ **b** $x = 7, y = -1$
  c $a = 3, b = -2$ **d** $v = 3, w = 2$
  e $p = 0, q = 3$ **f** $x = 5, y = 2$

2 $2x + y = 12, y - x = 15; x = -1, y = 14$
  $2x + y = 12, 3x - 4y = 7; x = 5, y = 2$
  $y - x = 15, 3x - 4y = 7; x = -67, y = -52$

3 a $x = 6, y = 2$ **b** $a = 8, b = -1$
  c $p = 8, q = -1$

4 a 17, 24 **b** 17, 23
  c 4 large, 1 small **d** 105°, 37.5°, 37.5°
  e £5 for a paperback, £10 for a hardback

A5.5

1 ai $x = 2, y = 3$ **aii** $x = 1, y = 4$
  aiii $x = 1, y = 1$ **aiv** $x = \frac{1}{4}, y = 1\frac{3}{4}$
  b The lines are parallel so they never intersect.

2 a $x = 3, y = 7$ **b** $x = 1, y = 1$ **c** $x = 3, y = -1$

3 a $x = 1, y = \frac{1}{2}$ **b** $x = 1, y = \frac{1}{2}$
  c They are the same

4 a 2, 0 **b** James is 3, Isla is 1

5 a The lines are parallel so they never intersect.
  b Not if both are lines, but you could have a curve and a line intersecting twice or more.

6 a $y = 4x + 3, y = 6x + 1$ **b** $x = 1, y = 7$ **c** (1, 7)

A5.6

1 a $x \leqslant 3$ **b** $2 \leqslant x \leqslant 8$ **c** $-5 < x < 12$

2 a $x \leqslant 2$ **b** $x > -1$ **c** $x \geqslant -1$
  d $1 \leqslant x < 5$ **e** $-7 < x \leqslant 0$

3 a 2, 1, 0, −1, −2, −3, ...
  b 0, 1, 2, 3, ...
  c −1, 0, 1, 2, 3, ...

d 1, 2, 3, 4
e −6, −5, −4, −3, −2, −1, 0

4 True

5 a $x \leqslant 7$ **b** $x > 11$ **c** $p \leqslant -16$
  d $x > -3$ **e** $x \leqslant \frac{2}{3}$ **f** $y < -3$
  g $x \leqslant 2$ **h** $x > -5$ **i** $x \leqslant 10$
  j $p \leqslant -3$ **k** $x > -18$ **l** $x \geqslant \frac{1}{2}$

6 a $6(x-2) > 12 + 2(x-2), x > 5$ **b** 6

7 $-1 < x \leqslant 2$

(number line from −2 to 3, open circle at −1, closed circle at 2)

A5 Summary

1 a

x	−1	0	1	2	3	4
y	−3	−1	1	3	5	7

b

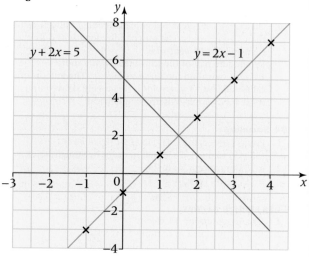

  c $x = 1.5, y = 2$

2 a $x = \frac{y+5}{3}$
  b $a = 9b - 2$

3 $x < 10$

4 $x = 3, y = 0.5$

N7 Check in

1 a 40 **b** 630 **c** 90 **d** 200

2 ai £49.50 **aii** 66 mm
  aiii 52.8 km **aiv** 4.4 hours
  bi 4.5 miles **bii** 43.5 minutes
  biii 10.5 kg **biv** £46.5

3 a 40 mph **b** 42.5 mph **c** 54 mph
  d 32 mph

N7.1

1 a £8.25 **b** £12.38 **c** £18.84 **d** £75.76

2 a £17.25 **b** £11.90 **c** £7.76 **d** £16.73

3 12 kg

4 75 kg

5 a 60p **b** £1.80

6 a 40p **b** £2.80

7 a 3p **b** 0.52 g

8 a £1.88 **b** £1.48 **c** £1.73
  d £2.04 **e** £1.86

9 £3.64

10 a Pack A = 1.21p per pin. Pack B = 1.15p per pin. Pack B is the better value.

11 Regular, at 0.7p per sheet, is better value than Super, at 0.77p per sheet.

N7.2

1 **a** $33 **b** $57.75 **c** $17.33 **d** $63.61
2 **a** €5.80 **b** €34.80 **c** €68.44 **d** €306.24
3 **a** £1.07 **b** £19.22 **c** £127.67 **d** £305.31
4 **a** £2 **b** £10 **c** £1.46 **d** £3.66
  e £31.71 **f** £2.90
5 **a** CAN$40.91 **b** $61.11
6 **a** €348 **b** $1250
7 1.760 pints
8 **a** 12.701 kg **b** 1016.047 kg
  c 170.324 kg **d** 17.293 kg
9 **a** 40.2 km to 3 sf **b** 6.21 miles to 3 sf
  c 24 900 miles to 3 sf **d** 7.24 km to 3 sf
10 **a** 157.5 litres **b** 11.1 gallons to 3 sf
  c 56.25 litres to 3 sf **d** 8.62 gallons to 3 sf

N7.3

1 **a** 25 g **b** 200 cm **c** 7.5 h **d** 625 kg
2 **a** 400 g **b** 20 sec **c** 30° **d** 48 h
3

Original number	Proportional change	Result
42	Decrease by $\frac{1}{4}$	31.5
110	Increase by $\frac{1}{5}$	132
250	Increase by $\frac{1}{10}$	275
450	Decrease by $\frac{2}{5}$	270
965	Increase by $\frac{1}{10}$	1061.5

4 90 g butter, 4.5 tsp sugar, 300 ml milk
5 **a** 280 g **b** 350 g **c** $233\frac{1}{3}$ g **d** 875 g
6 **a** £126 **b** £252 **c** £525 **d** £75.60
7 **a** £495 **b** £510 **c** £954 **d** £658.60
8 Andrew's pay increases by a larger amount (£10.80) than Bella's (£10.40)
9 £203.34 to 2 dp

N7.4

1 **a** 5.6 **b** 73 **c** 0.85 **d** −35
  e 110 **f** 38
2 **a** 141 kg **b** 8.4 m² **c** £13.63 **d** 45 sec
3 **a** 4.745 m, 4.755 m **b** 12.55 s, 12.65 s
  c 149.5 cm, 150.5 cm **d** 24.45 kg, 24.55 kg
  e 8.065 g, 8.075 g **f** 4.325 s, 4.335 s
4 **a** 2.545 m, 2.555 m **b** 1.65 s, 1.75 s
  c 1.548 m/s to 4 sf **d** 1.454 m/s to 4 sf
5 **a** 56.3 m², 72.3 m² **b** 40.3 cm², 41.6 cm²
  c 1.09 m², 1.11 m² **d** 8.70 mm², 9.30 mm²
  e 14.0 m², 14.1 m² **f** 99.7 cm², 99.9 cm²
6 **a** 36.765, 38.645 **b** 0.3, 0.5
  c 0.3618, 0.3864 **d** 0.1075, 0.1485 to 4 sf

N7.5

1 **a** $2\sqrt{3}$ **b** $2\sqrt{5}$ **c** 5 **d** $3\sqrt{7}$
2 **a** 2π **b** 3π **c** $2+2\pi$ **d** 10π
  e 18π **f** 16π
3 **a** $4+\sqrt{3}$ **b** $3+\sqrt{2}$ **c** $7+2\sqrt{7}$ **d** $14+\sqrt{17}$
4 **a** 32π **b** $7+2\pi$ **c** $28+4\sqrt{2}$ **d** 62π
5 **a** 2 **b** 5 **c** $3+3\sqrt{3}$ **d** $8+2\sqrt{3}$
6 **a** 5.66 **b** 3.24 **c** 4.24 **d** 106.10
7 **a** 20.9 **b** 44.7 **c** 12.3 **d** 32.2
8 **a** $2\sqrt{5}$ **b** $5\sqrt{5}$ **c** $3\sqrt{2}$ **d** $7\sqrt{2}$
9 **a** 3.146 **b** 3.162 **c** 11.18 **d** 4.897
10 **a** 3.146 **b** 3.162 **c** 11.18 **d** 4.899
11 **a** $14+7\sqrt{3}$ **b** $17+7\sqrt{3}$ **c** $15-3\sqrt{3}$ **d** 20

N7 Summary

1 **a** £350.96
2 **a** €546 **b** £78
3 4.57×10^{19}
4 **a** 100.5 m
  b 10.515 seconds
  c 9.5577746 m/s
  d 9.4536817 m/s

G5 Check in

1 **a** 49 **b** 52 **c** 34 **d** 48
  e 45 **f** 80 **g** 5 **h** 7
2 **a** 17.5 cm² **b** 15 m² **c** 50 m²
3 60 m³, 94 m²

G5.1

1 **a** 5 cm **b** 17 cm **c** 13 cm **d** 10.3 cm
  e 10.8 cm **f** 9.90 cm **g** 12.5 cm **h** 7.3 cm
2 3, 4, 5
  8, 15, 17
  5, 12, 13
3 **a** 8 cm **b** 19.8 cm **c** 10 cm **d** 9 cm
  e 7.5 cm **f** 5.4 cm **g** 5.5 cm **h** 7.1 cm
4 **a** 6, 8, 10; 10, 24, 26; 9, 12, 15
  b 6, 8, 10 is 3, 4, 5 doubled; 10, 24, 26 is 5, 12, 13 doubled; 9, 12, 15 is 3, 4, 5 trebled
5 **a** 5 **b** 17 **c** 13 **d** $\sqrt{106}$
  e $2\sqrt{29}$ **f** $7\sqrt{2}$

G5.2

1 **a** 8.6 cm **b** 7.9 cm
2 9.6 cm
3 11.3 cm
4 5.7 cm
5 16.7 cm
6 13.6 cm
7 5.4 m

G5.3

1 **a** (4, 5.5) **b** (0.5, 1) **c** (2, 3) **d** (1.5, 2)
2 **a** 5 units **b** 8.6 units **c** 10.0 units
  d 10.6 units
3 J (2, 4, 5), K (3, 1, 5), L (−1, 2, 5), M(4, −3, 0)
  N (0, 2, 0), P (−3, 2, −1), Q (0, 0, 2), R (2, 0, −3)
4 **ai** (1, 3, 3.5) **aii** (3.5, 3, 2.5)
  aiii (1.5, 3.5, 1.5) **aiv** (3, 2.5, 4.5)
  b ai Length of AB = 3.61 (2 dp)
    aii Length of CD = 4.24 (2 dp)
  c BC = $\sqrt{3}$ = 1.72 (2 dp)
    If rounded, $\sqrt{2}$ = 1.41 so BC = 1.73 (2 dp)

G5.4

1 **a** 25.1 cm **b** 239 mm **c** 50.3 cm
  d 47.1 cm **e** 75.4 mm **f** 132 mm
  g 82.9 cm
2 **a** 75.4 mm **b** 145 cm **c** 330 mm
  d 3.77 cm **e** 22.6 cm **f** 393 cm
3 **a** 50.3 cm² **b** 4540 mm² **c** 201 cm²
  d 177 cm² **e** 452 mm² **f** 1390 mm²
  g 547 cm²
4 **a** 452 mm² **b** 1660 cm² **c** 8660 mm²
  d 1.13 cm² **e** 40.7 cm² **f** 12 300 cm²
5 18.8 m
6 7.0 cm to 1 dp
7 **a** 13.9 cm² **b** 3.79 cm²

G5.5

1 a 39.3 cm² b 81.4 cm² c 905 mm²
   d 127 cm² e 402 mm² f 373 cm²
   g 422 cm² h 103 cm²
2 a 226 cm² b 8.31 cm² c 65.3 m²
   d 195 cm² e 142 mm² f 4.51 cm²
3 a 25.7 cm b 37.0 cm c 123 mm
   d 46.3 cm e 82.3 mm f 79.2 cm
   g 84.3 cm h 33.5 cm
4 a 61.7 cm b 11.8 cm c 33.2 m
   d 57.3 cm e 48.8 mm f 8.71 cm
5 a 6.28 m²
   b 20 flowers; he has 0.28 m² left
6 113 m
7 2510 cm²

G5.6

1 a 27 cm³ b 343 cm³
   c 64 cm³ d 0.125 m³
2 a 155.9 cm³ b 86.6 cm³
   c 304.8 cm³ d 125.7 mm³
3 a 384 cm² b 600 cm²
   c 216 cm² d 6 cm²
4 a 75.4 cm³, 100.5 cm²
   b 628.3 cm³, 408.4 cm²
   c 201.1 cm³, 201.1 cm²
   d 160.8 cm³, 257.6 cm²
5 a 1 cm × 1 cm × 80 cm, 2 cm × 2 cm × 20 cm,
      4 cm × 4 cm × 5 cm
   bi 112 cm² bii 322 cm²
6 a 1 cm × 1 cm × 24 cm, 2 cm × 2 cm × 6 cm; 56 cm²;
      98 cm²
   b 1 cm × 1 cm × 64 cm, 2 cm × 2 cm × 16 cm,
      4 cm × 4 cm × 4 cm; 96 cm²; 258 cm²

G5.7

1 a 400 m b 630 m c 4200 cm
   d 12 m e 80 km f 45 m
   g 50 mm h 300 mm i 60 cm
   j 70 mm k 4 cm l 18 050 m
2 a 2.6 m² b 70 000 m² c 4.5 cm²
   d 1200 cm² e 80 cm² f 4.5 cm²
   g 8 400 000 mm² h 300 m²
   i 20 000 cm² j 1 000 000 m²
3 a 3000 mm³ b 200 cm³ c 4800 cm³
   d 3000 cm³ e 10 m³ f 50 m³
4 a $\pi r + 2h + 2r$ b $2hr + \frac{1}{2}\pi r^2$
5 a $a^2 + 2ab + a\sqrt{a^2+b^2}$ b $\frac{1}{2}a^2b$
6 a The formula has only two dimensions; it should have
      three.
   b Area

G5 Summary

1 6.25 cm
2 15.42 cm (Accept 15.42 to 15.43)
3 302 g (Accept 301 or 302)
4 25 000 cm²

Case study 7: Radio maths

1 a 35 GHz = 35 000 000 000 = 35 000 000 kHz
      = 3.5 × 10⁷ kHz
   b 300 Hz = 0.000 300 MHz = 3.0 × 10⁻⁴ MHz
2 Waves **b** has the greatest volume because it has the
   greatest amplitude and wave **c** has the smallest volume
   because it has the smallest amplitude. The volume of
   wave **a** is between that of waves **b** and **c**. Wave **c** has the

highest pitch because it has the greatest frequency and
wave **b** has the lowest pitch because it has the lowest
frequency. The pitch of wave **a** is between that of waves
b and **c**.
3 ai 93.2 MHz
   aii 93 200 000 Hz = 9.32 × 10⁷ Hz
   aiii 0.0932 GHz = 9.32 × 10⁻² GHz
   b For Maths FM:
      wave speed = frequency × wavelength
                 = 9.32 × 10⁷ × 3.22
                 = 300,104,000 m/s
      For Maths AM:
      Wavelength = $\frac{\text{wave speed}}{\text{frequency}} = \frac{300\,104\,000}{930\,000} = 322.7$ m
   c frequency (kHz) × wavelength = 300 104
      so frequency = wavelength = $\sqrt{300104}$
                               = 547.8 kHz
      This is in the AM frequency band.

4 ai 40 mins aii 20 mins
   b Yes. A third of the show (33%) is speech based.

D5 Check in

1 a 0.55 b 0.04 c 0.72 d 0.625
   e 0.6 f 0.34 g 0.9 h 0.75
   i 0.18 j 0.17 k 0.192 l 0.12
2 a $\frac{1}{6}$ b $\frac{4}{5}$ c $\frac{2}{9}$ d $\frac{3}{4}$
   e $\frac{13}{15}$ f $\frac{11}{12}$ g $\frac{5}{9}$ h $\frac{8}{45}$

D5.1

1 a $\frac{4}{10}$ b $\frac{6}{10}$ c $\frac{7}{10}$ d $\frac{4}{10}$ e $\frac{8}{10}$ f $\frac{9}{10}$
2 241
3 102
4 a $\frac{13}{28}$ b $\frac{15}{28}$ c $\frac{11}{14}$ d $\frac{3}{14}$
5 ai $\frac{9}{16}$ aii $\frac{3}{8}$ aiii $\frac{13}{16}$ aiv $\frac{9}{16}$ av $\frac{5}{16}$ avi $\frac{1}{8}$
   b the black triangle
6 ai $\frac{1}{4}$ aii $\frac{3}{4}$ aiii 1 aiv 0 b 64

Coin/Dice	1	2	3	4	5	6
Head	H1	H2	H3	H4	H5	H6
Tail	T1	T2	T3	T4	T5	T6

   bi $\frac{1}{12}$ bii $\frac{1}{12}$ biii $\frac{1}{12}$

2 a

Red/blue	1	2	3	4	5	6
1	1, 1	1, 2	1, 3	1, 4	1, 5	1, 6
2	2, 1	2, 2	2, 3	2, 4	2, 5	2, 6
3	3, 1	3, 2	3, 3	3, 4	3, 5	3, 6
4	4, 1	4, 2	4, 3	4, 4	4, 5	4, 6
5	5, 1	5, 2	5, 3	5, 4	5, 5	5, 6
6	6, 1	6, 2	6, 3	6, 4	6, 5	6, 6

   bi $\frac{1}{36}$ bii $\frac{1}{36}$ biii $\frac{1}{12}$ biv $\frac{1}{18}$
3 $\frac{1}{6}$
4 a $\frac{1}{5}$ b $\frac{1}{20}$ c $\frac{3}{20}$ d $\frac{1}{10}$ e $\frac{1}{20}$
5 a $\frac{1}{100}$ b $\frac{1}{50}$ c 0 d $\frac{1}{25}$ e $\frac{3}{50}$
6 a

10p/2p	Head	Tail
Head	HH	HT
Tail	TH	TT

   b $\frac{1}{4}$

7 $\frac{1}{4}$

D5.3

1   First choice   Second choice

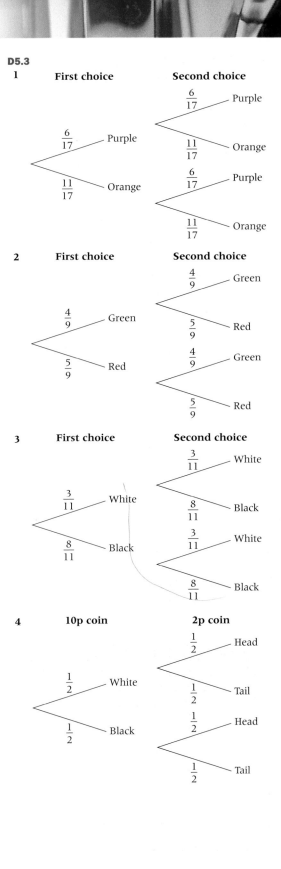

2   First choice   Second choice

3   First choice   Second choice

4   10p coin   2p coin

5   Bag   Coin

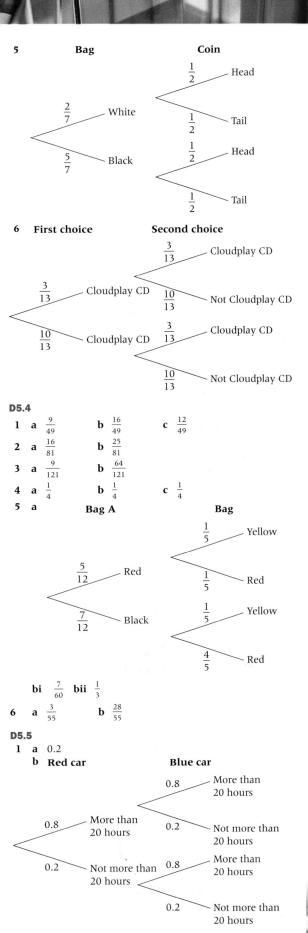

6   First choice   Second choice

D5.4

1 a $\frac{9}{49}$ b $\frac{16}{49}$ c $\frac{12}{49}$

2 a $\frac{16}{81}$ b $\frac{25}{81}$

3 a $\frac{9}{121}$ b $\frac{64}{121}$

4 a $\frac{1}{4}$ b $\frac{1}{4}$ c $\frac{1}{4}$

5 a   Bag A   Bag

   bi $\frac{7}{60}$ bii $\frac{1}{3}$

6 a $\frac{3}{55}$ b $\frac{28}{55}$

D5.5

1 a 0.2
   b Red car   Blue car

445

ci 0.64 **cii** 0.32 **ciii** 0.96

2 a First choice   Second choice

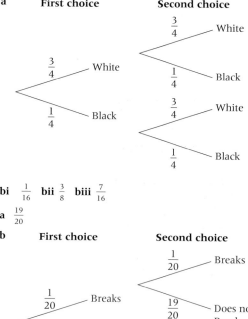

$\frac{3}{4}$ White

$\frac{1}{4}$ Black

$\frac{3}{4}$ White

$\frac{1}{4}$ Black

$\frac{3}{4}$ White

$\frac{1}{4}$ Black

bi $\frac{1}{16}$ **bii** $\frac{3}{8}$ **biii** $\frac{7}{16}$

3 a $\frac{19}{20}$

b First choice   Second choice

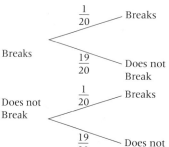

$\frac{1}{20}$ Breaks

$\frac{1}{20}$ Breaks

$\frac{19}{20}$ Does not Break

$\frac{1}{20}$ Breaks

$\frac{19}{20}$ Does not Break

$\frac{19}{20}$ Does not Break

ci $\frac{361}{400}$ **cii** $\frac{38}{400}$ **ciii** $\frac{39}{400}$

4 a First spin   Second spin

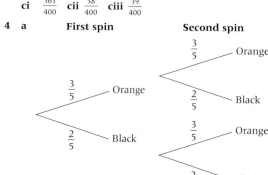

$\frac{3}{5}$ Orange

$\frac{3}{5}$ Orange

$\frac{2}{5}$ Black

$\frac{3}{5}$ Orange

$\frac{2}{5}$ Black

$\frac{2}{5}$ Black

bi $\frac{4}{25}$ **bii** $\frac{16}{25}$ **biii** $\frac{12}{25}$

5 a 0.9

b Grey plane   Orange plane

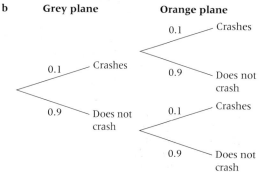

0.1 Crashes

0.1 Crashes

0.9 Does not crash

0.9 Does not crash

0.1 Crashes

0.9 Does not crash

c 0.01 **d** 0.18

6 a First choice   Second choice

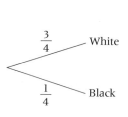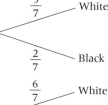

$\frac{3}{4}$ White

$\frac{5}{7}$ White

$\frac{2}{7}$ Black

$\frac{1}{4}$ Black

$\frac{6}{7}$ White

$\frac{1}{7}$ Black

b i $\frac{1}{28}$ **ii** $\frac{3}{7}$ **iii** $\frac{13}{28}$

D5 Summary
1 a 40 times
   b 0.6
2 a 0.2, 0.4, 0.4
   b 0.48
   c 0.44

G6 Check in
1 $a = 72°$ $b = 34°$ $c = 105°$
2 a Circle radius 3 cm **b** Arc radius 5 cm
3

G6.1
1 a 050°   **b** 320°
2 a Bearing of 070°   **b** Bearing of 155°
   c Bearing of 340°   **d** Bearing of 260°
3 a 316° **b** 265° **c** 068°
4 a 284° **b** 263° **c** 117°
5 b 170°
6 aii 328° **bii** 357°

G6.2
1 a Triangle with sides 8 cm, 4 cm, 7 cm
   b Triangle with sides 3 cm and 4 cm, and 30° angle
   c Triangle with sides 10 cm, 7.5 cm, 6 cm
   d Triangle with sides 8 cm and 2 cm, and 90° angle
   e Triangle with sides 6 cm, 9 cm, 5 cm
   f Triangle with two 45° angles and a 4 cm side
2 a 4 + 3 < 9, two short sides will never meet.
   b 4 + 5 = 9 so triangle is a straight line.
3 a Yes   **b** Yes   **c** Yes   **d** No
   e No   **f** No   **g** Yes   **h** Yes
4 a Triangle with sides 7 cm, 7 cm, 5 cm
   b Triangle with sides 5 cm, 5 cm, 7 cm
5 a Equilateral triangle with sides 5 cm
   b Rhombus with sides 5 cm
   c Rhombus with sides 3.5 cm
6 a Triangle with sides 5 cm, 12 cm, 13 cm
   b Right-angled triangle
7 a Triangle with sides 3 cm, 4 cm, 5 cm
   b Rectangle with sides 3 cm, 4 cm
   c Rectangle

8 a, b

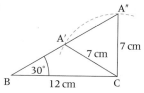

c No, SSA triangles are not unique since you can construct two different triangles with the given side and angle measurements.

G6.3

1 Angle bisectors
2 **a** Equilateral triangle with sides 5 cm
  b Angles bisectors of triangle in part **a**
  c Bisectors meet at a point and cut the midpoints of the opposite sides.
3 **a** Equilateral triangle with sides 4 cm
  b Angles bisector of triangle in part **a**
4 **a** Perpendicular bisector of 6 cm line
  b Perpendicular bisector of 9 cm line
  c Perpendicular bisector of 5.6 cm line
  d Perpendicular bisector of 10 cm line
  e Perpendicular bisector of 11.2 cm line
5 **a** Equilateral triangle with sides 5 cm
  b Perpendicular bisectors of sides of triangle in part **a**
  c Perpendicular bisectors and angle bisectors of equilateral triangles are the same.
6 **d** Both sets of lines meet at common points but not the same points.
7 **c** 90°, 135°, 180°, 225°, 270°, 315°

G6.4

1 Perpendiculars from points to lines
2 Perpendiculars from points on lines

G6.5

These sketches not drawn to scale.

1

2

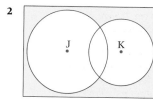

3

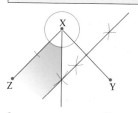

4

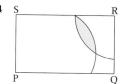

5

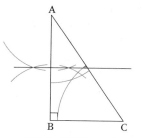

6

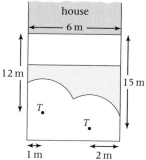

G6 Summary

1

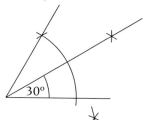

2

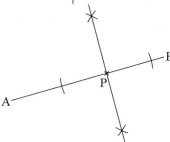

3 a, b

Scale: 1 cm represents 10 metres

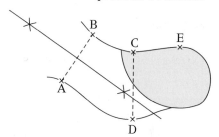

A6 Check in

1	**ai** 9	**aii** 4
	bi 27	**bii** −8
	ci 18	**cii** 8
	di 30	**dii** −10
	ei 24	**eii** 14
	fi 60	**fii** −20

gi 63   **gii** 8
hi 15   **hii** −30

2 a 20 mph   **b** 80 mph
    c 90 mph   **d** 24 mph

3 ai 3   **aii** (0, 4)   **aiii** upwards
    bi −4   **bii** (0, 10)   **biii** downwards
    ci 4   **cii** (0, 5)   **ciii** upwards
    di 2   **dii** $(0, 7\frac{1}{2})$ **diii** upwards
    ei 0   **eii** (0, 7)   **eiii** horizontal
    fi $\frac{1}{2}$ **fii** (0, 2)   **fiii** upwards

4 a $\frac{3}{2}$ **b** −1   **c** $-\frac{1}{2}$

    d $\frac{1}{3}$ **e** $\frac{3}{5}$

A6.1

1 a 60 km   **b** 1 hours 30 minutes
    c 60 km/h   **d** 11:30 am and 12 noon
    e 30 km/h

2 Claire and Christina: Claire's vertical line means she travelled a distance in no time, and Christina's line sloping backwards means she has travelled backwards in time.

3 a

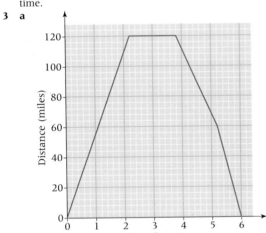

b

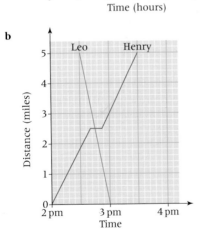

4

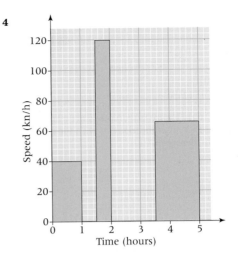

A6.2

1 A3, B1, C4, D2

2 Growth spurt at around 5, steadies down until next growth spurt at about 13–17 years (puberty) then steadies down again.

3

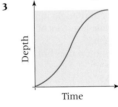

4 a

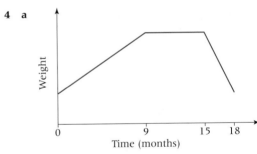

    b

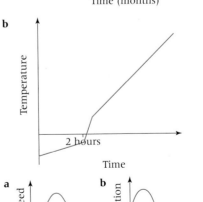

5 a   **b**

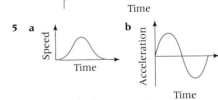

A6.3

1 6.2

2 **a** $x(x+4)-77=0$, length and width are 7 cm and 11 cm

b $h(h+1)(h+2)-990=0$; 9 cm by 10 cm by 11 cm

3 **a** $x=-5$ or 3

b Factorisation is more efficient

4 **a** 8.8   **b** 4.6   **c** 9.51   **d** 4.28

5 **a** $x(x+1)(x+2)=85\ 140$; 43, 44, 45

b $x(x+2)=57$; 6.6 cm

c $3x^3=500$, 5.5 mm

6 0 or 6.03

A6.4

1 **a**

Straight line	Parabola
$y=3x-2$	$y=x^2-2$
$3x+2y=8$	$y=10+x^2$
$y=x$	$y=x^2+2x+1$

2 **b**

x	−4	−3	−2	−1	0	1	2	3	4
x^2	16	9	4	1	0	1	4	9	16
$y=x^2-2$	14	7	2	−1	−2	−1	2	7	14

c

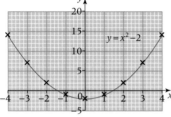

x	−4	−3	−2	−1	0	1	2	3	4
$y=x^2+3$	19	12	7	4	3	4	7	14	19

aiii

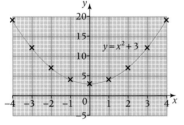

aiv (0, 3)

bi

x	−4	−3	−2	−1	0	1	2	3	4
$y=2x^2$	32	18	8	2	0	2	8	18	32

biii

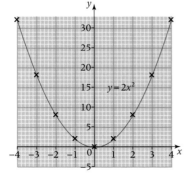

biv (0, 0)

ci

x	−4	−3	−2	−1	0	1	2	3	4
$y=3x^2-1$	47	26	11	2	−1	2	11	26	47

ciii

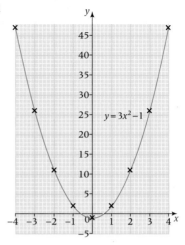

civ (0, −1)

di

x	−4	−3	−2	−1	0	1	2	3	4
$y=x^2+x$	12	6	2	0	0	2	6	12	20

diii

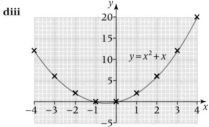

div (−0.5, −0.25)

4 False, since $4^2-5 \neq 10$.

5 **a** graph will be other way up

b

x	−3	−2	−1	0	1	2	3
$y=10-x^2$	1	6	9	10	9	6	1

c

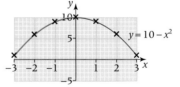

6 **a**

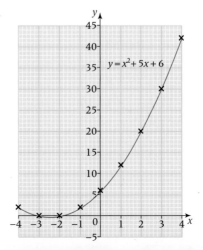

b $(-2, 0)$, $(-3, 0)$
c $x = -2$ or -3
d The coordinates give the solutions of the equation, as x-axis is the line $y = 0$

A6.5

1 From left to right: $y = x^3$, $y = 3 - 2x$, $y = x^2 - x - 6$, $x = \frac{1}{2}$.
Remaining graphs:

2 **b**

x	-3	-2	-1	0	1	2	3
y	-26	-7	0	1	2	9	28

c

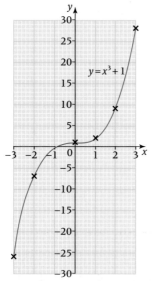

di 4.375   **dii** 1.125

3 **a**

x	-2	-1	0	1	2	3
x^3	-8	-1	0	1	8	27
$x^3 - 4$	-12	-5	-4	-3	4	23

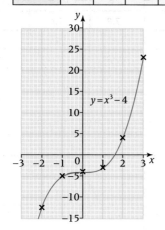

b

x	-2	-1	1	2	3
$\frac{1}{x}$	$-\frac{1}{2}$	-1	1	$\frac{1}{2}$	$\frac{1}{3}$
y	-1	-2	2	1	$\frac{2}{3}$

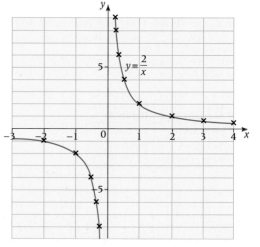

c

x	-2	-1	0	1	2	3
x^3	-8	-1	0	1	8	27
$x + 1$	-1	0	1	2	3	4
y	-9	-1	1	3	11	31

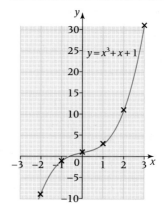

d

x	-2	-1	1	2	3
$\frac{3}{x}$	$-\frac{3}{2}$	-3	3	$1\frac{1}{2}$	1
y	$-\frac{1}{2}$	-2	4	$2\frac{1}{2}$	2

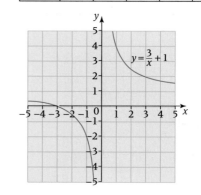

4 a

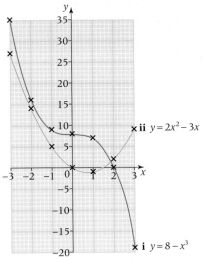

ii $y = 2x^2 - 3x$

i $y = 8 - x^3$

b (1.9, 1.4)

5

x	$-\frac{1}{2}$	$-\frac{1}{3}$	$-\frac{1}{4}$	$\frac{1}{2}$	$\frac{1}{3}$	$\frac{1}{4}$
$\frac{1}{x}$	-2	-3	-4	2	3	4
y	-4	-6	-8	4	6	8

See Q3b for graph

A6.6

1 **a** $x = -1.5$ or 2.5 **b** $x = -0.6$ or 1.6
  c $x = -1$ or 3
2 **a** $x = -1.2$ or 3.2 **b** $x = 0$ or 2
  c $x = 1$ **d** $x = -0.3$ or 3.3
  e $x = -1.6$ or 2.6 **f** $x = -1$ or 3
3 **a** $x = -2.6$ or 2.6 **b** $x = -1.3$
  c $x = 0$ or 0.5 **d** $x = 1.7$

A6.7

1 **a** $x = -3$ or 1 **b** $x = -2.6$ or 1.6
  c $x = -3.4$ or 1.4 **d** $x = -1$ or 0
2 **a** $y = 3$ **b** $y = 0$ **c** $y = 2x + 1$
  d $y = 4$ **e** $y = 2$ **f** $y = 10x - 4$
3 **a** $x = 1.1$ **b** $x = 1.2$ **c** $x = -1.5$
  d $x = -1.6$, 0 or 0.6 **e** $x = -2.5$ **f** $x = 0.8$

A6.8

1 $\left(-\frac{1}{4}, -4\frac{1}{8}\right)$

2 **a**

x	-3	-2	-1	0	1	2	3
x^2	9	4	1	0	1	4	9
y	7	3	1	1	3	7	13

b

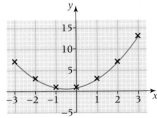

c Minimum is 0.75 at $x = -0.5$

3 a

Time (x)	0	1	2	3	4	5
$20x$	0	20	40	60	80	100
$4x^2$	0	4	16	36	64	100
Height (y)	0	16	24	24	16	0

b

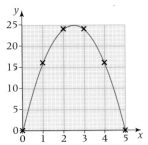

ci 25 m, 2.5 seconds
cii 0.7 seconds, 4.3 seconds
ciii 3.2 seconds

A6 Summary

1 **a** $x = 2.1$
2

B
D
C
A
F
E

Case study 8: Business
Original table:

	January (£)	February (£)	March (£)
Quantity of standard	18	16	24
Quantity of deluxe	12	14	9
Standard card sales	45.9	40.8	61.2
Deluxe card sales	43.2	50.4	32.4
TOTAL INCOME	89.1	91.2	93.6
Materials used	7.8	7.6	9
Wages	42	44	42
Craft fair fees	10	10	10
Advertising	5	5	5
TOTAL EXPENDITURE	64.8	66.6	66
NET CASH SURPLUS/DEFICIT	24.3	24.6	27.6
CASH BALANCE BROUGHT FORWARD		24.3	48.9
CASH BALANCE TO CARRY FORWARD	24.3	48.9	76.5

Fair fees £15:

	January (£)	February (£)	March (£)
Quantity of standard	18	16	24
Quantity of deluxe	12	14	9
Standard card sales	45.9	40.8	61.2
Deluxe card sales	43.2	50.4	32.4
TOTAL INCOME	89.1	91.2	93.6
Materials used	7.8	7.6	9
Wages	42	44	42
Craft fair fees	15	15	15
Advertising	5	5	5
TOTAL EXPENDITURE	69.8	71.6	71
NET CASH SURPLUS/DEFICIT	19.3	19.6	22.6
CASH BALANCE BROUGHT FORWARD		19.3	38.9
CASH BALANCE TO CARRY FORWARD	19.3	38.9	61.5

Material costs increase by 40%:

	January (£)	February (£)	March (£)
Quantity of standard	18	16	24
Quantity of deluxe	12	14	9
Standard card sales	45.9	40.8	61.2
Deluxe card sales	43.2	50.4	32.4
TOTAL INCOME	89.1	91.2	93.6
Materials used	10.92	10.64	12.6
Wages	42	44	42
Craft fair fees	10	10	10
Advertising	5	5	5
TOTAL EXPENDITURE	67.92	69.64	69.6
NET CASH SURPLUS/DEFICIT	21.18	21.56	24
CASH BALANCE BROUGHT FORWARD		21.18	42.74
CASH BALANCE TO CARRY FORWARD	21.18	42.74	66.74
Materials cost standard	0.42		
materials cost deluxe	0.28		

20% discount:

	January (£)	February (£)	March (£)
Quantity of standard	18	16	24
Quantity of deluxe	12	14	9
Standard card sales	36.72	32.64	48.96
Deluxe card sales	34.56	40.32	25.92
TOTAL INCOME	71.28	72.96	74.88
Materials used	7.8	7.6	9
Wages	42	44	42
Craft fair fees	10	10	10
Advertising	5	5	5
TOTAL EXPENDITURE	64.8	66.6	66
NET CASH SURPLUS/DEFICIT	6.48	6.36	8.88
CASH BALANCE BROUGHT FORWARD		6.48	12.84
CASH BALANCE TO CARRY FORWARD	6.48	12.84	21.72
discounted price standard	2.04		
discounted price deluxe	2.88		

Scenario with £8.00 for materials and £40 for wages.

Constraints:
$0.3s + 0.2d \leqslant 8$
$s + 2d \leqslant 40$
The aim is to maximise profit £P, where
$P = 1.25s + 1.40d - 15$
Gives maximum profit £24 when $s = 20$ and $d = 10$

G7 Check in

1   **a** $x = 6y$ **b** $x = 5y$ **c** $x = 10y$

     d $x = \frac{2}{y}$ **e** $x = \frac{5}{y}$ **f** $x = \frac{8}{y}$

2   **a** $a = 8.1$ units   **b** $b = 9.8$ units

     c $c = 8.5$ units   **d** $d = 13.7$ units

G7.1

1   **a**   3.63 cm   **b**   11.3 cm   **c**   10.8 cm

     d   3.02 cm   **e**   6.38 cm   **f**   3.79 cm

     g   30.5 cm   **h**   14.2 cm   **i**   74.5 cm

     j   6.06 cm   **k**   7.00 cm   **l**   12.3 cm

2   Right-angled isosceles triangles; They are the same as the other shorter side in each triangle.

G7.2

1   **a**   6.89 cm   **b**   8.60 cm   **c**   7.88 cm

     d   4.62 cm   **e**   4.77 cm   **f**   39.7 cm

     g   10.6 cm   **h**   9.56 cm   **i**   34.4 cm

     j   13.5 cm   **k**   11.8 cm   **l**   6.15 cm

     m   14.2 cm   **n**   7.88 cm   **o**   5.12 cm

G7.3

1.
   a. 26.4° b. 25.7° c. 56.2° d. 67.4°
   e. 17.6° f. 23.6° g. 69.2° h. 55.1°
   i. 19.8° j. 35.7° k. 27.7° l. 40.1°
   m. 73.0° n. 41.0° o. 45°   p. 30°
   q. 60°

G7.4

1.
   a. $a = 8$ cm, $b = 14.4$ cm
   b. $c = 9.57$ cm, $d = 16.0$ cm
   c. $e = 13.2$ cm, $f = 10.8$ cm, $g = 13.1$ cm
2. a. 28.9° b. 35.5° c. 61.2°
3. a. 15.2 cm b. 19.5°
4. a. 8.5 cm b. 58.4°
5. a. 15.2 cm b. 6.9 cm

G7.5

1. 49.4 cm²
2. 63.8 cm²
3. 31.5 cm²
4. 91.2°
5. 10.3 cm
6. 14.7 cm
7. 38.9°
8. 7.5 km
9. a. 130° or 050° b. 20.8 km or 5.4 km
10. 6.6 m
11. Yes, both give roughly the same height for phone mast.

G7 Summary

1. 230
2. a. 45.6° b. 8.1 cm

Index